Y0-CAH-618

M. C. Escher, "Stars" (1948), © M. C. Escher Heirs, % Cordon Art, Baarn, Holland

Crystal Chemistry and Refractivity

HOWARD W. JAFFE

Professor of Geology, Emeritus
University of Massachusetts at Amherst

DOVER PUBLICATIONS, INC.
MINEOLA, NEW YORK

Copyright

Published in Canada by General Publishing Company, Ltd., 30 Lesmill Road, Don Mills, Toronto, Ontario.
Published in the United Kingdom by Constable and Company, Ltd., 3 The Lanchesters, 162–164 Fulham Palace Road, London W6 9ER.

Bibliographical Note

This Dover edition, first published in 1996, is an unabridged, corrected republication of the work first published by Cambridge University Press, Cambridge, England, 1988. For this edition the author has corrected a number of errors as well as providing a new frontispiece ("Stars" by M. C. Escher) and commentary, "About the Frontispiece," which appears on page vi.

Library of Congress Cataloging-in-Publication Data

Jaffe, Howard W.
Crystal chemistry and refractivity / Howard W. Jaffe.
p. cm.
"An unabridged, corrected republication of the work first published by Cambridge University Press, Cambridge, England, 1988"—T.p. verso.
Includes bibliographical references (p. –) and indexes.
ISBN 0-486-69173-X (pbk.)
1. Crystallography. I. Title.
QD921.J28 1996
549′.18—dc20 96-4258
CIP

Manufactured in the United States of America
Dover Publications, Inc., 31 East 2nd Street, Mineola, N.Y. 11501

To
Elizabeth Boudreau Jaffe

About the Frontispiece

Maurits Cornelis Escher (1898–1972) was one of Europe's greatest graphic artists. Although he was neither crystallographer nor mineralogist, Escher's illustrations of geometric solids are greatly admired by many scientists who work with minerals. I am one of these.

Escher saw his "Stars" as a sky full of geometric forms that float like brilliantly faceted jewels through a black outer space. The central object of "Stars" is formed by interpenetration of three octahedral frameworks, outlined only by their 36 (3×12) edges. He chose "two chameleons as denizens of this framework because they are able to cling by their legs and tails to the beams of their cage as it swirls through space." Wide-eyed and green-scaled, the two chameleons sail through space in their octahedral cage, one with his long tongue extended, the other with his mouth closed.

Surrounding the central star are smaller satellite forms such as interpenetrated or twinned dodecahedra, tetrahedra, cube-octahedra, and cube-dodecahedra. Yet smaller "stars" include the single, regular cube, dodecahedron, octahedron, tetrahedron and icosahedron.

Escher didn't tell us why he chose chameleons to ride through space inside a triple octahedral cage. Like many a great artist, he left something to the viewer's imagination.

Contents

Preface *page* ix
Acknowledgments xi

PART I: ***Principles of crystal chemistry and refractivity*** 1

1. Atoms, electrons, orbitals, and minerals 1
2. The excited atom: spectra and quanta, ionization potential, electronegativity, and chemical bonding 15
3. The crystal chemistry of the covalent bond 27
4. The crystal chemistry of the ionic bond 42
5. Pauling's second rule of electrostatic valency in ionic or coordination compounds 61
6. External (nontranslational) and internal (translational) symmetry 71
7. Crystal field theory 89
8. Polyhedral distortion 95
9. Diadochy and isostructural crystals 100
10. Density, volume, unit cells, and packing 108
11. Refractivity and polarizability 118

PART II: ***Descriptive crystal chemistry*** 147

12. Silicates: classification and formulation 147
13. Sorosilicates – dumbbell silicates – pyrosilicates 177
14. Cyclosilicates 186
15. Inosilicates and phyllosilicates 201
16. Tektosilicates 232
17. The crystal chemistry of the borate minerals 251
18. The crystal chemistry of oxides and some fluoride minerals 265

19. Structures built of complex anions: sulfates, tungstates, phosphates, carbonates, and fluocarbonates 305
20. Sulfides, arsenides, and related compounds 321

Mineral index 327
Subject index 330

Preface

Crystal chemistry books are often written by chemists, physicists, and mathematical crystallographers, who assume that their readers have a sound background in chemistry, physics, mathematics, and elementary crystallography. This author, a geologist with more than forty years of research and teaching experience in mineralogy and crystal chemistry, has had the task of teaching advanced mineralogy and crystal chemistry to first-year graduate students studying for M.S. and Ph.D. degrees in the geological sciences. Most of the incoming graduate students, who have arrived with undergraduate degrees from a representative cross section of American colleges and universities, do not have this adequate background in the aforementioned basic subjects. This is perhaps because chemists teach elementary chemistry courses using generally nonmineralogical examples; and mineralogists teaching elementary mineralogy, in attempting to cover the entire mineral kingdom in a one-semester course, emphasize hand-specimen identification and external symmetry, with the focus on crystal morphology. Incoming graduate students thus require, and are grateful to receive, a review of chemistry, physics, and internal crystallographic symmetry, needed to further their education in crystal chemistry and enable them to comprehend journal articles from the contemporary mineralogical literature.

The book is divided into two parts: Part I (Chapters 1–11) takes the reader through principles of basic chemistry, physics, and internal crystallographic symmetry into more sophisticated areas of crystal chemistry and refractivity. Part II (Chapters 12–20) is addressed to graduate students and researchers in geology or in materials science who would like to know more about atomic arrays, atomic bonds, and the chemical, optical, and other physical properties they express. Extensive data are given for unit cell dimensions, interatomic distances, coordination polyhedra, electronegativity differences, optical constants, and the variation of velocity of propagation of light with direction and atomic polymerization patterns. These data will be useful to the research scientist working with naturally occurring and synthetic materials.

For about twenty years, the author's graduate and some undergraduate students have been required to construct ball or packing models of minerals or synthetic compounds, using the raw crystallographic data supplied in appropriate journals and textbooks. In addition to learning the fundamental concepts of crystal chemistry as a graphic rather than as an abstract wave function summation, the students were encouraged to relate the known optical properties to the structure of the model they built: the discipline of refractivity. Because of a paucity of published data on this subject, this book attempts to present the principles of refractivity and polarizability that relate to coordination chemistry and chemical bonding in diverse but representative crystal structures.

Just as the field geologist measures the attitude of bedding, foliation, and lineation of rocks as a means of deciphering the complex geological history of a region, so too can the mineralogist and crystal chemist measure, in crystals, the foliation and lineation of polymerized units that exercise control over the path and magnitude of the electric field or electric vector of light.

No attempt has been made to duplicate the voluminous data on crystal structures covered in the many volumes of *Structure Reports* (1928 to date), Wyckoff's *Crystal Structures* (1968), and the excellent series of *Short Course Notes* of the Mineralogical Society of America (1974 to date). Instead, about sixty representative examples of silicates, oxides, borates, carbonates, sulfates, phosphates, and sulfides were selected for analysis of the basic principles of crystal chemistry they illustrate.

Geneva, Switzerland
February 1986

Howard W. Jaffe

Acknowledgments

The author wishes to thank The Institute of Mineralogy, Section of Earth Sciences, University of Geneva, Switzerland, and its former head, Professor Marc B. Vuagnat, for the use of its superb international mineralogical library and other facilities.

Margarita Shannon and David Elbert worked out the mathematical formula for the direct calculation of the unit cell volume of a trigonal–rhombohedral mineral, using the cell edge, a_{rh}, and the interaxial angle, α_{rh} (Chapter 10).

Perceptive reviews by Dr. Brian Mason, Smithsonian Institute, Washington, D.C., Professor Robert T. Dodd, SUNY, Stony Brook, NY, Professor Roger Burns, Massachusetts Institute of Technology, Cambridge, MA, and Mr. David Elbert, University of Massachusetts, Amherst, as well as the editorial assistance of Peter-John Leone and Michael Gnat, Cambridge University Press, have greatly contributed to the improvement of the book. Thanks are also due the following students for locating errors during trial use of the book manuscript: John Burr, Marshall Chapman, Vincent Dellorusso, Karin Olson, Virginia Peterson, Jennifer Thompson, Suizhou Xue, and Yang Yu.

The author is especially grateful to Mrs. Marie Litterer, University of Massachusetts, for the superb drafting of the many complex illustrations.

Many of the packing models illustrated in this book were built by the author; others were built by students in the advanced mineralogy and optical geochemistry courses at the University of Massachusetts and have become part of a valuable teaching collection. These latter include the following:

sphene (Carl Francis, Fig. 12.3)
datolite (Richard Jackson, Fig. 12.9)
staurolite (Kurt Hollocher, Fig. 12.15)
chloritoid (David Elbert, Figs. 12.16, 12.17)
lawsonite (Katherine Caldwell, Fig. 13.1)
epidote (Robert Tracy, Fig. 13.5)
cordierite (Henry Berry, Fig. 14.3)
wadeite (David Devoe, Fig. 14.4)
azurite (George Cody, Fig. 19.9)

PART I
Principles of crystal chemistry and refractivity

1 Atoms, electrons, orbitals, and minerals

A crate of oranges hastily filled at the orchard can be more efficiently packed if vigorously shaken a few times to eliminate waste space. In a similar way, atoms loosely collected or disordered in space can become more energetically stable by bonding together into an ordered crystal structure. Evaluation of the manner in which the various atoms become ordered into stable chemical compounds, both natural and synthetic, and their resultant chemical and physical properties are the principal objectives of the science of crystal chemistry.

For example, atoms of silicon, Si, and oxygen, O, may combine or become ordered into different crystal structures under different conditions of growth. At low pressures and moderate temperatures of formation, each Si atom will surround itself with four O atoms to form the SiO_4 tetrahedron. Each such tetrahedron will then share each of its four corner O atoms with adjoining Si atoms to build a giant three-dimensional polymer or framework of tetrahedra that characterizes the common mineral quartz. Each Si is thus said to be coordinated with four O atoms, and each O atom by two Si atoms to yield the formula $Si^{IV}O_2^{II}$.

At very high pressures and temperatures of formation, such as those induced by an impacting meteorite, six O atoms can be compressed about the Si atom in an octahedral array to form the mineral stishovite. Now, each Si atom is coordinated by six O atoms and each O by three Si atoms to yield the formula $Si^{VI}O_2^{III}$. Quartz, the low pressure, low coordination form of SiO_2, has the more open structure and possesses a density of 2.65 g/cm^3 and a mean index of refraction of 1.547; stishovite, the high pressure, high coordination form of SiO_2, has a density of 4.28 g/cm^3 and a mean index of refraction of 1.815.

Similarly, elemental carbon, C, will, at low pressures of growth, surround itself with three additional C atoms arrayed in an equilateral triangular coordination, with resulting polymerization into sheets of C atoms; whereas during growth at the high pressures that obtain deep in the earth, each C atom will surround itself by four additional C atoms arrayed at the corners of a tetrahedron to build the three-dimensional framework that typifies the mineral diamond. Graphite has the formula C^{III}; diamond, C^{IV}. Graphite has a density of 2.23 and is opaque, reflecting incident light, whereas diamond has a density of 3.51 and a very high index of refraction, 2.418. The causes for such contrasts in physical properties will be further explored and explained in Chapter 2.

The student of crystal chemistry must therefore become familiar with the nuclear and electronic structures and properties of atoms as well as with the organization chart of atoms known as the periodic table or periodic classification of the elements (Fig. 1.1).

Atoms, elements, and nuclides

From the viewpoint of the chemist, the atom may be regarded as a spherical bundle of energy or matter of small mass that retains its identity in chemical combination. The atom consists of *nucleons* (nuclear particles) and extranuclear *electrons.* Nucleons consist primarily of positively charged *protons* and uncharged or neutral *neutrons,* along with subordinate numbers of other particles such as *mesons,* which may have electrical charge of +, −, or 0. The number of protons is equal to the *atomic number* (*Z*); and the number of neutrons, to the *neutron number* (*N*). The sum of protons and neutrons is equal to the *relative atomic mass number,* or *atomic weight* (*A*); thus $Z + N = A$. The atomic mass is relative because it is generally not given in a quantity such as grams but as a number that uses elemental carbon, C = 12.000, as a reference standard. Thus, elemental hydrogen, H, with one proton and no neutrons in its

Figure 1.1. Periodic classification of the elements.

←Rrepesentative→		*←Transition*							*→*	*←Representative*							*→*
s^1	s^2	d^1	d^2	d^3	d^4	d^5	d^6	d^7	d^8	d^9	d^{10}	p^1	p^2	p^3	p^4	p^5	p^6
H 1 1.00797																H 1 1.00797	He 2 4.0026
Li 3 6.939	Be 4 9.0122											B 5 10.811	C 6 12.01115	N 7 14.0067	O 8 15.9994	F 9 18.9984	Ne 10 20.123
Na 11 22.9898	Mg 12 24.312											Al 13 26.9815	Si 14 28.086	P 15 30.9738	S 16 32.064	Cl 17 35.453	Ar 18 39.948
K 19 39.102	Ca 20 40.08	Sc 21 44.956	Ti 22 47.90	V 23 50.942	Cr 24 51.996	Mn 25 54.9381	Fe 26 55.847	Co 27 58.9332	Ni 28 58.71	Cu 29 63.54	Zn 30 65.37	Ga 31 69.72	Ge 32 72.59	As 33 74.9216	Se 34 78.96	Br 35 79.909	Kr 36 83.80
Rb 37 85.47	Sr 38 87.62	Y 39 88.905	Zr 40 91.22	Nb 41 92.906	Mo 42 95.94	Tc 43 (99)	Ru 44 101.07	Rh 45 102.905	Pd 46 106.4	Ag 47 107.870	Cd 48 112.40	In 49 114.82	Sn 50 118.69	Sb 51 121.75	Te 52 127.60	I 53 126.9044	Xe 54 131.30
Cs 55 132.905	Ba 56 137.34	La* 57 138.91	Hf 72 178.49	Ta 73 180.948	W 74 183.85	Re 75 186.2	Os 76 190.2	Ir 77 192.2	Pt 78 195.09	Au 79 196.967	Hg 80 200.59	Tl 81 204.37	Pb 82 207.19	Bi 83 208.980	Po 84 (210)	At 85 (210)	Rn 86 (222)
Fr 87 (223)	Ra 88 (226)	Ac** 89 (227)															

*Lanthanides	Ce 58 140.12	Pr 59 140.907	Nd 60 144.24	Pm 61 (145)	Sm 62 150.35	Eu 63 151.96	Gd 64 157.25	Tb 65 158.924	Dy 66 162.50	Ho 67 164.930	Er 68 167.26	Tm 69 168.934	Yb 70 173.04	Lu 71 174.97
**Actinides	Th 90 232.038	Pa 91 (231)	U 92 238.03	Np 93 (237)	Pu 94 (242)	Am 95 (243)	Cm 96 (247)	Bk 97 (249)	Cf 98 (251)	Es 99 (254)	Fm 100 (253)	Md 101 (256)	No 102 (253)	Lw 103 (257)

simple nucleus, has atomic number 1, neutron number 0, and atomic mass 1.00797 or about one-twelfth of the mass of the reference standard C = 12. From the foregoing discussion, it might be assumed that all of the mass of the atom is contained in the nucleus if $Z + N = A$. This is close to the truth, inasmuch as a proton has about 1837 times the mass of an electron. The following mass numbers are accepted for atomic particles:

proton = 1.00797	μ meson = 0.215
neutron = 1.00894	π meson = 0.280
electron = 0.00055	

In the neutral atom, the number of electrons must equal the number of protons in order to maintain electrical neutrality. Although, at this writing, 106 chemical elements have been identified and periodically classified, there exist more than 1500 variants of these, which are classified as nuclides. A *chemical element* is an atomic species possessing a specific number of protons in its nucleus; a *nuclide* is simply any atomic species that has a specified atomic mass, in which case there is no restriction on the number of either protons or neutrons. The same chemical element Z may exist in species with different numbers of neutrons, and these are classed as *isotopes* (the **"p"** is to remind you that isotopes contain equal numbers of protons). Table 1.1 lists the isotopes of oxygen in

Table 1.1. *Classification of nuclides as isotopes (=Z), isotones (=N), and isobars (=A)*[a]

Isotopes of oxygen, Z = 8, A = 15.9994[b]

${}^{16}_{8}O_{8}$ ${}^{17}_{8}O_{9}$ ${}^{18}_{8}O_{10}$

Isotones of manganese, iron, and nickel, N = 30

${}^{55}_{25}Mn_{30}$ ${}^{56}_{26}Fe_{30}$ ${}^{58}_{28}Ni_{30}$

Isobars of argon, potassium, calcium, A = 40

${}^{40}_{18}Ar_{22}$ ${}^{40}_{20}Ca_{20}$ ${}^{40}_{19}K_{21}$

Isotope	Mass	% Natural abundance	Product
${}^{16}O$	15.9949	99.759	15.956352
${}^{17}O$	16.9914	0.037	0.0062897
${}^{18}O$	17.99916	0.204	0.0367183
		100.00	$A = 15.9993600$

[a] Surrounding the elemental symbol are the atomic mass number (upper left), atomic (proton) number (lower left), and neutron number (lower right). This system is used throughout this book.

[b] Oxygen has atomic mass $A = 15.9994$, based on the mass of carbon (C = 12.000).

which atomic nuclei all contain eight protons ($Z = 8$) combined with either eight, nine, or ten neutrons ($N = 8$, 9, or 10) to yield the oxygen isotopes ^{16}O, ^{17}O, and ^{18}O.

Reversing the situation, we find that other elements or nuclides may have equal numbers of neutrons combined with unequal numbers of protons, and these are classed as *isotones* (the "**n**" is to remind you that these contain equal numbers of neutrons). Thus Table 1.1 lists the isotones ^{55}Mn, ^{56}Fe, and ^{58}Ni, all containing thirty neutrons. There also exist nuclides that contain unequal numbers of both protons and neutrons but have equal integral mass number: these are classed as *isobars* (the "**a**" is to remind you that these have equal atomic mass numbers). In Table 1.1, isobars are represented by ^{40}Ar, ^{40}K, and ^{40}Ca.

Nuclear structure and stability

Of the more than 1500 nuclides, both natural and synthetic, that have been identified, only 269 are stable, that is, do not decay by radioactive processes. Although all 106 of the chemical elements contain a percentage of radioactive nuclides, those with atomic numbers $Z = 1$–82 (H through Pb) contain only a small percentage of these. Those elements with atomic numbers $Z = 83$–106 consist entirely of radioactive nuclides; or, to put it differently, those elements with $Z > 82$ are all unstable and decay naturally by a variety of radioactive processes. Thus, our 269 stable nuclides must come from elements with $Z = 1$–82. It is common to classify the stable nuclides into four groups characterized by the even and odd proton and neutron constitution of their nuclei. It is obvious from Table 1.2 that nuclides having even numbers of both protons and neutrons have exceptionally high stability and abundance, whereas those having an odd–odd combination are rare in nature. The other two groupings having even–odd and odd–even combinations are of intermediate abundance. This tells us that the atomic nucleus also has a structure, and that protons and neutrons are not merely an irregular or disordered aggregate of

Table 1.2. *Classification of stable nuclides on the basis of proton and neutron configuration*

Nuclear content		No. of	
Protons (Z)	Neutrons (N)	nuclides	Example
Even	-Even (e–e)	160	$^{16}_{8}O_8$
Even	-Odd (e–o)	55	$^{9}_{4}Be_5$
Odd	-Even (o–e)	49	$^{27}_{13}Al_{14}$
Odd	-Odd (o–o)	5	$^{14}_{7}N_7$
		269	

both. We need not look too far for the reasons for this, when we note that inside the nucleus repulsive proton–proton forces far exceed attractive forces. At short range, protons and neutrons are strongly interactive, and their nuclear pairing $(n + p)$ creates cohesive forces that greatly exceed magnetic, electric, or gravitative forces. Here, the aforementioned mesons also play an important role. As noted earlier, μ mesons and π mesons are of different mass and may also have electric charge +, −, or 0. These mesons appear to help overcome the electrostatic proton–proton repulsion in the nucleus and further contribute to nuclide stability.

Where atomic nuclei become either oversaturated or undersaturated with protons or neutrons, they tend to expel the excess particles by means of natural radioactive processes. Where neutrons are in excess, the number of protons is increased by a *β-decay process,* whereby a neutron decays to a proton and ejects an electron; this process is written

$$n \longrightarrow p + e^- \text{ ejection}$$

(e.g., ${}^{87}_{37}\mathrm{Rb}_{50} \rightarrow {}^{87}_{38}\mathrm{Sr}_{49}$ + energy). If protons are in excess, the reverse occurs, and we have an *electron capture process:*

$$p \longrightarrow n + e^+ \quad \text{positron ejection}$$

(e.g., ${}^{40}_{19}\mathrm{K}_{21} \rightarrow {}^{40}_{18}\mathrm{Ar}_{22}$ + energy). ^{87}Rb and ^{40}K constitute small percentages of radioactive isotopes associated with their abundant stable family members ^{86}Rb and ^{39}K. Radioactive ^{40}K and ^{87}Rb contribute much of the heat in the earth's crust, inasmuch as they are concentrated in potassium feldspar, sodium–calcium feldspar, and micas, all principal minerals of the more common rocks that make up the earth's crust. Over the 4.6×10^9-yr history of the solar system these radioactive nuclides, along with those of uranium, U, and thorium, Th, have been producing large, but decreasing, amounts of heat at known rates of atomic disintegration. Each radioactive nuclide has a known *half-life* or time interval in which one-half of the atoms present decay. Radioactive decay also plays an important role in crystal chemistry directly, particularly in those cases where decay of unstable nuclides, such as ^{238}U, ^{235}U, and ^{232}Th produce copious quantities of alpha particles that bombard and gradually destroy the crystal structures of the minerals in which they reside. Minerals whose structures are so damaged or destroyed are classed as *metamict minerals.* They are discussed further in a subsequent chapter dealing with radioactive minerals.

At this point, the important conclusion may be made that nuclear structure of atoms controls the abundance of the chemical elements in the universe, whereas electronic structure of atoms controls the species of minerals that are permitted to grow or crystallize. Protons and neutrons control nuclear stability, but electrons control crystal stability via processes of chemical bonding. Accordingly, a more detailed examination of the electronic structure of atoms is essential to our comprehension of modern concepts of crystal chemistry.

Electronic structure

Matter has both corpuscular and undulatory properties, and it is possible to consider the electron either as a particle or as a wave. An *electron* is perhaps best described as a continuous distribution of negative electrical charge carried by waves. In these waves, *electron density* ψ^2 is proportional to the *intensity* of the waves. The point that represents the *energy* of the electron is guided by the waves that surround it. In the early treatment of the atom, attributed to Niels Bohr, the electron was perceived as a negative particle revolving around the nucleus in a three-dimensional racetrack or solar system. To a first approximation, the comparison between the atomic nucleus and its electron cloud with the sun and the solar system is impressive. Table 1.3 illustrates the remarkable correlation between the ratio of the diameter and mass of the atomic nucleus and its electron cloud and that of the sun and its planetary solar system, and serves to emphasize the concentration of mass in the atomic nucleus and the vast amount of space between the nucleus and the outer limits of the electron cloud.

The analogy falters, however, when we note that the elliptical orbits of the planets about the sun all lie more or less in one plane. For the atom, this is contrary to the concepts of modern quantum physics and the wave mechanical treatment of the electron developed by Schrödinger and his associates fairly early in this century. In the Schrödinger atom, the electron is assumed to be smeared out in an electron cloud or orbital of specific shape or dimension, denser in some regions than in others. Wave mechanics actually predicts the *probability of distribution of electron density* of different orbitals or energy levels ascribed to particular quantum states. It tells us that, at any given instant in time, a wave has, at a given position in space, a specific amplitude. This amplitude, at every position in space, at all times, completely describes the wave. Surfaces are drawn enclosing the amplitude of the wave function, which outline the electron cloud, or *atomic orbital.* Once again, these surfaces or orbitals indicate the restricted areas where, with high probability, the electron will be found. Thus the electron may be considered as a particle concentrated in

Table 1.3. *Comparison of diameters and mass percentage of an average atomic nucleus with its electron cloud and of the sun with the solar system*

Object	Diameter	Mass %
Atomic nucleus	10^{-12} cm	99.95
Electron cloud	10^{-8} cm	0.05
Sun	10^{9} m	99.85
Solar system	10^{13} m	0.15

the denser regions of the electron cloud or as a wave of negative charge smeared out over the confines of the electron cloud or orbital.

In some cases, the one treatment, and in some cases, the other, provides better insight for the explanation of various atomic phenomena. We may note that ψ is the orbital wave function in space and time, ψ^2 is the probability of finding the electron at a particular point, and $\psi^2 R^2$ is the probability of finding the electron in a shell of a given thickness and radius; $4\pi R^2 \psi^2$ describes the number of electrons in a spherical shell of a given thickness at a radius R from the nucleus.

To solve for the wave function and determine the probable location of the electron in its orbital, Schrödinger showed that four quantum numbers or parameters had to be determined: the *principal quantum number n*, which establishes the radial distance of the electron from the nucleus; the *azimuthal* or *angular momentum quantum number l*, which gives the angular momentum of the electron in orbit and helps define the shape of the orbital; the *magnetic quantum number* m_l, which determines the orientation of the angular momentum in space; and the *spin quantum number s*, which determines the sense of rotation of the electron or "spin up or spin down," in the magnetic field of the spinning electron.

The principal quantum number n has integer values and determines the position of the horizontal rows of the conventional periodic table (Fig. 1.1). The azimuthal quantum momentum number l may take the values 0, 1, 2, 3, 4, which are equivalent to the letter designations *s, p, d, f, g,* . . . . An *s* orbital has

Figure 1.2. Orbitals and quantum numbers.

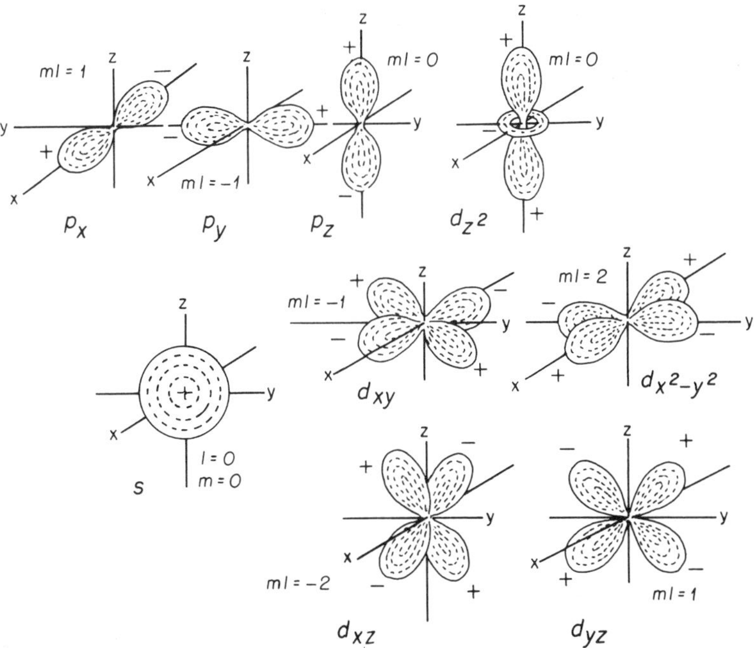

no (i.e., 0) angular momentum, because it is spherical; a *p* orbital is quasi-dumbbell-shaped and axially directed; a *d* orbital has various shapes, three possible shapes being double dumbbells of electron density, axially directed, a fourth, axially directed, and a fifth, axially directed but quasi-dumbbell-shaped with a central doughnut or node of electron density (Fig. 1.2).

The magnetic quantum number m_l restricts the orientation and number of each type of orbital, so there are one *s* orbital, three *p* orbitals, five *d* orbitals, and seven *f* orbitals in a particular quantum shell. Each of these orbitals may be occupied by a maximum of two electrons, provided that their spins are opposed. Each orbital level, *s*, *p*, *d*, and *f*, may be regarded as an energy level when given in combination with the principal quantum number (Fig. 1.3). Electrons introduced into an atom will occupy the lowest energy levels first in the sequence 1*s*, 2*s*, 2*p*, 3*s*, 3*p*, 4*s*, but then 3*d*, before 4*p*, and so on. No two electrons in the same atom can be described by the same wave function ψ unless their spins are opposed. Further, in filling orbitals, each quantum state must be occupied singly before it can be occupied doubly; thus where five *d* orbitals are being filled, we must place one electron in each before a second electron with opposite spin can be introduced. These so-called selection rules (Hund's rules)

Figure 1.3. Orbitals and energy levels.

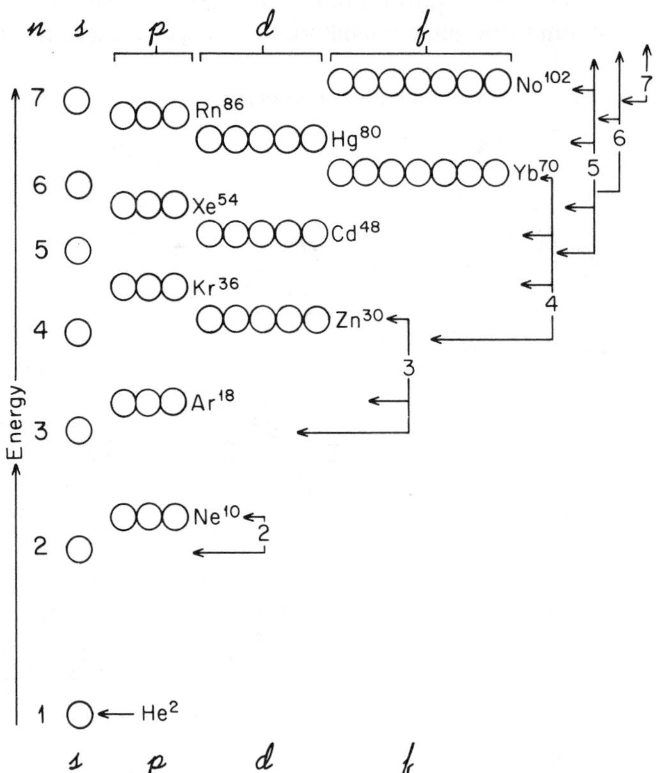

just alluded to express the tendency for electrons to remain as far apart as is possible. We emphasize that we are now considering the *unexcited, neutral,* or *ground state* atoms.

Periodic classification

We are now sufficiently well informed to analyze and use the periodic table or periodic classification of the elements, for which several formats exist. We will make use of both the standard row and column, or so-called long form, format (Fig. 1.1) as well as the less familiar but more elegant spiral format (see Fig. 1.4), introduced by the American Chemical Society. After some study, it will become apparent that the periodic classification attributed to Mendeleev (for whom element $Z = 101$ is named mendelevium, Md) organizes the elements according to their quantum numbers, and, in so doing, also organizes them into vertical groups according to similarities in their chemical properties. This leads to an arrangement with seven horizontal rows or periods representing the principal quantum numbers, $n = 1, \ldots, 7$, also designated as the *shells* K–Q. Within each row or shell, the elements are arranged in order of increasing atomic number, Z. The vertical columns are called *groups* or *subshells* and contain the orbitals s, p, d, and f, giving one, three, five, and seven sites, respectively, in which we can place two, six, ten, and fourteen electrons. Ignoring, for the moment, the f orbitals, we can divide our table into eighteen vertical *groups* by successively adding $2 + 6 + 10 = 18$ single and opposed spin pairs of electrons into the s, p, and d orbitals. The seven f orbitals, in which can be placed fourteen electrons building up fourteen different elements, are grouped separately at the bottom of the long form table. Before going further, we note that spectroscopists and chemists use different nomenclature for the periodic classification of the elements. From Table 1.4 it can be seen that there is overlap in the energies of elements of the third and fourth principal quantum levels, and the $4s$ level is lower in energy than the $3d$ in the neutral atom. Thus the krypton shell of the chemist, the fourth horizontal row of the table, contains or includes the third level of the spectroscopists' M shell. The important feature to note is that filling of the fourth principal quantum level that begins with potassium (K, $Z = 19$) proceeds in accordance with increasing energy levels in the sequence $4s$, $3d$, $4p$. Sequential filling of the eighteen electron sites, two electrons in s orbitals, ten in d orbitals, and six in p orbitals, will build the neutral atom structure of the eighteen elements following K (see Table 1.5).

By placing one electron in the lowest energy level, the spherical $1s$ orbital level, we form hydrogen, coded $_1$H $1s^1$, sequentially giving the atomic number, the element symbol, the orbital energy level, and the number of electrons contained. A second s electron with opposed spin placed in the $1s$ orbital completes the K (or He) shell, forming the gas helium $_2$He $1s^2$. The next electron is placed in the $2s$ orbital, a sphere of electron density with a radius larger than that of $1s$, and this gives $_3$Li $1s^2 2s^1$ for the element lithium. A fourth electron yields beryllium, $_4$Be $1s^2 2s^2$, completing the $2s$ subshell.

The fifth electron is placed in a dumbbell-shaped p orbital to form boron, $_5$B $1s^2 2s^2 2p^1$. The sixth electron must be placed in a different p orbital (single before double occupancy) in forming carbon, $_6$C $1s^2 2s^2 2p^2$. The seventh electron is placed in the third p orbital, giving nitrogen, $_7$N $1s^2 2s^2 2p^3$ with one electron in *each* of the three $2p$ orbitals. The eighth electron is, with spin opposed, placed in the first p orbital to form oxygen, $_8$O $1s^2 2s^2 2p^4$; the ninth electron forms fluorine $_9$F $1s^2 2s^2 2p^5$, and a tenth electron closes out the L (or Ne) shell with the rare gas neon, $_{10}$Ne $1s^2 2s^2 2p^6$.

For each periodic row after the first, $n = 1$, an eight-electron complement of s^2p^6 completes the energy level with the formation of a *rare* or *noble* gas. Several of the elements have been given special group names because their similarity in electronic structure leads to similarities in chemical behavior. In addition to the rare or noble gases of Group VIII (i.e., those with a full complement of s^2p^6 outer electrons), we have the *alkali* elements or *alkalies* of Group IA, which are those with an outer electron configuration of s^1; these are followed by the *alkaline earth* elements of Group IIA, which are those with an outer electron configuration of s^2 and the *halogens* of Group VIIB (alternatively labeled VIIA by other schemes), which consist of those elements having an outer electron configuration of s^2p^5. Those sequences of elements containing outer electrons in d orbitals, d^1–d^{10}, form horizontal sequences classed as *transition elements* or *transition metals.* Two other sequences, represented by those elements having outer electrons in f orbitals, f^1–f^{14}, are classed as *lanthanides* or *rare earth elements* and *actinides.* Here, the elegance of the arrangement of the spiral

Table 1.4. *Classification of electron shells*

Spectroscopists' nomenclature			Chemists' nomenclature		
Shell	Electron complement	No. of electrons	Shell	Electron complement	No. of electrons
K	$1s^2$	2	He	$1s^2$	2
L	$2s^2 2p^6$	8	Ne	$2s^2 2p^6$	8
M	$3s^2 3p^6 3d^{10}$	18	Ar	$3s^2 3p^6$	8
N	$4s^2 4p^6 4d^{10} 4f^{14}$	32	Kr	$3d^{10} 4s^2 4p^6$	18
O	$5s^2 5p^6 5d^{10} 5f^{14} (5g^{18})$[a]	50	Xe	$4d^{10} 5s^2 5p^6$	18
P	$6s^2 6p^6 6d^{10}$ . . .[b]	72	Rn	$4f^{14} 5d^{10} 6s^2 6p^6$	32
Q	$7s^2 7p^6$ . . .		Eka-Rn[c]	$5f^{14} 6d^{10} 7s^2 7p^6$	32

[a] Nine g orbitals with eighteen electrons are theoretically possible.
[b] Eleven h orbitals with twenty-two electrons are theoretically possible.
[c] The rare gas with atomic number 118 has not been discovered, but chemists have provisionally assigned it the shell name of Ekaradon.

periodic table (Fig. 1.4) beautifully illustrates these sequences and the manner in which transition-metal elements, lanthanides, and actinides interrupt the alkali s^1 to rare gas p^6 sequences shown as the quasi-circular portion of the table. The spiral table also serves to emphasize that the simplest element, H, is the nucleus out of which all of the other elements are formed. Thus, the fusion of four hydrogen atoms of atomic mass $1.0079 \times 4 = 4.0316$ yields He, with atomic mass 4.0026. The small difference in masses, 0.0290, represents the energy released, which is enormous when millions of atoms participate. This process is called *hydrogen burning,* which, in addition to forming He and, ultimately, the other elements, also supplies the energy to keep the sun both hot and luminous.

We return now to the filling sequence of the periodic table. After the completion of the second period with neon, Ne, with its complement of $1s^22s^22p^6$ electrons, a third period, the M shell, is opened with a third spherical $3s$ orbital located at a considerable increase in radial distance from the closed K and L shells. Here, successive s orbitals or clouds of high electron density are separated

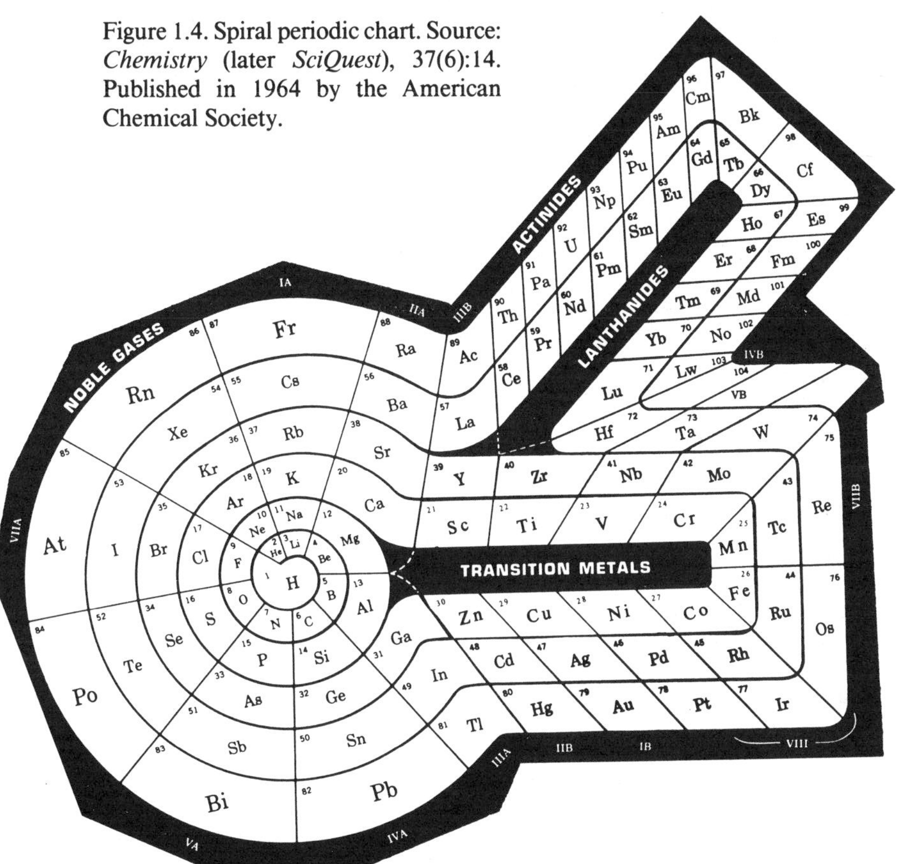

Figure 1.4. Spiral periodic chart. Source: *Chemistry* (later *SciQuest*), 37(6):14. Published in 1964 by the American Chemical Society.

from one another by *nodes* or regions in which the electron density is zero. Filling of the third periodic row leads to the rare gas argon, $_{18}Ar$ $1s^2 2s^2 2p^6 3s^2 3p^6$, which we can further abbreviate as $_{18}Ar(Ne)3s^2 3p^6$, stating that argon has a ten-electron core of neon inside the outermost $3s^2 3p^6$ shell. Our new, shortened code then gives, in parentheses, the symbol of the nearest previously filled rare gas core followed by the outer electrons that succeed it. We will now fill the entire fourth period of the table by adding outer electrons to the Ar core, going from elements $Z = 19$ to $Z = 36$, or K to Kr. The orbital filling sequence is represented schematically in Table 1.5.

Elements of the fifth period, rubidium, Rb, to xenon, Xe, are filled in the same sequence as the fourth row, but they now begin with a 5*s* orbital and closing with a 5*p* orbital of six electrons, with a ten-electron 4*d* sequence of transition elements lying between *s* and *p* orbitals. After xenon, a sixth period is opened with the formation of cesium, Cs, with $6s^1$, followed by barium, Ba, $6s^2$, and then lanthanum, La, $6s^2 5d^1$.

Now the 5*d* filling sequence is interrupted by the introduction of the seven 4*f* orbitals. When these are filled with their fourteen electrons, they give rise to the lanthanides or rare earth elements, from cerium, Ce, to lutetium, Lu. Note that although this completes the spectroscopists' N shell, it lies within the chemists' radon shell. The latter is completed with resumption of filling of the 6*p* orbitals with six electrons to close the shell with the rare gas radon, Rn. Now the pattern of the sixth period is begun anew. The seventh period opens with a 7*s* orbital

Table 1.5. *Sequence of filling of orbitals from $_{19}K$ to $_{36}Kr$*

Element		Z	4*s*	3*d*	4*p*	Electron code
Potassium	K	19	′			$(Ar)4s^1$
Calcium	Ca	20	″			$(Ar)4s^2$
Scandium	Sc	21	″	′		$(Ar)4s^2 3d^1$
Titanium	Ti	22	″	′ ′		$(Ar)4s^2 3d^2$
Vanadium	V	23	″	′ ′ ′		$(Ar)4s^2 3d^3$
Chromium	Cr	24	′	′ ′ ′ ′ ′		$(Ar)4s^1 3d^5$
Manganese	Mn	25	″	′ ′ ′ ′ ′		$(Ar)4s^2 3d^5$
Iron	Fe	26	″	″ ′ ′ ′ ′		$(Ar)4s^2 3d^6$
Cobalt	Co	27	″	″ ″ ′ ′ ′		$(Ar)4s^2 3d^7$
Nickel	Ni	28	″	″ ″ ″ ′ ′		$(Ar)4s^2 3d^8$
Copper	Cu	29	′	″ ″ ″ ″ ″		$(Ar)4s^1 3d^{10}$
Zinc	Zn	30	″	″ ″ ″ ″ ″		$(Ar)4s^2 3d^{10}$
Gallium	Ga	31	″	″ ″ ″ ″ ″	′	$(Ar)4s^2 3d^{10} 4p^1$
Germanium	Ge	32	″	″ ″ ″ ″ ″	′ ′	$(Ar)4s^2 3d^{10} 4p^2$
Arsenic	As	33	″	″ ″ ″ ″ ″	′ ′ ′	$(Ar)4s^2 3d^{10} 4p^3$
Selenium	Se	34	″	″ ″ ″ ″ ″	″ ′ ′	$(Ar)4s^2 3d^{10} 4p^4$
Bromine	Br	35	″	″ ″ ″ ″ ″	″ ″ ′	$(Ar)4s^2 3d^{10} 4p^5$
Krypton	Kr	36	″	″ ″ ″ ″ ″	″ ″ ″	$(Ar)4s^2 3d^{10} 4p^6$

and the formation of the rare elements francium, Fr, radium, Ra, and actinium, Ac, with orbitals containing $7s^1$, $7s^2$, and $6d^1$ electrons.

Filling of the $6d$ subshell is now interrupted by the appearance of $5f$ orbitals, and fourteen new elements, the actinides, are formed. These contain the naturally radioactive Th and U, as well as the rare radioactive elements from neptunium, Np, through lawrencium, Lw ($Z = 103$), largely spontaneous fission products. Elements 104–106 have been discovered by different investigators in the United States and the USSR, and the names proposed for these remain in dispute. Elements 107–118 have not yet been discovered, and the seventh period remains incomplete, although chemists have long ago proposed the name "Ekaradon" for this shell.

This completes our discussion of the nuclear and electronic structure of the neutral atoms and their arrangement in a periodic classification. Although periodic irregularities do exist, periodic classification of the elements constitutes a great creative scientific achievement. In Table 1.5, a periodic irregularity may be noted in the *absence* of a $3d^4$ configuration in going from $_{23}$V, $3d^3$ to $_{24}$Cr, $3d^5$. Note also the absence of $3d^9$ in going from $_{28}$Ni, $3d^8$ to $_{29}$Cu, $3d^{10}$. In both of these irregularities the Cr and Cu atoms carry one rather than two electrons in their $4s$ orbitals. Continued study of and regular reference to periodic classification offers great rewards.

Summary

Crystal chemistry evaluates the influence of the packing of atoms in minerals on their physical and chemical properties.

The atom is the smallest amount of matter that retains its identity in chemical combination. It consists of a nucleus of protons, neutrons, and mesons, which contains the greater part of its mass and of a cloud of electrons surrounding the nucleus, whose behavior is governed by four parameters or quantum numbers.

There are 106 chemical elements, based on the number of protons in the nucleus. Of these, elements 1–82 are relatively stable and 83–106 naturally radioactive.

The stable elements form 269 nuclides (atomic species with specified atomic mass). These may be isotopes (equal protons, differing neutrons), isotones (differing protons, equal neutrons), or isobars (differing protons and neutrons, equal atomic mass).

The 269 stable nuclides may have even numbers of protons and neutrons: these are the most abundant. Nuclides with odd numbers of protons and neutrons are rarest in nature. Nuclides with odd–even or even–odd numbers of protons and neutrons are intermediate in abundance.

Atomic nuclei having excess or lack of protons and neutrons are unstable and decay radioactively, thus producing heat.

The number of electrons must equal the number of protons in a neutral or ground state atom.

Electrons are distributed in shells with different energy levels and shapes of orbitals according to their quantum numbers.

Electrons govern the chemical behavior of atoms, and how they may be packed into minerals.

Nuclear structures determine the abundance of the elements.

Elements may be arranged into a periodic table according to their number of protons and the structure of their electron cloud. In this table, affinities between families of elements are evident.

Bibliography

Benfy, T., ed. (1964). Spiral periodic chart. In *Chemistry. Amer. Chem. Soc.*, Washington, D.C.

Cotton, F. A., and Wilkinson, G. (1966). *Advanced inorganic chemistry,* 2d ed. Interscience, New York.

Coulson, C. A. (1961). *Valence,* 2d ed. Oxford Univ. Press, London.

Dickerson, R. E., Gray, H. B., and Haight, G. P., Jr. (1974). *Chemical principles,* 2d ed. W. A. Benjamin, Menlo Park, Calif.

Gray, H. B. (1965). *Electrons and chemical bonding.* W. A. Benjamin, Menlo Park, Calif.

Greenwood, N. N., and Earnshaw, A. (1984). *Chemistry of the elements.* Pergamon Press, New York.

Krebs, H. (1968). *Grundzüge der anorganischen Kristallchemie.* Ferdinand Enke Verlag, Stuttgart.

Mason, Brian, and Moore, C. B. (1982). *Principles of geochemistry,* 4th ed. Wiley, New York.

Nassau, Kurt (1983). *The physics and chemistry of color.* New York.

Periodic Table of the Elements (1979). Sargent-Welch Scientific Co., Skokie, Ill.

Pauling, L. (1960). *The nature of the chemical bond,* 3d ed. Cornell Univ. Press, Ithaca, N.Y., pp. 28–63.

Rankama, K., and Sahama, Th. G. (1950). *Geochemistry.* Univ. of Chicago Press, Chicago, Ill.

Sanderson, R. T. (1960). *Chemical periodicity.* Reinhold, New York.

Sebera, D. K. (1964). *Electronic structure and chemical bonding.* Blaisdell, Waltham, Mass.

2 The excited atom: spectra and quanta, ionization potential, electronegativity, and chemical bonding

In the undisturbed, neutral, or ground state of an atom, all electrons will occupy the lowest possible energy levels in the sequence $n = 1, 2, 3, 4, \ldots$ and $l = 0, 1, 2, 3, \ldots$, equivalent to *s*, *p*, *d*, *f*, ... Electrons can change their energy level only by "jumping" from one quantum state to another under the influence of some external added energy source. When sufficient energy is thus absorbed, an electron may leave its *s* orbital and "leap" to a higher vacant *p* orbital site; this energy will be released when the excited electron (or a neighboring electron in the *p* site) drops down or returns to its original *s* orbital. This release of energy, often radiant visible energy, is detectable as emission spectral lines of specific wavelength and frequency. With a spectroscope or a spectrograph, it is possible to observe and measure the wavelength or frequency of each spectral line and to determine the specific quantum states or orbital levels involved in a given electronic transition. Thus, careful study of emission spectral lines has provided independent verification of the quantum theory of atomic and electronic structure, in that each line can be catalogued in terms of principal, azimuthal or angular, and spin quantum numbers. An omnipresent serendipity has also given the crystal chemist a very valuable method for the chemical analysis of minerals, rocks, and other inorganic materials.

Atoms can be excited by various means, including radiation or heat from an ordinary light bulb, an ultraviolet lamp, a candle, a Bunsen flame, or a higher energy electric carbon arc. Atoms can also be excited by the close approach of, and/or collision with, other atoms, which may give rise to chemical bonding of atoms and formation of crystals. Some atoms are very difficult to excite and are thus reluctant to emit spectral lines or combine with other atoms. This is particularly true of the rare or noble gas atoms with electronic configuration $1s^2$ and $2–7s^2p^6$, He, Ne, Ar, Kr, Xe, and Rn (see Fig. 1.1). Completed s^2p^6 orbitals endow the rare gases with high stability, allowing them to remain as gases. Conversely, elements of the alkali group, $n \ldots s^1$, Li, Na, K, Rb, and Cs, all have a single electron outside of a core of a rare gas element. This electron will be very loosely bound to the atomic nucleus, and can be excited to leave its orbital with only a small input of external energy. Such a loosely bound electron contained in an orbital lying outside of a rare gas core is called a *valence electron.* Moving to different groups in our periodic tables (see Fig. 1.1 or 1.4), we can observe an increase in the number of valence electrons with the vertical group number (e.g., K $4s^1$, Ca $4s^2$, Sc $4s^2 3d^1$, Ti $4s^2 3d^2$, V $4s^2 3d^3$, . . .). On reaching the element zinc, Zn, $4s^2 3d^{10}$, the completed $3d$ subshell binds these electrons securely so that Zn effectively has only the two s electrons that are valence electrons. Spectroscopy has shown that electrons in s orbitals are easy to excite, and yield spectral lines of high sensitivity, permitting their detection even when present in very minute quantities. Elements with p orbital valence electrons are very difficult to excite and have low spectral sensitivity, and those with d and f orbital valence electrons are of intermediate excitation and detection level.

When common salt, NaCl (halite), is held in a candle or a Bunsen flame, the small thermal energy source excites the $3s^1$ electron to jump or ascend to an

Table 2.1. *Correlation of the periodic variation of spectral line multiplicity with maximum valence of atoms*

4s, 3d electrons	s^1	s^2	s^2d^1	s^2d^2	s^2d^3
Princ. quantum #4	K	Ca	Sc	Ti	V
Valence electrons	1	2	3	4	5
Electron spin, s	$\pm\frac{1}{2}$	± 1	± 2	± 3	± 4
Sum of spins, S	$\frac{1}{2}$	0,1	$\frac{1}{2}, 1\frac{1}{2}$	0, 1, 2	$\frac{1}{2}, 1\frac{1}{2}, 2\frac{1}{2}$
Multiplicity M = 2S + 1	2	1,3	2,4	1, 3, 5	2, 4, 6
Spectral region	Red	Green	Green	Green	Blue
Maximum multiplicity pattern	||	|||	||||	|||||	||||||
Wavelength in Å	7699 7664	5270 5265 5261	5087 5085 5083 5081	5014 5007 4999 4991 4981	4481 4876 4865 4851 4832 4827

empty $3p$ orbital: when this electron falls back to home base, a pair of spectral lines is emitted, constituting the sodium, Na, doublet at wavelengths $\lambda =$ 5890 Å and 5896 Å (Jaffe, 1949). Although the separation is only 6 Å (1 Å $= 10^{-8}$ cm), this line pair is easily resolved with the simplest of spectroscopes. The wavelengths cited lie in the yellow region of the visible spectrum of electromagnetic radiation and color the flame a bright yellow. Because many Na atoms will be excited, some may be induced to leap to levels higher than $3p$ and may return via a longer route in a manner that will cause additional Na doublets to appear in different parts of the spectral region: the latter will be less sensitive to spectral detection.

The way in which spectra have been used to verify quantum theory is best illustrated by the multiplicity M of spectral lines, caused by the clockwise and counterclockwise spin of electrons. Spectroscopists discovered that the simple formula relating multiplicity to electron spin S is $2S + 1 = M$, where S is the sum of the possible spins, s, and M is the multiplicity of lines emitted. Spin quantum numbers may have a value of $+\frac{1}{2}$ or $-\frac{1}{2}$ (or $\pm\frac{1}{2}$) per electron. Thus, an alkali element with one valence electron, s^1, can have the electron spinning either way, giving $s = \pm\frac{1}{2}$. Accordingly, S can only equal $\frac{1}{2}$ and $2S + 1 = 2$, indicating that alkali elements should emit their spectral lines in pairs or doublets, even though only one electron is involved per atom. If we move across a row of the periodic table, using principal quantum number 4 as an example, we see that spectral multiplicity, $2S + 1 = M$, will consistently be one number greater than the number of valence electrons for the elements K, Ca, Sc, Ti, and V (Table 2.1).

It has been shown that the multiplicity term $2S + 1 = M$ leads to an emission spectral pattern of singlets and triplets for the divalent calcium, Ca, atom (Table 2.1 and Fig. 2.1). However, at the short-wavelength end of the visible region, we note the appearance of a doublet pattern (3968.5 Å and 3933.7 Å) emitted by Ca. Inasmuch as doublet spectral patterns or multiplicities imply the presence

Figure 2.1. Atomic (Ca), ionic (Ca$^+$), and molecular (CaO) spectra of calcium.

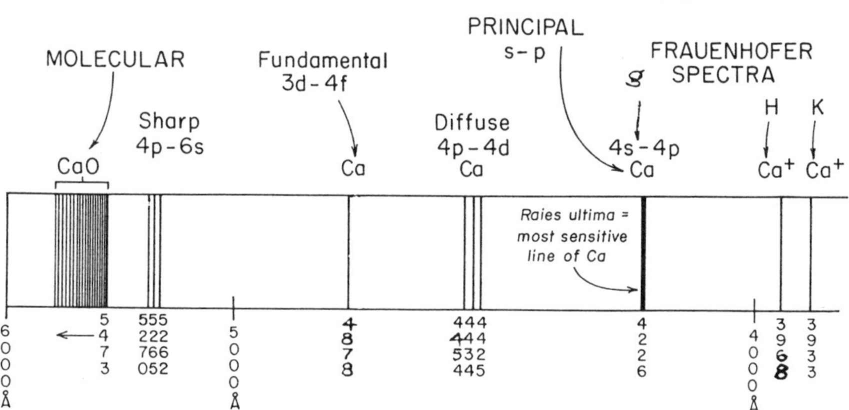

of an atom with one valence electron, $S = \pm\frac{1}{2}$, one must conclude that such a pattern is emitted not by the neutral Ca atom but by the singly ionized Ca^+ ion.

When enough energy is supplied (e.g., higher temperature or higher voltage), Ca will give up one of its $4s^2$ valence electrons to become Ca^+, with a $4s^1$ electron structure, and this univalent ion then behaves like a univalent alkali element and emits doublet spectra. Because of the higher energy absorbed, the spectral lines are shifted to shorter wavelengths, causing the Ca^+ doublet to appear in the violet region of the visible spectrum. The appearance of the Ca^+ ion doublet in stellar spectra has been used to measure stellar temperatures. Low temperature red stars contain only Ca atom spectra (singlets and triplets) and CaO band and molecular spectra; higher temperature stars, such as the sun, show weak lines of the Ca^+ ion doublet; and high temperature blue stars emit intense doublet spectra of Ca^+.

The energy required to remove an s electron from the neutral Ca atom to infinity is termed the *ionization potential* I (Table 2.2). With still higher energy input, additional electrons may be removed at energy increments labeled I_2, I_3, and so on, referred to as the first, second, third, and so on, ionization potentials for a given chemical element. It is generally the first ionization potential, I_1, that is most important in determining whether an element will ionize. The ease

Table 2.2. *Some crystal chemical concepts*

Ionization potential (I or I_P): Energy or potential required to move an electron from its normal quantum level to infinity. Energy is acquired as $X \rightarrow X^+ + e^-$. Also a measure of electronegativity, power of a free cation to attract anions, nuclear screening, and polarizing power. I_P is usually given in electron volts (ev): 1 ev = 23 kcal/mole.

Electron affinity (E or E_A): Energy released in adding an electron from ∞ to a neutral atom, $X + e^- \rightarrow X^-$. Usually expressed in ev or kcal.

Electronegativity (χ): Power of an atom in a molecule to attract electrons. $\chi = (I_P + E_A)/125$ in kcal/mole. I_P refers to the first ionization potential. $\Delta\chi$ of two atoms in a molecule is directly proportional to the percentage of ionic bonding and inversely proportional to the percentage of covalent bonding. A pure ionic bond is impossible: a pure covalent bond is possible. Why? (See Chapter 4.)

Polarizing power (ze/r^2): Charge divided by radius squared is equal to electric field strength. Small, highly charged cations can readily deform or polarize large anions.

Polarizability: Dipole moment induced in an ion by a unit electrical field. A measure of the deformability of an ion. Large ions are readily deformed; most cations are not.

Ionic potential (ϕ) = (ze/r): Charge divided by radius of ion. A measure of electrostatic charge on the surface of an ion. High potential (ϕ) results in a tendency to form complex anions.

Isoelectronic: Atoms or ions having the *same* electronic or extranuclear structure are said to form an isoelectronic series. Cl^-, neutral Ar, K^+, Ca^{2+}, Sc^{3+}, Ti^{4+}, V^{5+}, Cr^{6+}, and Mn^{7+} are isoelectronic because all have the structure $1s^2$-$2s^2$-$2p^6$-$3s^2$-$3p^6$, that of argon in the ground state.

or difficulty with which an element can be made to give up a valence electron is, once again, a periodic function.

Here, a few observations are important: an element having its valence electrons in s orbitals, which have low ionization potential, will ionize readily. Such elements include the alkalies and alkaline earth elements of Groups IA (s^1) and IIA (s^2) of each principal quantum row ($n = 1, \ldots, 7$). Thus, the alkali elements, Li, Na, K, Rb, and Cs, will have ionization potentials lower than those of any other element, because all have a single valence electron in an s orbital. Accordingly, these alkali elements will ionize readily to form *ionic bonds* in crystal growth.

Alkaline earth elements of Group IIA, Be, Mg, Ca, Sr, and Ba, have valence electrons in s orbitals and slightly higher ionization potentials; they will each lose both s^2 electrons to form divalent or $2+$ ions, but will ionize less readily than the alkali elements. The reason for the higher energy of ionization is that the two electrons with opposed spin in s^2 orbitals are paired and fill the s^2 subshell. Removal of an electron from a completed shell or subshell always requires higher energy, indicating that completed electron shells represent a condition of greater stability than do uncompleted shells.

As we move across the periodic table to higher group numbers having s^2d^n valence electrons, the ionization potentials increase, informing us that d electrons are more firmly bound than are s electrons. When a d^{10} shell is completed, as for example at atomic number 30, Zn has electronic structure $4s^23d^{10}$, and at this point the ionization potential takes a rapid jump to a higher value caused by the completion of the $3d$ or M shell. Data in Table 2.3 indicate that for Zn the values for I_1, I_2, and I_3 are, respectively, 9.392, 17.960, and 39.701 ev (electron volts). The high value for I_3 indicates why Zn never achieves a valency higher than 2 in the excited atom. Here, the reader might assume that two d electrons leave first, because they arrived last, but it is the two s electrons that are removed, because the completed $3d^{10}$ shell will now have a higher orbital energy. After this, any additional electrons must be placed in the $4p$ orbitals, first singly and then doubly, and I_1 potentials will increase more or less regularly from gallium, Ga, $4p^1$ to krypton, Kr, $4p^6$. Electrons in p orbitals are the most difficult of all to remove, and the rare gas Kr, $4p^6$, has potential $I_1 = 13.996$ ev. All of the rare or noble gases (He, $1s^2$; Ne, $2s^2p^6$; Ar, $3s^23p^6$; Kr, $4s^24p^6$; Xe, $5s^25p^6$; and Rn, $6s^26p^6$) have relatively high I_1 values and tend to avoid transferring or sharing their valence electrons with other elements; they remain as gases.

In summary, we see that ionization energies increase and spectral detection sensitivities decrease in the periodic group sequence s, d or f, and p, that such energies increase rapidly and suddenly where shells or subshells become closed. Careful study of Table 2.3 will show how ionization energy requirements restrict the number of valence electrons used by the various elements in chemical bonding.

Leaving ionization potentials for the moment, and reversing our trek across the periodic table, we note that Kr, $4p^6$, has no place to add another electron, whereas bromine, Br, $4p^5$, would readily accept a sixth p electron to become *isoelectronic* with Kr and, in so doing, increase its stability. Selenium, Se, $4p^4$, has room for a pair of p electrons.

Just as energy is absorbed when atoms lose an electron, energy is released when atoms gain an electron. This energy is the *electron affinity* (E_A, see Table

Table 2.3. *First five ionization potentials of the elements (in electron volts)*

Z	Element	I	II	III	IV	V
1.	H	13·595	—	—	—	—
2.	He	24·581	54·403	—	—	—
3.	Li	5·390	75·6193	122·420	—	—
4.	Be	9·320	18·206	153·850	217·657	—
5.	B	8·296	(25·149)	37·920	259·298	340·127
6.	C	11·2564	24·376	47·871	64·476	391·986
7.	N	14·529	29·593	47·426	77·450	97·863
8.	O	13·614	35·108	54·886	77·394	113·873
9.	F	17·418	35·012	{62·689 62·646	87·139	114·214
10.	Ne	21·559	40·955	(63·450)	97·024	126·260
11.	Na	5·138	47·290	71·628	98·880	138·367
12.	Mg	7·644	15·031	80·117	109·294	141·231
13.	Al	5·984	18·823	28·441	119·957	153·772
14.	Si	8·149	16·339	33·459	45·130	166·725
15.	P	10·484	19·720	30·156	51·354	65·007
16.	S	10·357	23·345	34·799	47·292	(72·474)
17.	Cl	13·014	23·798	39·649	53·450	67·801
18.	A	15·755	27·619	40·705	(59·793)	(75·002)
19.	K	4·339	31·620	45·793	(60·897)	(82·6)
20.	Ca	6·111	11·868	50·881	(67·181)	84·385
21.	Sc	6·5384	12·797	24·753	73·911	(91·847)
22.	Ti	6·818	13·573	27·467	43·236	99·842
23.	V	6·743	14·651	29·314	(48·464)	65·198
24.	Cr	6·764	16·493	30·950	(48·580)	(73·093)
25.	Mn	7·432	15·636	33·690	(*53*)	(76·006)
26.	Fe	7·868	16·178	30·643	(*56*)	(*79*)
27.	Co	7·875	17·052	33·491	(*53*)	(*82*)
28.	Ni	7·633	18·147	35·165	(*56*)	(*79*)
29.	Cu	7·724	20·286	37·079	(*59*)	(*83*)
30.	Zn	9·391	17·959	39·701	(*62*)	(*86*)
31.	Ga	5·997	20·509	30·702	(64·157)	(*90*)
32.	Ge	7·889	15·930	34·215	45·700	(93·434)
33.	As	9·813	18·628	28·344	50·122	62·612
34.	Se	9·750	21·512	31·979	{(42·898) (47·3)	73·103
35.	Br	11·843	21·799	35·887	(50·186)	(59·7)
36.	Kr	13·996	24·565	36·940	(*52*)	(*65 66*)
37.	Rb	4·176	{27·499 27·560	39·664	(*53*)	(*71*)
38.	Sr	5·692	11·027	(43·6)	(57·017)	(*72*)
39.	Y	6·377	12·233	20·514	(61·8)	(76·849)
40.	Zr	6·835	13·126	22·980	34·330	(*82 83*)
41.	Nb	6·881	14·316	25·038	38·251	50·534
42.	Mo	7·097	16·151	27·133	(46·38)	(61·152)

* After LAKATOS, BOHUS and MEDGYESI (1959) except those given in italics which are either from LANDOLT-BÖRNSTEIN (1950) or FINKELNBURG and HUMBACH (1955). LAKATOS *et al.* list numerous sources (up to mid-1958) for their data. Values in brackets are uncertain or have been obtained through extrapolation and not direct measurement.

2.2) and is a much smaller quantity than the ionization energy. The power, or the tendency, of an atom in a molecule to gain or attract electrons is measured by its *electronegativity* χ, most commonly expressed in arbitrary units (Table 2.4). Electronegativity values also increase in the periodic sequence *s*, *d* or *f*, *p*. Thus a high ionization potential will be matched by a high electronegativity. Stating this differently, those elements with a low electronegativity will readily

Table 2.3. *(cont.)*

Z	Element	I	II	III	IV	V
43.	Tc	7·276	15·258	29·537	*(43)*	*(59)*
44.	Ru	7·364	16·758	28·459	*(47)*	*(63)*
45.	Rh	7·461	18·072	31·049	*(46)*	*(67)*
46.	Pd	8·334	19·423	32·921	*(49)*	*(66)*
47.	Ag	7·574	21·481	34·818 *(36)*	*(52)*	*(70)*
48.	Cd	8·991	16·904	37·766	*(55)*	*(73)*
49.	In	5·785	18·865	(28·025)	(54·4 (57·9)	—
50.	Sn	7·342	14·628	30·494	40·724*	72·263
51.	Sb	8·639	16·526 18·593	25·317	44·146	55·691)
52.	Te	9·007	18·593	30·616	37·816	60·270
53.	I	10·454	19·094	*(31)*	*(42)*	*(52)*
54.	Xe	12·127	21·204	32·114	*(45)*	*(57)*
55.	Cs	3·893	25·070 23·37	(33·97)	*(46)*	(62)
56.	Ba	5·210	10·001	*(37)*	*(49)*	*(62)*
57.	La	5·614	11·433	19·166	*(52)*	*(66)*
58.	Ce	6·54 6·91	12·31	19·870	36·714	*(70)*
59.	Pr	(5·76)	—	—	—	—
60.	Nd	(6·31)	—	—	—	—
61.	Pm	—	—	—	—	—
62.	Sm	5·6	11·4	—	(36·5)	—
63.	Eu	5·67	11·24	—	—	—
64.	Gd	6·16	(12)	—	—	—
65.	Tb	(6·74)	—	—	—	—
66.	Dy	(6·82)	—	—	—	—
67.	Ho	—	—	—	—	—
68.	Er	—	—	—	—	—
69.	Tm	—	—	—	—	—
70.	Yb	6·22	12·10	—	—	—
71.	Lu	6·15	14·7	—	—	—
72.	Hf	7·003	(14·874)	*(21)*	*(31)*	—
73.	Ta	7·883	(16·2)	*(22)*	*(33)*	*(45)*
74.	W	7·982	(17·7)	*(24)*	*(35)*	*(48)*
75.	Re	7·875	(16·6)	—	—	—
76.	Os	8·732	(17)	—	—	
77.	Ir	9·1	—	—	—	
78.	Pt	8·962	18·558	*(29)*	*(41)*	*(55)*
79.	Au	9·223	20·045 20·452	*(30)*	*(44)*	*(58)*
80.	Hg	10·435	18·751	34·210	*(46)*	*(61)*
81.	Tl	6·106	20·423	29·822	50·708	*(64)*
82.	Pb	7·415	15·028	31·929	42·310	68·792
83.	Bi	7·287	16·684	25·556	45·304	56·0
84.	Po	8·426	—	—	—	—
85.	At	—	—	—	—	—
86.	Rn	10·746	—	—	—	—

* Various values of the fourth ionization potential of Sn, ranging from 39·4 to 46·4 V, have been given in the literature. The value given here is clcse to that 41 V recommended by AHRENS (1956).

transfer an electron to a bonding partner possessed of a high electronegativity; the latter will acquire the electron and its negative charge.

A positively charged atom is called a *cation;* a negatively charged atom, an *anion.* Their union constitutes the *ionic bond.* From Tables 2.3 and 2.4, it follows that those elements having a maximum difference in electronegativity $\Delta\chi$ or in ionization potential ΔI will have the strongest tendency to form ionic bonds: for example, for Na^+, $\chi = 0.9$; for F^-, $\chi = 4.0$ and $\Delta\chi = 3.1$.

Linus Pauling, the great American chemist who developed a still widely used table of electronegativities, or the electronegativity scale, also developed formulae for estimating the percentage of ionic bonding associated with different values of $\Delta\chi$. Thus, for NaF, the value $\Delta\chi = 3.1$ indicates that the compound has 91% ionic bonding. This leads to the concept that bonding of atoms in crystals is often only partly ionic and partly something else. Here, we may briefly categorize the principal chemical bond types:

the ionic bond, involving *electron transfer*
the covalent bond, involving *electron sharing*
the metallic bond, involving *electron mobility*
the hydrogen bond, involving *electron orientation*
the van der Waals bond, involving *electron synchronization*

Actually, Pauling developed his concept of electronegativity by determining that the energy holding bonded atoms together in simple chemical compounds is always equal to or greater than the energy of a covalent bond between the constituent atoms. The excess energy is due to the ionic energy of the bond. The line of reasoning is as follows. The simple compound H_2 is fully covalent, and a spherical *s* orbital of each atom is overlapped so that one electron from each H atom, $1s^1$, is shared (Fig. 2.2). In this *s–s* overlap, each atom acquires the electronic structure of helium, He, $1s^2$. The bond dissociation energy *D* equals 104.2 kcal/mole. In Cl_2, also fully covalent, a dumbbell-shaped orbital of each chlorine, Cl, atom, $3s^2 3p^5$, is overlapped with that of the other so that now the sharing of one electron each from both Cl atoms is diagrammed as in Figure 2.2. In this *p–p* overlap, each Cl atom acquires the electronic structure of argon, Ar, $3s^2 3p^6$. The bond dissociation energy *D* equals 58.0 kcal/mole. Thus, *D* for the compound HCl should be equal to

$$\tfrac{1}{2}H_2 + \tfrac{1}{2}Cl_2 = 162.2/2 = 81.1 \text{ kcal/mole}$$

Figure 2.2. Orbital overlap in H_2 and Cl_2.

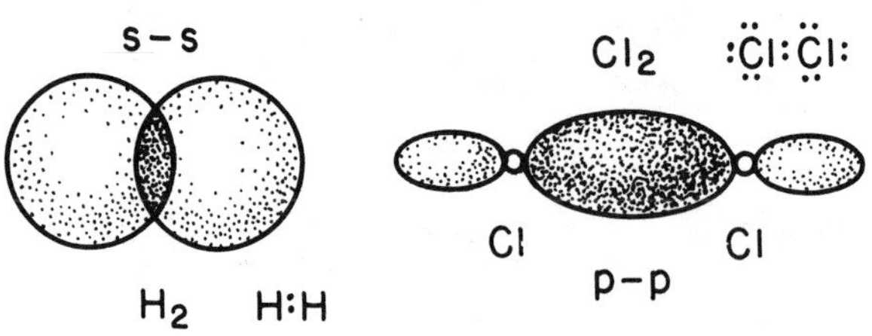

Table 2.4. *Electronegativities of the elements*

Element	PAULING and others as listed by FYFE (1951)	LITTLE and JONES (1960)	Element	PAULING and others as listed by FYFE (1951)	LITTLE and JONES (1960)
H	2·1	2·1	Sb	III 1·8	1·82
He	—	—		V 2·1	
Li	1·0	0·97	Te	2·1	2·01
Be	1·5	1·47	I	2·6	2·21
B	2·0	2·01	Xe	—	—
C	2·5	2·50	Cs	0·7	0·86
N	3·0	3·07	Ba	0·85	0·97
O	3·5	3·50	La	0·85	1·08
F	4·0	4·10	Ce	1·05	1·08
Ne	—	—	Pr	1·1	1·07
Na	0·9	1·01	Nd	—	1·07
Mg	1·2	1·23	Pm	—	1·07
Al	1·5	1·47	Sm	—	1·07
Si	1·8	1·74	Eu	—	1·01
P	2·1	2·06	Gd	—	1·11
S	2·5	2·44	Tb	—	1·10
Cl	3·0	2·83	Dy	—	1·10
A	—	—	Ho	—	1·10
K	0·8	0·91	Er	—	1·11
Ca	1·0	1·04	Tm	—	1·11
Sc	1·3	1·20	Yb	—	1·06
Ti	IV 1·6	1·32	Lu	—	1·14
	III 1·35		Hf	(1·3)	1·23
V	IV 1·6	1·45	Ta	(1·4)	1·33
	V 1·8		W	IV (1·6)	1·40
	II 1·5			VI 2·1	
Cr	III 1·6	1·56	Re	—	1·46
	VI (2·1)		Os	(2·1)	1·52
	II 1·4		Ir	2·1	1·55
Mn	III (1·5)	1·60	Pt	2·1	1·44
	VII (2·3)		Au	2·3	1·42
Fe	II 1·65	1·64	Hg	I 1·8	1·44
	III 1·8			II 1·9	
Co	1·7	1·70	Tl	I 1·5	1·44
Ni	1·7	1·75		III 1·9	
Cu	I 1·8	1·75	Pb	II 1·6	1·55
	II 2·0			IV 1·8	
Zn	1·5	1·66	Bi	1·8	1·67
Ga	1·6	1·82	Po	2·0	1·76
Ge	1·7	2·02	At	2·4	1·90
As	2·0	2·20	Rn	—	—
Se	2·3	2·48	Fr	0·7	0·86
Br	2·8	2·74	Ra	0·8	0·97
Kr	—	—	Ac	(1·0)	1·00
Rb	0·8	0·89	Th	1·1	1·11
Sr	1·0	0·99	Pa	(1·4)	1·14
Y	1·2	1·11	U	1·3	1·22
Zr	1·4	1·22	Np	—	1·22
Nb	(1·6)	1·23	Pu	—	1·22
Mo	IV (1·6)	1·30	Am	—	(1·2)
	V (2·1)		Cm	—	(1·2)
Tc	—	1·36	Bk	—	(1·2)
Ru	2·05	1·42	Cf	—	(1·2)
Rh	2·1	1·45	Es	—	(1·2)
Pd	2·0	1·35	Fm	—	(1·2)
Ag	1·8	1·42	Md	—	(1·2)
Cd	1·5	1·46	No ?	—	(1·2)
In	1·6	1·49			
Sn	II 1·65	1·72			
	IV 1·8				

Source: Reprinted with permission from L. H. Ahrens, *Physics and chemistry of the earth,* © 1964, Pergamon Books, Ltd.

The determined value for $D_{HCl} = 103.2$, and

$$103.2 - 81.1 = 22.1 \text{ kcal/mole}$$

The value $22.1 = \Delta$, the extra ionic contribution or *ionic resonance energy*, and the formula

$$\Delta = D(A - B) - \tfrac{1}{2}D(A - A) + 1/D(B + B)$$

is applicable.

Using bond energies from dielectric moments or measurements, Pauling quantified this excess ionic resonance energy Δ by assigning values of electronegativity χ for each chemical element such that $\Delta\chi$ measures the ionic contribution to bonding between two elements. After assigning a value of $\chi = 2.05$ for H, Pauling derived values of $\chi = 3.0$ for Cl, $\chi = 3.5$ for O, and $\chi = 4.0$ for F, from simple H compounds. He then derived values for χ for other elements by a substitution scheme.

Other investigators have obtained electronegativity values which vary slightly from those of Pauling, and some of these give lower percentages of ionic bonding than those calculated by Pauling, who cautioned that his electronegativity scale was only semiquantitative. Nonetheless, very similar electronegativity values were obtained by Mulliken, who averaged the first ionization potential and the electron affinity of an atom according to the formula $I_1 + E_A/125 = \chi$, when the values for electron affinity and ionization potential are expressed in kcal/mole. The element fluorine, F, has an ionization potential of 17.42 ev; and 1 ev = 23.06 kcal/mole, giving 401.71 kcal/mole for F. The electron affinity value for F is 83.5 kcal/mole. Their sum divided by 125 yields 3.88 for the electronegativity of F, essentially the same value as that obtained by Pauling, who rounded it to 4.0.

It is significant that the following different equations or methods used to estimate the percentage of ionic bonding all give similar, although not identical, results.

For HCl, with $\chi_H = 2.1$ and $\chi_{Cl} = 3.0$, we obtain Pauling's relation

$$\text{Percent ionic bonding} = \Delta/D$$

where $\Delta = 23(\chi_A - \chi_B)^2 = 18.63$ and D = bond dissociation energy = 103.2 = 18.1%, and Hannay and Smith's relation

$$\text{Percent ionic bonding} = 16(\Delta\chi) + 3.5(\Delta\chi)^2 = 17.2\%$$

From dielectric properties and dipole moments, μ:

$$\mu_{obs}/\mu_{calc} \quad \text{for } H^+ - Cl^- = 17.0\%$$

$$(\mu_{obs} = 1.03 \text{ debyes}/\mu_{calc} = 6.07 \text{ debyes})$$

From the foregoing discussion, it should be evident that the chemical bonds joining atoms into crystal structures need not be solely covalent, ionic, or metallic, but are a mixture of these types. In addition to Pauling's criterion of differences in electronegativity, we have other means of evaluating the nature of the chemical bond in crystals.

Measurement of other physical and chemical properties of crystals also provide equally, if not more, valuable criteria for evaluation of chemical bond type. These properties include luster, reflectivity, magnetism, electrical and thermal conductivities, hardness, color, refractivity and index of refraction, coordination number, sums of radii of bonded atoms, solubility of crystals in nonpolar and polar solvents, and the distortion of atomic orbitals in atoms or ions in response to the location of charged atoms or ions in different crystal structure sites.

Thus, although Pauling's electronegativity differences provide the primary and best theoretical method of predicting chemical bond types, measurement of the physical and chemical properties cited provides more direct evidence to confirm those inferences.

Correlation of the physical and chemical properties of minerals with chemical bond type will be developed further in the ensuing chapters.

Summary

When an atom is excited by the addition of external energy, the energy absorbed causes electrons to leap to a higher energy state; when they fall back, they emit the added energy as spectral lines characteristic of the particular element.

Elements with the most loosely bound valence electrons (e.g., *s* electrons) are easiest to excite.

The multiplicity M of spectral lines is related to the spin quantum number S of the electron ($M = 2S + 1$).

Ionization potential is the energy required to remove an electron from a neutral atom to infinity.

Elements with low ionization potential are most apt to form ionic bonds.

Elements that attract electrons have electron affinity, expressed as electronegativity.

Those elements with low ionization potential also have low electronegativity.

Elements with low electronegativity combine with those having high electronegativity to form ionic bonds most readily.

Although bonds between elements may be ionic, covalent, metallic, or van der Waals, they are most commonly mixtures of more than one bond type.

Percent ionic bonding can be estimated from differences in electronegativity. In addition, bond character can be inferred from physical properties such as luster, hardness, transparency, and electrical conductivity.

Bibliography

Ahrens, L. H. (1953). The use of ionization potentials, Part 2. Anion affinity and geochemistry. *Geochim. et Cosmochim. Acta,* 3: 1–29.

(1964). The significance of the chemical bond for controlling the geochemical distribution of the elements. Part 1, Appendix A. *Physics and chemistry of the earth,* Pergamon, New York, vol. 5, pp. 3–54.

and Taylor, S. R. (1961). *Spectrochemical analysis,* 2d ed. Pergamon Press, New York.

Aller, L. H. (1961). *The abundance of the elements.* Interscience, New York.

Fyfe, W. S. (1964). *Geochemistry of solids.* McGraw-Hill, New York, chaps. 6, 7.

Hannay, N. B., and Smyth, C. P. (1946). The dipole moment of hydrogen fluoride and the ionic character of bonds. *J. Am. Chem. Soc.,* 68: 171–3.

Harrison, G. R. (1939). *M.I.T. wavelength tables.* Wiley, New York.

Herzberg, G. (1944). *Atomic spectra and atomic structure.* Dover, New York.

(1950). *Molecular spectra and molecular structure,* 2d ed. Van Nostrand, New York.

Jaffe, H. W. (1947). Reexamination of sphene (titanite). *Am. Mineral.,* 32: 637.

(1949). Visual arc spectroscopic detection of halogens, rare earths and other elements by use of molecular spectra. *Am. Mineral.,* 34: 667.

Lakatos, B., Bohus, J., and Medgyesi, Gy. (1959). A new way for the calculation of the degree of polarity of chemical bonds. *Acta Chim. Hungar.,* 20: 1; 21: 293.

Little, E. J., and Jones, M. M. (1960). A complete table of electronegativities. *J. Chem. Ed.,* 37: 231.

Merrill, P. W. (1963). *Space chemistry.* Univ. of Michigan Press, Ann Arbor, Mich.

Mulliken, R. S. (1934). A new electronegativity scale; together with data on valence states and on valence ionization potentials and electron affinities. *Jr. Chem. Phys.,* 2: 782.

Pauling, L. (1960). *The nature of the chemical bond.* 3d ed. Cornell Univ. Press, Ithaca, New York, chap. 3.

Pearse, R. W. B., and Gaydon, A. G. (1950). *The identification of molecular spectra,* 2d ed. Wiley, New York.

Peterson, M. J., and Jaffe, H. W. (1953). Visual arc spectroscopic analysis. U.S. Dept. of the Interior, Bur. Mines, Bull. 524, Washington, D.C.

Peterson, M. J., Kauffman, A., and Jaffe, H. W. (1947). The spectroscope in determinative mineralogy. *Am. Mineral.,* 32: 322.

Slavin, M. (1940). Prism versus grating for spectrochemical analysis. Proc. 7th Summer Conf. on Spectroscopy, p. 51.

3 The crystal chemistry of the covalent bond

This chapter considers a variety of crystal structures in which the atoms are bonded by the sharing of one or more electrons. Wave functions ψ of atom pairs are added in such a way that electron density charge clouds overlap to produce a new charge cloud with even greater electron density than the sum of its parts.

Pure or extreme covalent bonds can be formed only between atoms with equal electronegativity, such that $\Delta\chi_{A-B} = 0$, equal numbers of valence electrons, and equal coordination numbers. Obviously, these conditions are fulfilled only in the combination of two atoms of the same species.

s–s, p–p, and *s–p* bonds

In the covalent bonding of two hydrogen atoms in H_2, two spherical s orbitals overlap to yield H_2 (Fig. 3.1A).

Using dots for electrons, we may use *Lewis notation* to indicate that the two H atoms are joined by a *single covalent bond* in which one electron from each atom is shared. For H_2, $H\cdot + \cdot H = H:H$, and thus each H atom is surrounded by two electrons rather than one, conferring on each H atom the outer or extranuclear electronic structure of helium, He:, the nearest rare gas. In both covalent and ionic bonding, linked atoms tend to attain the electronic structure of the nearest rare gas and, in so doing, achieve the high stability associated with s^2p^6 completed shells of rare gas atoms. This is known as the *octet rule.*

Similarly, when chlorine, Cl, bonds to Cl in Cl_2, two dumbbell-shaped p orbitals overlap to attain the electronic structure of the nearest rare gas, in this case argon, Ar (see Fig. 3.1B).

In Lewis notation for Cl_2,

$$:\ddot{\underset{\cdot\cdot}{Cl}}\cdot + \cdot\ddot{\underset{\cdot\cdot}{Cl}}: = :\ddot{\underset{\cdot\cdot}{Cl}}:\ddot{\underset{\cdot\cdot}{Cl}}:$$

and each Cl atom now has eight rather than seven electrons in its outermost orbital, which is the same as, or *isoelectronic* with, Ar, ${}_{18}Ar = ({}_{10}Ne) + 3s^2 3p^6$, the nearest rare gas.

It has already been shown that the bonding of one atom of H with one of Cl to give HCl has an additional ionic component Δ, and the bond is thus no longer entirely covalent. Nevertheless, the small electronegativity difference $\Delta\chi_{H-Cl} = 2.1 - 3.0 = 0.9$ indicates an ionic contribution of only 19%, and the bonding is dominantly covalent. The electronic overlap of H and Cl is shown in Figure 3.1C. Here, in combination, H has the extranuclear electronic structure of He, $1s^2$, and Cl, the structure of Ar, $3s^2 3p^6$, obeying the octet rule.

σ, π, and single, double, and triple covalent bonds

All three examples cited earlier – *s–s* for H_2, *p–p* for Cl_2, and *s–p* for HCl – involve single covalent bonds of the type classed as σ bonds; these are symmetrical and rotatable about the bond axis.

In compounds such as the oxygen molecule O_2, two unpaired electrons in each atom are available for bond formation. The orbital configuration of oxy-

Figure 3.1. Covalent bonds: (A) *s–s* (H_2); (B) *p–p* (Cl_2); (C) *s–p* (HCl); (D) *p–π*(O_2).

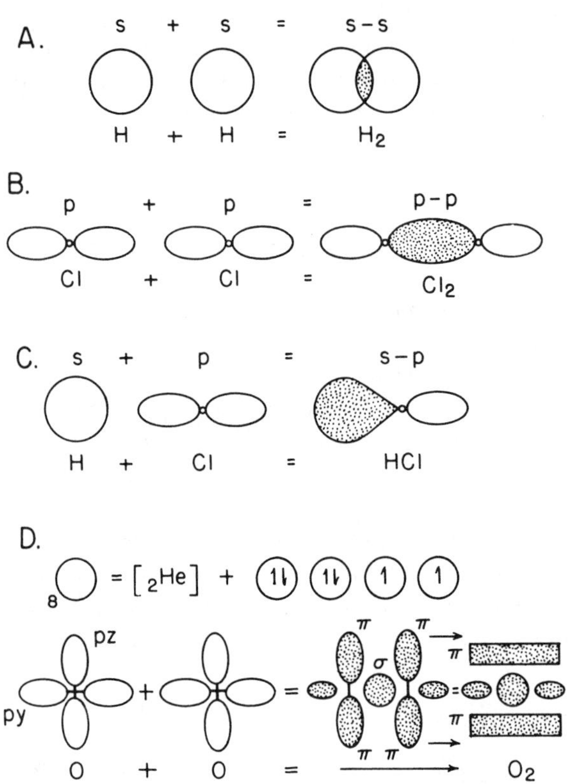

gen is shown in Figure 3.1D. Both the p_y and p_z orbitals may be used to form a *double bond.* In Figure 3.1D, the p_y orbitals along the y bond axis form a σ bond (rotatable), whereas the p_z orbitals combine by "sideways" overlap to form a π bond (nonrotatable or of restricted rotation). The latter is finally further combined into the "stretched rubber-band" configuration shown in Figure 3.1D to become a delocalized orbital that extends throughout the molecule; it is a molecular orbital. Note that the *double bond* molecule consists of a σ bond and a π bond and has drawn the atoms closer together than would a σ bond alone. It should also be apparent that the double bond is stronger than a single bond. Later we will show that the closer the atoms can be brought together, the greater will be the hardness or scratch resistance of the mineral. For this reason alone, minerals containing essential amounts of small atoms with low atomic number, such as boron B, and beryllium, Be, all have great hardness.

Triple covalent bonds also occur when three electrons from each atom are available for bonding. These are usually formed from one σ bond and two π bonds, as in the nitrogen molecule N_2:

$$_7N = [_2He] + \underset{2s^2}{(\uparrow\downarrow)}\ \underset{2p^3}{(\uparrow)(\uparrow)(\uparrow)}$$

$$:\dot{\underset{\cdot}{N}}\cdot + \cdot\dot{\underset{\cdot}{N}}: = :N:::N: \qquad \text{or} \qquad N{\equiv}N$$

Here, the three mutually perpendicular orbitals, p_x, p_y, and p_z, will form a triple bond, and each nitrogen atom in N_2 will attain the stable structure of $_{10}Ne$. The shortening of the interatomic distance in going from a single to a double to a triple bond and the consequent increase in bond strength are illustrated in Table 3.1.

The atomic radius for singly bonded carbon, C, 0.77 Å, is obtained by the convention of halving the interatomic distance, and this value is expected to be preserved with close correspondence in many different crystal structures such as diamond, C^{IV}, and silicon carbide, $Si^{IV}C^{IV}$.

Minerals such as diamond, C^{IV}, and graphite, C^{III}, having identical chemical

Table 3.1. *Bond energy, interatomic distance, and atomic radius in single, double, and triple covalent bonds*

Bond	Bond energy (kcal/mole)	Interatomic distance (Å)	Atomic radius (Å)
C—C (single)	83	1.54	0.77
C=C (double)	146	1.33	0.665
C≡C (triple)	200	1.20	0.600
N—N (single)	38	1.48	0.74
N=N (double)	100	1.24	0.62
N≡N (triple)	226	1.10	0.55

composition but different crystal structure are classed as *polymorphs*. The coordination numbers, CN IV for C in diamond and CN III for C in graphite, inform us that the C atoms are more closely packed in diamond than in graphite, and that the former is the high pressure polymorph of carbon.

Promotion and hybridization of covalent bond orbitals

The sp^3 hybrid bond orbital

In the ground or unexcited state, ${}_6C$ has the electronic configuration

$${}_6C = ({}_2He) + \underset{2s}{⑪} \; \underset{2p_x}{①} \; \underset{2p_y}{①} \; \underset{2p_z}{○}$$

indicating that each C atom has four electrons available for bond formation outside of the closed He $1s^2$ shell. In the excited state, these electrons may be used in different ways. Whenever four valence electrons are available, as in C, silicon, Si, tin, Sn, and germanium, Ge, a stronger bond can be formed between atom pairs if all four electrons have equal energy. Thus, in diamond, the four

Figure 3.2. Hybrid bond orbitals, sp^3, and tetrahedra in diamond.

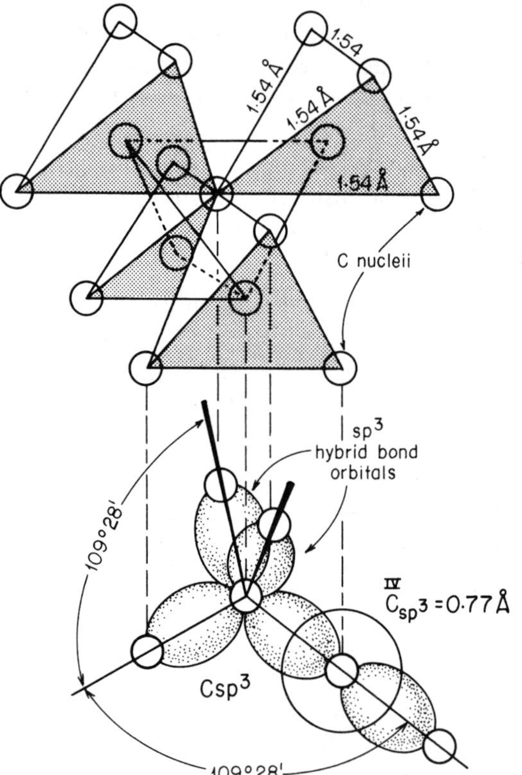

valence electrons are made equal in energy by unpairing one of the $2s$ electrons of C and promoting it to the empty p_z orbital. Thus $_{6}\mathrm{C} = (_{2}\mathrm{He}) + 2s^2 2p^2$ is excited to the configuration $2s^1 2p^3$ and is rewritten as the C_{sp^3} hybrid, as follows:

$_{6}\mathrm{C} = (_{2}\mathrm{He}) +$ [①①①①]

The four new or hybridized orbitals, now equal in energy, will project four semi-dumbbell-shaped globes of electron density or electron clouds from the C nucleus to the four corners of a regular tetrahedron. This is the classic sp^3 hybrid bond orbital (Fig. 3.2), in which lines drawn from the nucleus through each of the four orbitals will subtend angles of $109°28'$, the regular tetrahedral angle. In diamond, each lobe of the sp^3 hybrid on C will overlap equally with a lobe of an adjoining C atom. This overlap will place paired electrons in each orbital, conferring on each C atom the $2s^2 2p^6$ electron structure of neon. In diamond, we are dealing with a framework extended tetrahedrally in three-dimensional space. Each C atom is at the center of a tetrahedron of C atoms and is at the same time a common corner to four other tetrahedra (Fig. 3.2). In diamond, $8/[\mathrm{C}^{\mathrm{IV}}]$, the eight carbon atoms of the unit cell or smallest repeat unit form a face centered cube (Fig. 3.3) with the atoms disposed as follows:

8 C on corners $\times \frac{1}{8} = 1$
6 C on face centers $\times \frac{1}{2} = 3$
4 C inside the cube $\times 1 = \underline{4}$
8 C atoms in 1 unit cube (see also Fig. 10.1)

If we replace the eight carbon atoms of diamond with four atoms of zinc, Zn, and four atoms of sulfur, S, we derive the structure of sphalerite, 4/ZnS, the principal ore mineral of zinc. Thus, for Figure 3.4, C atoms numbered 2 and 4 in Figure 3.3 represent Zn and those numbered 1 and 3 represent S. The unit

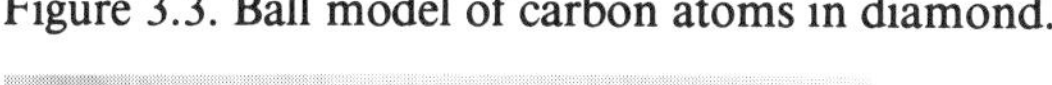
Figure 3.3. Ball model of carbon atoms in diamond.

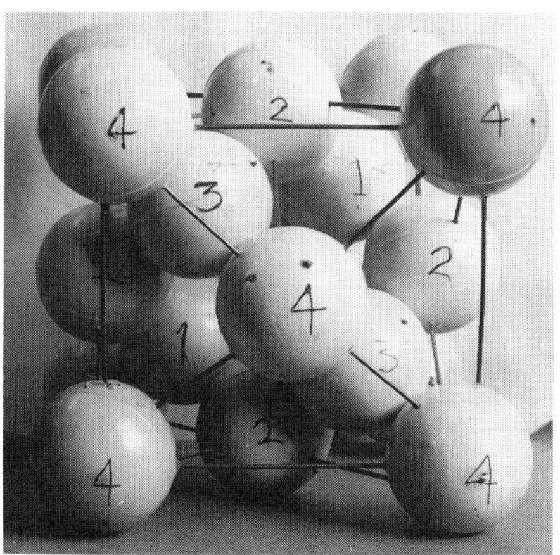

cell will dispose its eight atoms as follows:

8 Zn at corners	$\times \frac{1}{8} = 1$	4 Zn
6 Zn on face centers	$\times \frac{1}{2} = 3$	
4 S inside the cell	$\times 1 = 4$	4 S

Although the orbital configuration and valence electron structure of Zn, $3d^{10}4s^2$, and S, $3s^2 3p^4$, would not suggest an sp^3 hybridization of either atom, both achieve this by pooling the $4s^2$ electrons on Zn with the $3s^2 3p^4$ electrons on S and dividing them equally among each atom. Whenever eight valence electrons are available for bonding of two atoms and the electronegativity difference is small, an sp^3 hybrid may be used by both atoms and it does not matter if they are supplied equally by each atom, as in diamond or silicon carbide. The electronegativity difference, Δ, for sphalerite from $Zn_{\chi 1.5} - S_{\chi 2.5}$ is 1.0, and the bond is predicted to be dominantly covalent but with 22% ionic character.

The elements in Group IVB of the periodic table all contain four valence electrons outside of closed electron shells, and thus C, Si, Ge, and Sn would be expected to form sp^3 hybrid bonds. This is borne out by the crystal structures of silicon carbide, silicon, germanium, and "gray" tin, in addition to diamond. From Table 3.2 we can see that B-group elements located an equal number of places on either side of Group IVB elements will combine to form sp^3 hybrid

Figure 3.4. Covalent and ionic packing drawings of sphalerite, ZnS.

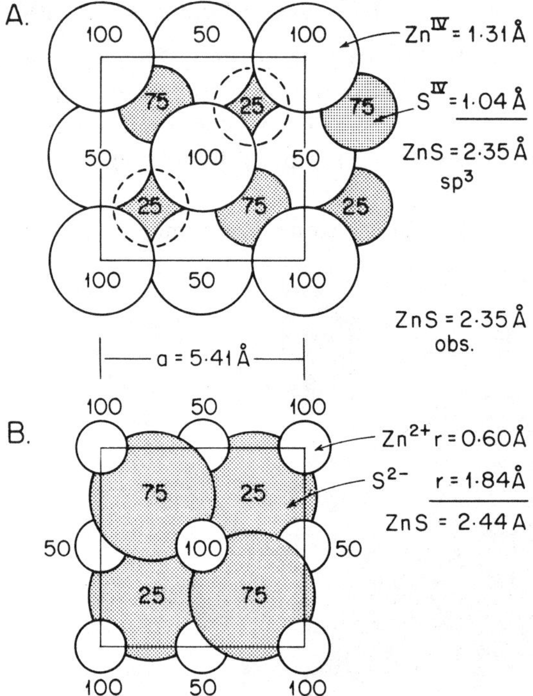

bonds. Further, those compounds having the same sum of atomic numbers have a virtually constant interatomic distance when using this sp^3 hybrid configuration. If the bonds were ionic, the interatomic distances would decrease for compounds having an equal sum of atomic numbers, because of contraction in the radii of cations with increasing charge. These relations are contrasted in Table 3.2, where it is shown that the ionic minerals villiaumite, NaF, and periclase, MgO, both have atomic sums of 20 but show a large difference in interatomic distance. Neither, of course, uses the sp^3 covalent hybrid bond, but rather the close-packed structure of halite, NaCl. The B-group compounds AgI, CdTe, InSb, and SnSn (gray tin) all have atomic number sums of 100, and all have equal interatomic distance sums of 2.80 Å.

It has been shown that the tetrahedral covalent or sp^3 radius of C (0.77 Å) was obtained by halving the distance between the centers of the bonded atoms. Using similar treatment for other covalent compounds, Pauling obtained radii for Si, S, and O from interatomic distances for Si—Si, S—S, and O—O. Then, by a substitution method, he derived a series of single bond tetrahedral covalent radii for use in compounds bonded by sp^3 hybrid orbitals. Radii used by the

Table 3.2. *IV–IV compounds of B-group elements bonded by sp^3 hybrid orbitals*

IB s^1p^0	IIB s^2p^0	IIIB s^2p^1	IVB s^2p^2	VB s^2p^3	VIB s^2p^4	VIIB s^2p^5	Rare gas s^2p^6
		$_5$B	$_6$C	$_7$N	$_8$O	$_9$F	$_{10}$Ne
		$_{13}$Al	$_{14}$Si	$_{15}$P	$_{16}$S	$_{17}$Cl	$_{18}$Ar
$_{29}$Cu	$_{30}$Zn	$_{31}$Ga	$_{32}$Ge	$_{33}$As	$_{34}$Se	$_{35}$Br	$_{36}$Kr
$_{47}$Ag	$_{48}$Cd	$_{49}$In	$_{50}$Sn	$_{51}$Sb	$_{52}$Te	$_{53}$I	$_{54}$Xe

	Valence electrons	Atomic nos.	$A - X$ (Å)	$\Delta\chi$
IV–IV tetrahedral sp^3 hybrid compounds				
CuCl	1 + 7	29 + 17 = 46	2.34	1.2
ZnS	2 + 6	30 + 16 = 46	2.35	1.0
GaP	3 + 5	31 + 15 = 46	2.35	0.5
AgI	1 + 7	47 + 53 = 100	2.80	0.8
CdTe	2 + 6	48 + 52 = 100	2.80	0.6
InSb	3 + 5	49 + 51 = 100	2.80	0.5
SnSn	4 + 4	50 + 50 = 100	2.80	0.0
VI–VI ionic compounds				
NaF	1 + 7	11 + 9 = 20	2.31	3.1
MgO	2 + 6	12 + 8 = 20	2.06	2.3

Note: all have eight electrons per atom pair and a small difference in electronegativity, $\Delta\chi$. Ionic compounds, e.g., $Na^{VI}F^{VI}$, $Mg^{VI}O^{VI}$, have large $\Delta\chi$, use higher CN, and use the close-packed NaCl structure.

tetrahedral compounds of Table 3.2 are given in Table 3.3. Here, it must be emphasized that radii differ with chemical bond type, and even the same atom or ion may have a different radius in different types of crystal structures. For this reason, different sets of radii have been derived for use in comparing covalent, ionic, and metallic minerals and for specific structure types as well.

At a short radial distance from the nucleus of atoms and ions, the electron density is high and is easily associated with or recognized as belonging to a specific atom. However, at the considerably greater radial distances occupied by valence electrons, we are dealing with much lower electron density, a considerable amount of relatively empty space, and the complication of orbital overlap. It is now difficult to say where one atom ends and the other begins. The method of halving the interatomic distance to derive radii is a convenience that is replete with pitfalls and contradictions. Atoms and ions are not beads on a string. It is only by comparison and evaluation of crystal chemical and associated physical and chemical properties of a solid that we gain insight into the nature of the chemical bonding operative in it, and can thus draw conclusions as to the size of atoms. Although the distance between atomic centers or nuclei, the interatomic distance, can be measured with high precision, the partition of this distance into discrete atomic radii is subject to speculation.

Although hydrogen in H_2 and oxygen in O_2 use *s*–*s* and *p*–*p* single covalent bonds, respectively, it is the sp^3 hybrid that is used by oxygen in the water molecule H_2O. Oxygen hybridizes its six $2s^22p^4$ valence electrons into an unusual sp^3 hybrid tetrahedral orbital in which two lobes contain paired electrons (lone pairs) and the other two lobes contain the single unpaired electrons (Fig. 3.5). Hydrogen atoms, each with one electron, occupy the unpaired lobes of the sp^3 oxygen hybrid, and the H_2O molecule becomes polar: the lone pair side negative, and the H_2 side positive. In water, the octet rule is preserved, because each H has the electronic structure of He and each O that of Ne.

Table 3.3. *Single bond tetrahedral covalent radii*

IB	IIB	IIIB	IVB	VB	VIB	VIIB
		B	C	N	O	F
		0.88	0.77	0.70	0.66	0.64
	Mg	Al	Si	P	S	Cl
	1.40	1.26	1.17	1.10	1.04	0.99
Cu	Zn	Ga	Ge	As	Se	Br
1.35	1.31	1.26	1.22	1.18	1.14	1.11
Ag	Cd	In	Sn	Sb	Te	I
1.52	1.48	1.44	1.40	1.36	1.32	1.28

Source: After Pauling (1960).

The sp^2 hybrid bond orbital

Returning now to the C atom, we can clearly see that the chemical bonding employed in graphite, C^{III}, is entirely different from that than in diamond. Although diamond, C^{IV}, is colorless, adamantine in luster, a transparent insulator with very high refractive index, and the hardest mineral known, graphite, C^{III}, in contrast, is black, of metallic luster, opaque, a conductor of electricity, and one of the softest minerals known. Thus, whereas diamond is used in expensive jewelry, graphite is used in inexpensive "lead" pencils.

The reasons for this remarkable difference in physical properties come down to the disposition of one electron per atom. Diamond uses an sp^3 hybrid bond orbital, hybridizing all four $2s^22p^2$ valence electrons, but graphite hybridizes only three of these electrons and leaves the fourth as a p valence electron outside of the hybrid (Fig. 3.6).

Now, three lobes or orbitals promoted to equal energy lie in a plane and project at angles of 120°, or an equilateral triangle. These sp^2 hybrid C lobes join together in graphite into hexagonal sheets because each lobe is shared (Fig. 3.6). These bonds are σ bonds, and are as strong as those in diamond. The fourth, unhybridized p orbital projects perpendicularly above and below the sp^2 sheets. It draws each pair of C atoms closer together and, in so doing, forms a π bond that becomes delocalized by sideways overlap and extends as a "stretched

Figure 3.5. Chemical bonding in the water (H_2O) molecule.

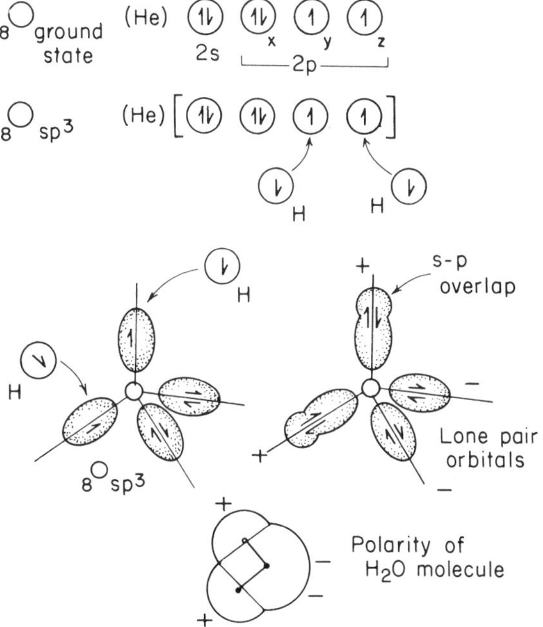

rubber-band" orbital through the entire crystal: it is a molecular orbital (Fig. 3.6). A continuous flow of excited *p* electrons through this delocalized orbital or band causes absorption of wavelengths of visible light, leading to opacity, metallic luster, and high electrical conductivity.

In addition to σ and π bonds, a third type of weak chemical bond is present – the weak van der Waals bond that holds the graphite sheets together. When we write with a "lead" pencil, we smear sheets of graphite on paper by breaking these weak van der Waals bonds.

The d^2sp^3 hybrid bond orbital

The most common sulfide mineral, pyrite, $Fe[S_2]$ (Figs. 3.7, 3.8), uses a d^2sp^3 hybrid bond orbital on the iron, Fe, atom and an sp^3 hybrid bond orbital on each S atom (Pauling 1960; Burns and Vaughan 1970), and its formula may be written as either $Fe^{VI}[S_2]^{VI}$ or $Fe^{VI}S^{IV}—S^{IV}$. Pyrite uses a variant of the familiar $Na^{VI}Cl^{VI}$ crystal structure: Fe atoms lie on the corners and face centers of the unit cube, and S_2 pairs of dumbbells lie on the twelve edges and in the

Figure 3.6. Chemical bonds, sp^2, π, and van der Waals bonds, in graphite, C.

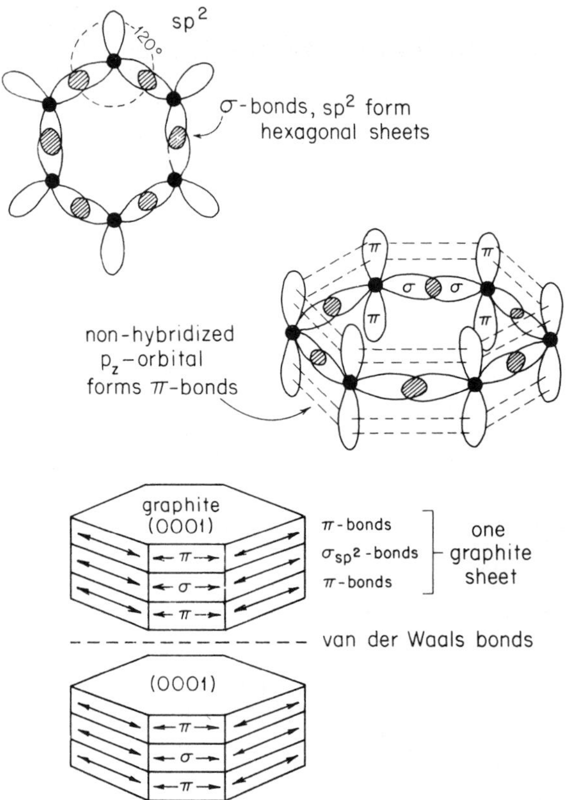

Figure 3.7. Packing models of pyrite with atoms given covalent radii. (A) projected along the *c* axis [(100) projection] and (B) projected along a threefold axis (111).

Figure 3.8. Packing drawings of the pyrite structure, comparing the ionic with covalent models. Sulfur atoms use sp^3 hybrid bonds in pairs, joining $S^I—S^I$, and Fe atoms use d^2sp^3 bonds joining Fe^{II} and S_2.

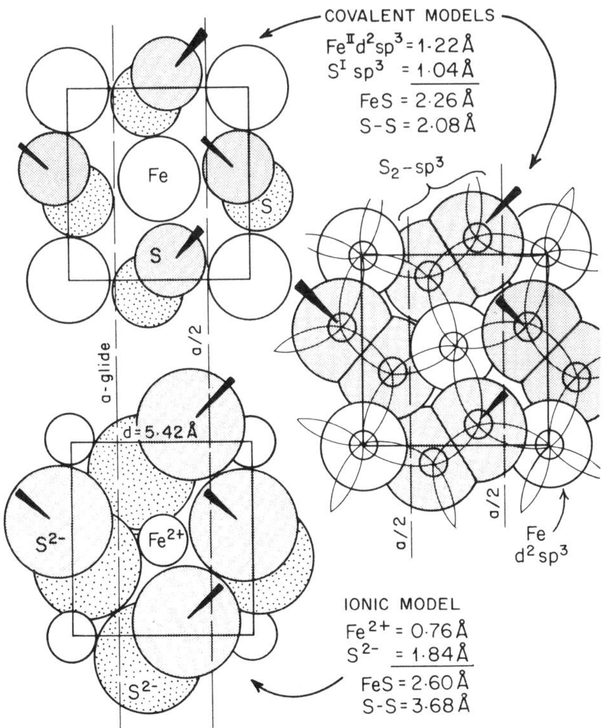

center of the unit cube. It is the different cant or opposed orientation of these S_2 dumbbells that lowers the cubic symmetry of pyrite from the face centered, $Fm3m$ space group of halite to the primitive, $Pa3$ space group. In pyrite, then, S atoms are joined in pairs by sp^3 hybrid bond orbitals by one lobe per atom, and the three remaining lobes overlap and bond with empty d^2sp^3 hybrid lobes of the Fe atom (Figure 3.8). Here, the six $3d$ valence electrons of Fe are paired and placed in three low energy $3d$ orbitals, resulting in a net low electron spin configuration. The two unused high energy $3d$ orbitals, the now empty $4s$ orbital, and three unused $4p$ orbitals are hybridized to equal energy and project six empty lobes toward the corners of a regular octahedron. Now, each Fe atom has loaned its two $4s$ electrons to S. Each Fe atom will be coordinated by VI S atoms, and each S atom will be coordinated tetrahedrally, or in CN IV, by three Fe atoms and one S atom. Thus, each $Fe_{d^2sp^3}$ hybrid has its six empty lobes occupied and overlapped by twelve valence electrons supplied by S atoms. Now, the plot thickens, as the twelve S valence electrons plus the two $4s$ electrons supplied by Fe are believed to join forces to form a fourteen-electron delocalized molecular orbital in which electrons can flow. This explains the metallic luster and electrical conductivity of pyrite. An alternative suggestion by Burns and Vaughan (1970) relates these properties to the formation of a delocalized d-π covalent bond between the paired $3d$ electrons of Fe and the unused d orbitals of S.

Here, we may note that the isostructural compounds, pyrite, FeS_2, and hauerite, MnS_2, although having identical crystal structures, have totally different physical and chemical properties and must therefore have different chemical bonding. $_{25}Mn$ and $_{26}Fe$ differ in the ground state by only one $3d$ valence electron. In both sulfide minerals, the S—S pairs have an identical interatomic distance of 2.16 Å (hence radii of 1.08 Å), but here the similarity ends. Pyrite is totally nonmagnetic or diamagnetic: its electrons fail to orient with an applied magnetic field, whereas hauerite has a paramagnetic response or susceptibility equivalent to five unpaired electrons, its total complement of d electrons. Thus, a magnetic criterion of bond type is apparent and informs us that in pyrite all six Fe $3d$ electrons are paired (low spin state), resulting in a net spin or Bohr magneton value $\mu_B = 0$ (Bohr magneton, $\mu_B = 9.273 \times 10^{-21}$ erg gauss^{-1}), whereas in hauerite all five $3d$ electrons of Mn must be unpaired, yielding $\mu_B = 5.92$. Paramagnetic susceptibility may be measured in units of Bohr magnetons by the formula $\mu_B = \sqrt{n(n+2)}$, where n is the unpaired electron spin per atom (Pauling, 1960). Thus, for the atoms and ions in question, we can derive

$$
\begin{array}{lll}
Fe^{++} \text{ ion}, & n = 4, & \mu_B = 4.87 \\
Mn^{++} \text{ ion}, & n = 5, & \mu_B = 5.92 \\
Fe^{II} \text{ low spin}, & n = 0, & \mu_B = 0
\end{array}
$$

Pyrite is opaque, metallic, conductive, and diamagnetic, whereas hauerite, although opaque or nearly so, is nonmetallic and strongly paramagnetic. Thus the Mn—S bond must be dominantly ionic, in stark contrast to the d^2sp^3 covalent bond of Fe—S. Thus, the identity of formula and crystal structure does not correctly predict the chemical bond. The electronegativity differences for Fe—S, $\Delta\chi = 0.7$ (=10% ionic bonding), and for Mn—S, $\Delta\chi = 1.1$ (=28% ionic bonding), fall short of confirming the dominant ionic character of the Mn—S bond. Evidently, electronegativity, like atomic and ionic radius, is not constant for different crystal structures.

A set of octahedral, d^2sp^3 covalent radii for use in pyrite-type structures is given in Table 3.4. These radii are derived by subtracting half the S—S distance from the metal–S distance (A—X). Because recent determinations of the S—S distance in pyrite (2.14 Å) and in sulfur (2.05 Å) differ from Pauling's assumption that both were 2.08 Å, the values in Table 3.4 may be expected to cover a range and, once again, must be considered to represent approximate values. Although the A—X distances are exact, the radii are not.

The dsp^2 hybrid bond orbital

Not all covalent bonds involving CN IV result from tetrahedral sp^3 hybrids. An interesting group of minerals and synthetic compounds dominated by oxides, chlorides, and sulfides of platinum, Pt, palladium, Pd, nickel, Ni, and copper, Cu, use dsp^2 square (or rectangular) coplanar hybrid bond orbitals in which four lobes of electron density project toward corners of a rectangle, almost a square. In cooperite, PtS (Fig. 3.9), and synthetic PdO, each metal atom pairs eight electrons in four *d* orbitals and loans its other two valence electrons to either S or O, enabling it to project four empty dsp^2 equal energy hybrid orbitals toward corners of a nearly square rectangle. These are then overlapped by the four S or O atoms at corners of the rectangle. There are now four S atoms around each Pt atom (or four O around each Pd) at corners of a square, and, in turn, four Pt atoms around each S (or four Pd around each O) at corners of a tetrahedron (Fig. 3.9). Note that whenever the bipositive Cu "ion" is coordinated by VI O anions, the four oxygens at corners of a square lie close to the Cu atom, whereas the other two are at a greater distance. The destabiliza-

Table 3.4. *Covalent d^2sp^3 octahedral atomic radii for use in AX_2, pyrite-type, and related structures*

Fe	Cu	Ni
1.17–1.24 Å	1.22–1.33 Å	1.21–1.39 Å
Ru	Rh	Pd
1.28–1.32	1.28–1.32	1.27–1.31
Os	Ir	Pt
1.30–1.34	1.28–1.32	1.27–1.31

Source: After Pauling (1960).

tion of the octahedron in favor of the square suggests that Cu much prefers the *dsp*^{2} hybrid square planar configuration to the ionic octahedron.

The sp linear bond orbital

Earlier it was shown that HCl uses an *sp* linear covalent bond. Only rarely are metallic atoms or ions found coordinated to only two nonmetallic atoms, as in cuprite, Cu_2O, in synthetic Ag_2O, and in the "secondary" uranium, U, minerals containing the uranyl ion $(UO_2)^{2+}$, as in carnotite, a principal uranium ore mineral in Colorado Plateau ores, $K_2(UO_2)_2(VO_4)_2 \cdot 3H_2O$.

In cuprite, Cu_2O, each O atom lies on the corner and center of a body centered cubic lattice, whereas each Cu atom is located on an interpenetrating face centered cubic lattice (see Fig. 18.26). Each Cu atom forms linear, CN II bonds with two O atoms, which are, in turn, each coordinated tetrahedrally by four Cu atoms. The linear O—Cu—O bonds interpenetrate throughout the cubic structure but are not cross-linked (see Fig. 18.26). Electronegativity differences, $\Delta\chi_{Cu-O} = 1.7$, are equivalent to roughly 50% ionic bonding. Cuprite has a deep red color, is barely transparent, and has index of refraction (n) of 2.705 (greater than that of diamond), strong internal reflections, and an adamantine luster. Thus, although by electronegativity difference it is 50% ionic, the physical and chemical properties of cuprite are much more those of a covalent rather than an ionic compound.

Although this completes our treatment of the covalent bond, we shall have many further occasions to evaluate the chemical bond in minerals in the ensuing chapters.

Figure 3.9. Rectangular ("square") planar, *dsp*^{2} hybrid bond orbitals on Pt atoms link to four S or O atoms in PtS or PtO (see also Fig. 18.17).

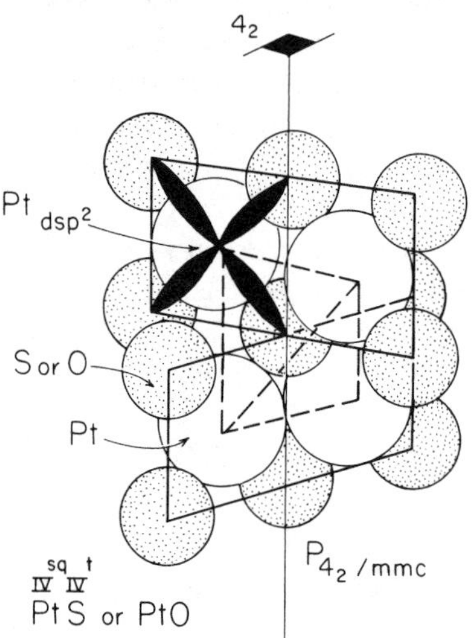

Summary

Covalent bonds are formed by the sharing of electrons. Pure covalent bonds form only between atoms of the same species, by the overlap of *s* or *p* orbitals, making single or σ bonds.

Double and triple bonds, stronger than single bonds, are formed of one σ and one or more π bonds. Quadruple bonds may form sp^3 hybrid orbitals, as in diamond, the hardest known compound, and in the water molecule.

In graphite, carbon is linked into sheets by sp^2 (σ) and π bonds; weak van der Waals bonds link the sheets.

In pyrite, Fe atoms use d^2sp^3 hybrid bonds, S atoms use sp^3 hybrid bonds. Hauerite, isostructural with diamagnetic pyrite, is strongly paramagnetic and must therefore be dominantly ionic.

Oxides, chlorides, and sulfides of Pt, Pd, Ni, and Cu use square coplanar dsp^2 hybrid bond orbitals. Linear *sp* covalent bonds occur in HCl, Ag_2O, Cu_2O, and secondary uranium minerals.

Bibliography

Burns, R. G., and Vaughan, D. J. (1970). Interpretation of reflectivity behavior of ore minerals. *Am. Mineral.,* 55: 1576–86.

Coulson, C. A. (1961). *Valence,* 2d ed. Oxford Univ. Press, London.

Fyfe, W. S. (1964). *Geochemistry of solids.* McGraw-Hill, New York.

Goodenough, J. B. (1972). Energy bands in TX_2 compounds with pyrite, marcasite, and arsenopyrite structures. *J. Solid State Chem.,* 5: 144–52.

Jellinek, F. (1970). Sulfides. *In Inorganic sulfur chemistry,* ed. G. Nickless. Elsevier, Amsterdam, pp. 667–748.

Pauling, L. (1960). *The nature of the chemical bonds,* 3d ed. Cornell Univ. Press, Ithaca, New York, chaps. 4, 5, 6, 7, 11, 14.

Prewitt, C. T., and Rajmani, I. (1974). Electron interactions and chemical bonding in sulfides. *Mineral. Soc. Am. Rev. Mineral.,* 1: 1–4.

Rajmani, V., and Prewitt, C. T. (1973). Crystal chemistry of natural pentlandites. *Can. Mineral.,* 12: 178–87.

Sebera, D. K. (1964). *Electronic structure and chemical bonding.* Blaisdell, Waltham, Mass.

4 The crystal chemistry of the ionic bond

The nature of the ionic bond

Because the covalent bond is directional and a function of spatially oriented electron density clouds, it has been classed as *homopolar.* By contrast, the ionic bond is nonpolar, requires that electric charge be smeared out evenly and symmetrically over the surface of an atom, and is thus classed as *heteropolar.* One atom A of low ionization potential transfers an electron (or more than one) to a bonding partner, atom X, and the cation A^+ and the anion X^- are formed. Atom A shrinks because of the loss of radial distribution and because the nucleus pulls in the remaining electrons more closely to shield its positive charge. Atom X obviously expands enormously, because it adds an outer electron orbiting at a large distance from the nucleus. Electrostatic Coulomb attractive forces will cause A^+ and X^- to approach closely until their valence electrons begin to overlap. At this point, repulsion will set in at a given distance for a given atom pair in a particular crystal structure. This equilibrium distance of closest approach of A^+ and X^- is classed as the *interatomic distance,* A—X, usually cited in ångstrom units, Å (1 Å = 10^{-8} cm).

The ionic model assumes that the A—X distance is equal to the sum of the radii of rigid spheres, A^+ and X^-, in tangent contact. Now, each A^+ cation will attempt to surround itself with the maximum number of X^- anions permitted by geometry, and, in turn, each X^- will surround itself by the maximum number of A^+ ions. The maximum number of X^- ions permitted to surround A^+ is solely a function of the radius of the A^+ cation divided by the radius of the X^- anion, the *radius ratio.* The number of cations A^+ around an X^- anion is a function of the total contribution of electric positive charge to the neutralization of the negative charge of the anion X^-. Fractional charge distribution contributed by each cation to near anion neighbors is equal to cation charge divided by cation coordination number, and is equal to the *electrostatic bond strength* (EBS).

We have just defined the two most important of Pauling's famous five rules for the stability of ionic compounds. We will discuss them in considerable detail later. In ideal ionic bonding, the highest coordination of A^+ around X^- and vice versa results in closest packing of spheres of unlike size and charge; ideal shielding of the electric charge on each partner, equivalent to perfect *ordering;* and a maximum lowering of potential and kinetic energy, contributing to high stability under the particular conditions of formation, such as temperature and pressure.

The ideal ionic bond requires that orbitals of A^+ and X^- not be overlapped. They are thus separated at equilibrium distance by an electron density equal to zero. This density is never achieved, and all so-called ionic minerals have a covalent component, expressed as a significant electron density between bonded atoms or ions.

The magnitude or the percentage of ionic bonding in a compound is predicted by the electronegativity difference, $\Delta\chi_{A-X}$, of bonded atoms, or almost equally well from the difference in the first ionization potential, $\Delta I_{P_{A-X}}$. The closest approach to ideal ionic bonding will occur between atoms of low I_P and high χ. Because low I_P is analogous to low χ, we will continue to use $\Delta\chi_{A-B}$ as our yardstick for prediction of bond type. The lowest electronegativities and I_P values are associated with the alkali elements of periodic group IA or s^1, and the single s^1 valence electron can be removed from Li, Na, K, Rb, and Cs at a very low input of energy of excitation.

Conversely, the atoms having the greatest affinity for gaining electrons in bonding are the halogens of periodic group VIIB, or s^2p^5, only one electron shy

Figure 4.1. Curve relating percentage of ionic bonding to electronegativity difference $\Delta\chi_{A-B}$. Reprinted from Linus Pauling, *The nature of the chemical bond,* 3d ed. Copyright © 1960 by Cornell University. Used by permission of the publisher, Cornell University Press.

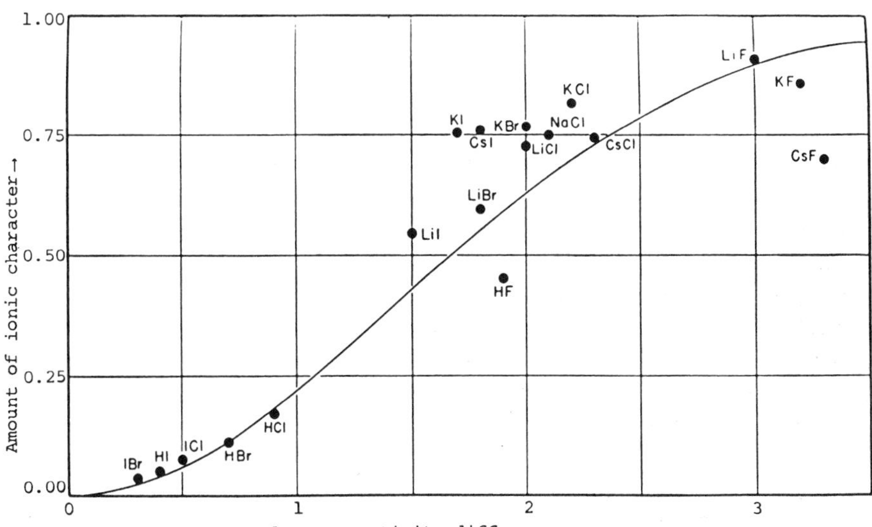

of a stable rare gas configuration. Of these, the electronegativity values fall off rapidly, where $\chi = 4.0, 3.0, 2.8$, and 2.6, for F, Cl, Br, and I, respectively. From the curve of Pauling (1960) relating the percentage of ionic bonding with $\Delta\chi_{A-X}$ (Fig. 4.1), it is apparent that ideal, or nearly ideal, ionic bonds are formed *only* between the alkali ions and fluorine, F, or A^+F^-. Thus, $\Delta\chi_{Na-F} = 3.1 = 91\%$ ionic bonding; $\Delta\chi_{Na-Cl} = 2.1 = 66\%$ ionic bonding; $\Delta\chi_{Na-Br} = 1.9 = 59\%$ ionic bonding; and $\Delta\chi_{Na-I} = 1.7 = 51\%$ ionic bonding. After F, $\chi = 4.0$, comes O with $\chi = 3.5$, and bonds to O with those elements having $\chi \leq 1.7$ will be predominantly ionic with a significant covalent contribution. The Si—O bond, the principal building block of the numerous silicate minerals, has $\Delta\chi = 1.7 = 51\%$ ionic bonding. The Si—O bond may thus be said to resonate between an Si—O ionic bond and an Si—O sp^3 hybrid covalent bond, to which is added an additional $d-\pi$ covalent bond (Brown and Gibbs 1969). The Si—O bond should not be considered to resonate from an ionic bond 50% of the time and a covalent bond 50% of the time, but rather as an ionic–covalent bond 100% of the time. Thus resonance implies that the bond is never wholly ionic nor wholly covalent, but is always a mixture.

The most appropriate generalization of the chemical bond type in minerals is as follows:

fluorides	ionic
chlorides, oxides, silicates	ionic–covalent
B-group metalloids	covalent
sulfides, arsenides	covalent–metallic
metals, alloys	metallic–covalent

Van der Waals and hydrogen bonds contribute to the ionic–covalent mixture in many minerals.

Because the ionic bond implies and requires sphericity of electron density, it should be obvious that certain valence configurations restrict and prevent formation of spherical ions. All elements of electronic valence configuration s^1 and s^2, alkali and alkaline earth elements of periodic groups IA and IIA will, on ionizing, yield their s^1 and s^2 valence electrons and assume a spherical electron density, as will the halogen, F $2s^22p^5$ of group VIIB on gaining an electron.

Those elements having *p*, *d*, or *f* valence electrons will have spherical electron density only when their threefold, fivefold, and sevenfold orbital energy levels, respectively, are either completely filled or half filled. It is this asymmetry, or departure from sphericity, that favors the formation of covalent bonds and complex anions. Thus, the valence electron structure of the B-group elements – $_5$B, $2s^22p^1$; $_6$C, $2s^2p^2$; $_7$N, $2s^22p^3$; $_{14}$Si, $3s^23p^2$; $_{15}$P, $3s^23p^3$; and $_{16}$S, $3s^23p^4$ – favors formation of covalent complexes with $_8$O, $2s^22p^4$, to produce complex anions of asymmetric electron density. Thus, $[CO_3]^{2-}$, $[NO_3]^{1-}$, $[SiO_4]^{4-}$, $[PO_4]^{3-}$, and $[SO_4]^{2-}$, to name a few, form complex anions that retain their identity and asymmetry in bonding to other elements. Thus, $[CO_3]^{2-}$ is a planar anion rather than a C^{4+} cation bonded to three O^{2-} anions, and $[PO_4]$ is a tetrahedral complex anion rather than a P^{4+} cation bonded to four O^{2-} anions.

We have already seen, in the previous chapter, that an S^{2-} anion does not exist in pyrite, Fe(S—S), but occurs as an $[S_2]^{2-}$ coupled anion complex using an sp^3 hybrid. The existence of the S^{2-} anion should always be regarded with suspicion, inasmuch as S has $\chi = 2.5$ and will, in any mineral, always form bonds that are more covalent than ionic.

Pauling's five rules for coordination (ionic) compounds

The principles that underlie the formation and stability of ionic or coordination compounds were codified by Pauling (1960) into five rules. We will now discuss them in detail and also consider them in subsequent chapters.

Rule 1. *Cation coordination and the radius ratio.* A polyhedron of spherical anions is grouped around each spherical cation such that the number of anions that may surround the cation is a function of the radius ratio $R = r_{cat}/r_{an}$. We assume that each cation is in tangent contact with each anion and lies in the center of the coordination polyhedron of anions that surround it.

Rule 2. *Electrostatic bond strength and the number of polyhedra with a common corner.* The number of cations that may surround a given anion is limited by the requirement that negative electric charge on anions be satisfied locally or over short range by *near cation neighbors.* A given anion will be the common corner to several coordination polyhedra, the exact number limited by the concept of EBS, equal to cation charge/cation CN. We assume that the charge of the cation is distributed evenly among all neighboring anions, and the sum of the fractional EBS contributions reaching an anion from all neighboring cations must equal the valency of the anion with sign reversed. Typical EBS values for some common cations are Li^+/CN VI $= +\frac{1}{6}$, Be^{2+}/CN IV $= +\frac{1}{2}$, Al^{3+}/CN IV $= +\frac{3}{4}$, Si^{4+}/CN IV $= +1$, and Nb^{5+}/CN VI $= +\frac{5}{6}$. Thus, EBS $= +2$ for O^{2-} and $+1$ for F^- or Cl^-. For low or α-quartz, $Si^{IV}O_2^{II}$, each Si^{4+} ion will occupy a tetrahedron of O^{2-} anions, and each Si will contribute $+4$/IV $= +1$ EBS units, limiting the CN of the O^{2-} anion to II. For halite, $Na^{VI}Cl^{VI}$, each Na^+ inside an octahedron, CN VI, contributes $+\frac{1}{6}$ EBS units to each Cl^-, requiring that there be six Na^+ around each Cl^-, and this leads to a so-called VI–VI structure. Most mineralogy textbooks fail to stress – indeed, they even ignore – the coordination or CN of the *anion:* this is as important as the CN of the cation for interpreting the structure of crystals. A detailed discussion and elaboration of this second rule is given in Chapter 5.

Rule 3. *The rule of polyhedral sharing.* Sharing of corners, edges, and faces of coordination polyhedra brings centers of positive charge closer together than they would be if edges were unshared, and this tends to decrease the stability of coordination compounds. Such sharing, however, is widespread in minerals; otherwise most of these would not be able to polymerize or link together to build extended crystal lattices. Pauling observed that whenever two cations approach one another across a common polyhedral edge, the two anions forming the common edge will draw closer together, reducing their interatomic distance, to better shield their negative charge clouds. This is known as the *shortening of shared edges* and provides proof of the presence of charged atoms in an ionic bond.

Conversely, those minerals in which shared polyhedral edges do not shorten provide evidence of a predominantly covalent bond. Many oxide and silicate minerals in which the O^{2-} anions are in close packing reveal O—O distances of 2.80 Å, twice the radius of O^{2-}, but any two O^{2-} anions forming a shared polyhedral edge show O—O interatomic distances of 2.50 Å, a considerable reduction. Shortening of shared polyhedral edges obviously distorts the geometry of CN polyhedra, and distorted polyhedra are actually the rule, rather than the exception, in minerals (see Chapter 8).

Rule 4. *Independence of polyhedra with small cations of high charge.* Cations with small radius and large charge tend not to share polyhedral elements with one another. This is an extension of Rule 2, where we have seen that in α-quartz, sharing of polyhedra is limited to corners because of the large EBS value of Si^{4+}/IV, which is equal to $+1$, or one-half the negative valence of the O^{2-} anion. Where the nominal electric charge exceeds 4^+, polyhedra containing these cations will remain independent of one another. Thus, EBS values for nominal C^{4+}, P^{5+}, and S^{6+} ions in CN III, IV, and IV, respectively, are $+1.33$, $+1.25$, and $+1.5$. Two such cations around an anion would overcharge the anion, and this restricts polyhedral sharing. Because the isolated tetrahedral $[PO_4]^{3-}$ groups must be joined together by other cations to form a phosphate mineral, it follows that they must be linked by cations with lower EBS. Thus, in berlinite, $[Al^{IV}P^{IV}O_4^{II}]$, isostructural with α-quartz, $[Si^{IV}O_2^{II}]$, each O^{2-} anion may be considered to receive an EBS contribution of $P^{5+}/IV = +1.25$ and $Al^{3+}/IV = +.75 = +2$.

This rule may be somewhat diminished by the fact that C^{4+}, P^{5+}, and S^{6+} cations do not really exist as such, but rather form complex anion groups or complex anions such as $[CO_3]^{2-}$, $[PO_4]^{3-}$, and $[SO_4]^{2-}$. These combine in minerals as though they were single or simple anions of nonspherical shape. Thus, for calcite, we may write the formula $Ca^{VI}C^{III}O_3^{III}$ or $Ca^{VI}[CO_3]^{VI}$. We may illustrate these alternatives in EBS diagrams as follows:

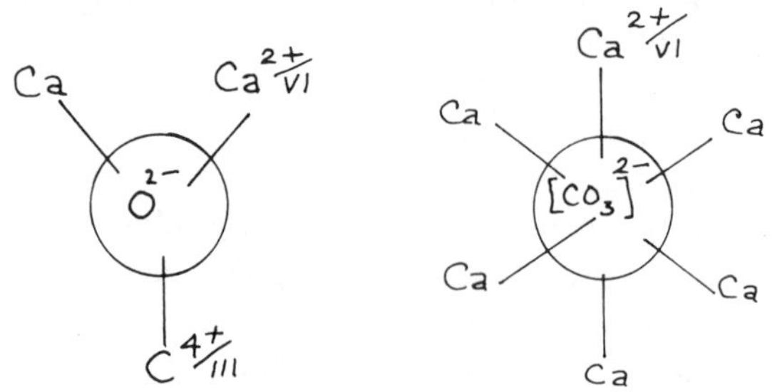

$$EBS = 0.33 + 0.33 + 1.33 = +2 \quad \text{or} \quad 0.33 \times 6 = +2$$

As can be seen here, there are two ways to do the arithmetic, but the second choice is the more realistic. In this connection, Pauling's rules may often be satisfied for covalent as well as for ionic compounds, as can be seen in the above. Here, although calcite is considered to be an ionic mineral, the $[CO_3]^{2-}$ group is itself a covalently bonded complex anion. The fact that no two $[CO_3]$, $[PO_4]$, or $[SO_4]$ polyhedral anion groups ever share corner oxygens with one another in any carbonate, phosphate, or sulfate mineral underscores the validity of this rule.

Rule 5. *The rule of parsimony.* This, the least important of Pauling's rules, states that a large number of different kinds of coordination polyhedra in a given mineral tends to decrease its stability, because this would build unnecessarily complex molecules. In most minerals, only two or three different types of polyhedra are found. Many exceptions to this rule may be cited, most notably the following:

Tourmaline

$$Na^{IX}Mg_3^{VI}Al_6^{VI}[Si_6^{IV}O_{18}][B^{III}O_3]_3[(OH), F]_4$$

and

Hornblende

$$Na_{0.5}^{X}Ca_2^{VIII}(Fe^{2+}, Al)_2^{VI}Fe_2^{2+VI}Mg^{VI}[Si_{6.5}^{IV}Al_{1.5}^{IV}O_{22}][(OH),F]_2$$

Tourmaline contains five different polyhedra, and hornblende seven, yet each exists over a wide range of temperature and pressure in the presence of H_2O and F-bearing fluids.

Radius ratios and coordination numbers in ionic compounds

Atoms widely separated in space or disordered can lower their potential and kinetic energies by bonding together to form ordered crystal structures. We began this book with an analogy of shaking a crate of oranges to bring them into a closer-packed and hence better-ordered state commensurate with lower energy and higher stability. A better analogy would be provided by a crate of tennis and Ping-Pong balls. If we vigorously shake the contents of this box, the tennis balls will pack so that each is surrounded or coordinated by twelve other tennis balls, whereas the much smaller Ping-Pong balls will occupy the interstices or voids formed between the packed tennis balls.

First, let us look at the packing of tennis balls alone. If we shake the box of tennis balls alone, each will wind up in CN XII. One of two arrangements of CN XII is possible: cubic close packing of spheres of equal radius, or hexagonal close packing of spheres of equal radius. *Both are equally close packed* (Fig. 4.2).

First we will pack six tennis balls in a plane around a central tennis ball. Next we will place another equilateral triangle of three tennis balls directly over the central ball of the first layer in close packing. Now we will place another equilateral triangle of three tennis balls directly under the central ball. There are now twelve spheres of equal radius around the central sphere, and we have a close-packed arrangement. However, we may orient the two triangular groups above and below our central sphere in parallel arrangement so that they form a trigonal prism and a hexagonal close-packed array (Fig. 4.2), or, on the other hand, we may orient the two triangular groups above and below the central sphere in opposition so that they form an octahedron, in which case we have a cubic close-packed arrangement (Fig. 4.3). In either case, the voids or interstices between the close-packed tennis balls, each in CN XII, will be of two types. One will be a void enclosed by six tennis balls, a CN VI site, and the other a void enclosed by four tennis balls, a CN IV site. These types will be oriented in alternating rows through the close-packed arrays (Fig. 4.3). It is important to see that the central tennis ball (Fig. 4.3B layers) is now surrounded by fourteen voids: six are octahedral, with three sites above and three below the central sphere; eight are tetrahedral, with four above and four below the same central sphere. In Figure 4.3, we have filled all fourteen sites with CN VI cations of large radius, and CN IV cations of smaller radius. This means that the central sphere, whether a tennis ball or a Cl^- anion, is a corner that is common to all fourteen polyhedra. If the central sphere is designated as a Cl^- anion, it should be

Figure 4.2. Cubic and hexagonal close-packing models and their geometries. Layers stack in sequence, C over B, over A.

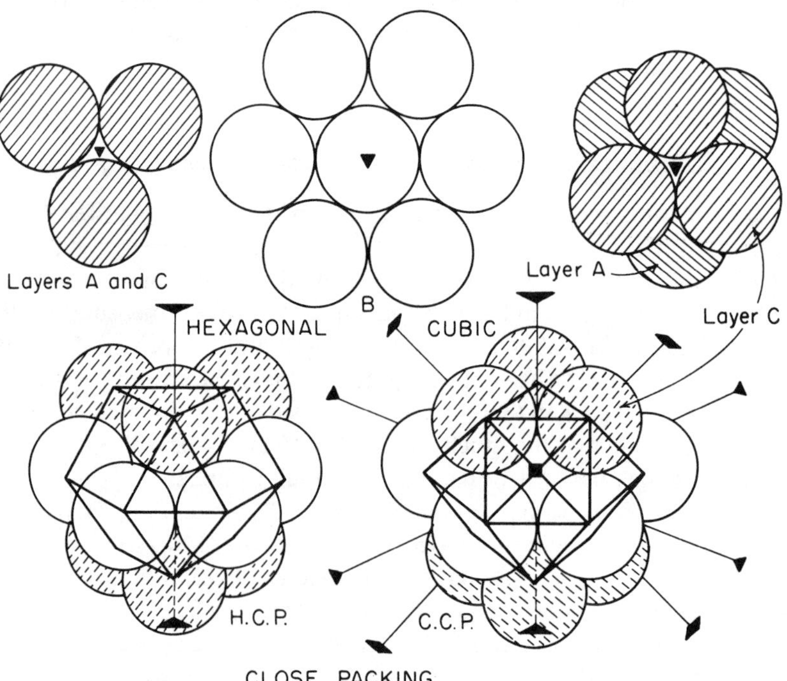

obvious that occupancy of all fourteen cation sites would convey a considerable excess of positive charge to the central Cl^- anion. Thus, in a close-packed array of anions, only a maximum of six of the fourteen sites will be occupied by cations in any given mineral. The limit on the number of sites that may be occupied by cations is strictly established by Pauling's second rule of electrostatic bond strength, which will be elaborated in considerable detail in the next chapter.

If the array of tennis balls is hexagonally closely packed, a CN IV site will lie directly over another such site, and a CN VI site will also lie directly over another identical site. If the array of tennis balls is cubic close packed, a CN IV site will lie directly over a CN VI site, and vice versa. To state this in another way, in HCP (hexagonal close packing) a tetrahedral site lies above a tetrahedral site, and an octahedral site lies over another of the same kind. In CCP (cubic close packing) the sites alternate positions so that a tetrahedral site lies over an octahedral site, and vice versa. These two arrangements favor the location of

Figure 4.3. Hexagonal close packing (HCP) resulting from placement of layer B over layer A (upper) stacks face sharing octahedra over one another along the threefold symmetry axis, whereas cubic close packing (CCP) with layer B over layer A (lower) stacks octahedra and tetrahedra in alternation along the threefold axis. Constraints on distribution of electric charge favor HCP in corundum, Al_2O_3 (four of six octahedral sites occupied), but favor CCP in periclase, MgO (six of six octahedral sites occupied). In both minerals tetrahedral sites are vacant.

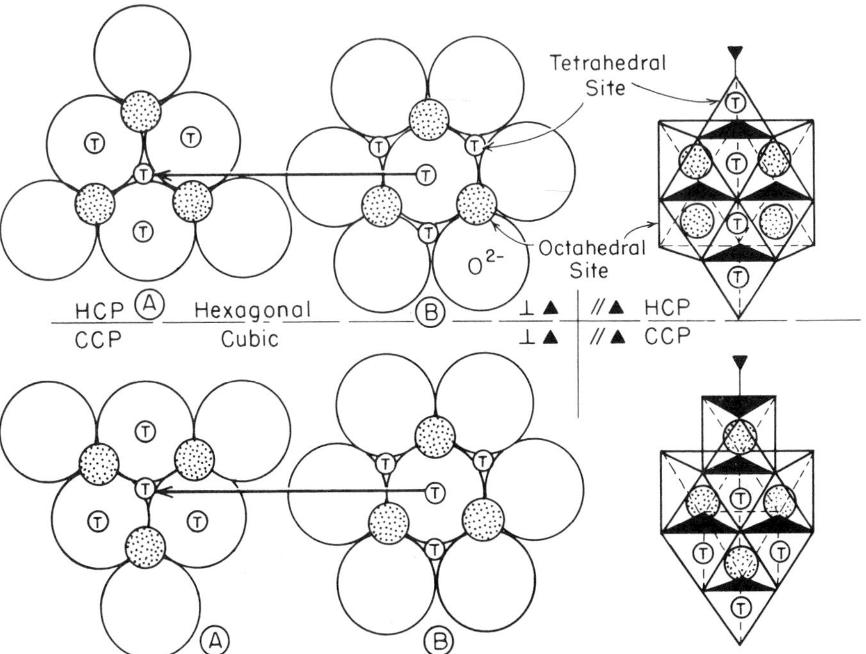

ions differently in different crystal structures, because one or the other arrangement will allow for a more geometrically equalized spacing of cations around an anion. Let us, for example, take the case of halite, NaCl, where our tennis balls will represent large Cl^- anions, and our Ping-Pong balls will represent the smaller Na^+ cations. In halite, each Cl^- is in CCP with XII Cl^- anions around each other. The Na^+ cations are too large to occupy the tetrahedral voids in the Cl^- array, so these will remain empty. The Na^+ cations all must thus occupy the CN VI octahedral sites or voids. Our central Cl^- anion will now be surrounded or coordinated as well by six Na^+ cations at corners of an octahedron: three above are opposed by three below. There are now VI Cl^- around each Na^+ and VI Na^+ around each Cl^-. If we change our Cl^- packing from CCP to HCP, we can still place VI Na^+ around each Cl^- anion, *but* our six points of positive charge, the Na^+ ions, will now be positioned directly above and below each other and the central Cl^- anion, outlining a trigonal prism rather than an octahedron of positive charge. The six points of positive charge will thus be closer together in HCP than in CCP. The CCP array is thus favored for NaCl because it permits more equidistant geometric placement of the six Na^+ cations. For this reason halite crystallizes in the cubic rather than in the hexagonal system. Halite is a VI : VI structure, because each ion is in CN VI.

By contrast, the mineral bromellite, BeO, crystallizes in the hexagonal crystal system. Each O^{2-} anion is in HCP, and now, because Be^{2+} is a very small ion, it will occupy only the tetrahedral, CN IV, sites or voids, leaving the CN VI sites empty. Electrical neutrality requires that only four bipositive Be ions be placed around each O^{2-} anion. The HCP of O^{2-} requires the placement of the four Be^{2+} ions at corners of a tetrahedron, with three below and one above the central O^{2-} anion. In this arrangement, all four BeO_4 tetrahedra point in the same direction, away from the viewer, and this gives rise to a polar structure. If the O^{2-} anions used CCP, the four Be^{2+} cations could not be placed geometrically equidistant, and one of the four would have to be placed too close to another. The hexagonal close-packed polar tetrahedral array is thus favored for bromellite, in contradistinction to halite. Bromellite is therefore a IV : IV structure, because cation and anion are each surrounded by four ions of the opposite species and charge.

Let us now return to our tennis and Ping-Pong balls to take a closer look at the radius ratios that control the site occupancies of the latter. We will start with a close-packed array of tennis balls. It is now immaterial whether we use CCP or HCP. The Ping-Pong ball has a radius of 1.9 cm, the tennis ball 3.2 cm. Thus the radius ratio is $R = r_{PB}/r_{TB} = 1.9/3.2 = 0.594$.

By simple geometry, we can predict that for $R = 0.594$, the Ping-Pong balls will occupy the six-fold CN sites and be in tangential contact with six tennis balls placed at corners of a regular octahedron (Fig. 4.4). Geometric relations show that it is solely the radius ratio R that governs the fit of small spheres in spherical voids or interstices of more or less close-packed large spheres. The maximum number of large spheres that may be in contact with a small sphere is controlled by the minimum radius ratio for a given geometry. The central cation must be in contact with all surrounding anions and must not "rattle" in the void, or else it must adopt a smaller CN polyhedron. The minimum radius ratios for the principal coordination polyhedra of ionic structures, and their geometric derivation, are shown in Table 4.1 and illustrated in Figure 4.4.

Note that it is immaterial whether the large spheres are tennis balls, $r =$ 3.2 cm, O^{2-} anions, $r = 1.40$ Å, or Cl^{1-} anions, $r = 1.81$ Å; it is solely the ratio of the small sphere to the large sphere that counts, $R = r_A/r_X$, where A is a small cation and X is a large anion. Note also that for all of these geometrically imposed limitations of occupancy of space we assume that we are dealing with rigid spheres of constant radius in the ionic model. Later it will become apparent that this treatment loses credibility because anions, like tennis balls, are indeed deformable and compressible. For now, we shall ignore or table this complexity, and deal only with our concept of spherical ions of constant radius in a given coordination polyhedron. We will now leave behind the useful tennis

Figure 4.4. Coordination polyhedra, their geometry, and limiting or minimum radius ratios, R. Large spheres are O^{2-} anions, $r = 1.40$ Å.

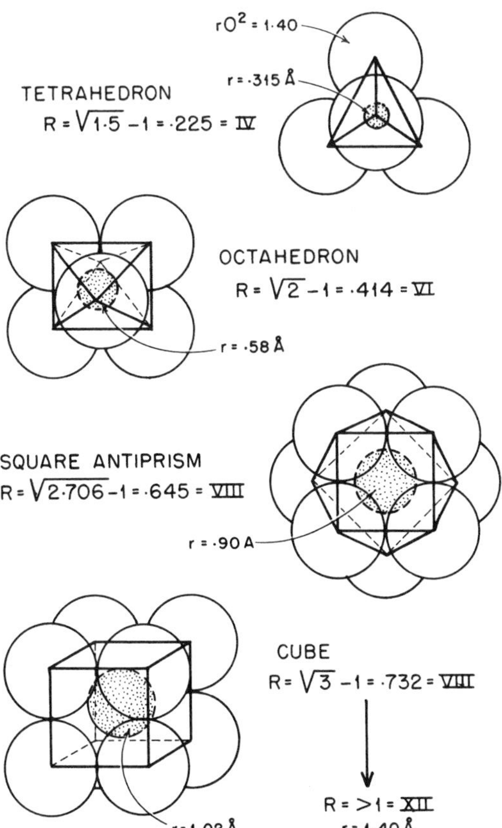

and Ping-Pong ball analogy in order to turn our attention to anions and cations and the derivation of their ionic radii.

Ionic radii

X-rays were discovered by Röntgen in 1895; the diffraction of X-rays by crystals was discovered by Friedrich, Knipping, and Laue in 1912; and the nature of the characteristic X-ray spectra was found by Mosely in 1913, shortly before his untimely death in World War I. Very soon thereafter, W. H. Bragg, W. L. Bragg, and others investigated and published the results of a number of crystal structure determinations based on X-ray analyses. Crystal chemistry was indeed born, and for the first time scientists had some concrete information on the arrangement and location of atoms and ions in crystals. It was not long before physicists, chemists, and geologists began to apportion the X-ray–determined interatomic distances into the radii of spherical ions in tangential contact.

Note that an ion in space has no fixed radius, because the electron cloud of the outermost electron falls off steadily with increasing distance from the nucleus. In crystals, however, in the ideal ionic case (e.g., Na^+F^-), two ions of opposite charge are attracted by Coulomb forces until the close approach of their electron clouds results in rapid repulsion. The balancing of attractive and repulsive forces at some equilibrium distance for a given pair of ions in a given crystal is equal to the A—X or interatomic distance, which, ideally, is the sum of the radii of each ion. The ions are believed to retain these radii in chemical combination with other elements in an additive sense. Recall, however, that electronegativity differences, $\Delta\chi$, for NaF and NaCl are 3.1 and 2.1, equivalent to bonds that are about 10% and 20% covalent, respectively. The orbitals are indeed overlapped, and the apportionment of interatomic distances, 2.31 Å (Na—F) and 2.76 Å (Na—Cl), into discrete radii for $Na^+ = 0.95$ Å, for $F^- =$ 1.36 Å, and for $Cl^- = 1.81$ Å, separated by a zero electron density, is a convenient distortion of the truth.

Table 4.1. *Minimum radius ratio and limiting geometry of regular coordination polyhedra*

Minimum radius ratio R	CN	Polyhedron	Geometry of limit
0.225	IV	Tetrahedron	$R = \sqrt{1.5} - 1$
0.414	VI	Octahedron	$R = \sqrt{2} - 1$
0.645	VIII	Square antiprism	$R = \sqrt{2.71} - 1$
0.732	VIII	Cube	$R = \sqrt{3} - 1$
1.0	XII	Cube–octahedron or hexagonal prism	$R = \sqrt{1}$

One of the first important derivations and compilations of the radii of ions was given by Wasastjerna in 1923. He apportioned interatomic distances on the basis of the molar refractivities of ionic salts in solution, on the assumption that molar refractivities were proportional to ionic volumes. He gave a value of 1.32 Å for the O^{2-} anion and 1.33 Å for the F^- anion. Goldschmidt used these anion values as a basis for the formulation of a fairly complete set of empirical ionic radii, which he published in 1926 (Table 4.2).

Pauling suggested that values of 1.40 Å for O^{2-} and 1.36 Å for F^- were more realistic, and in 1927 published a revised set of ionic radii based on an inverse variation of effective nuclear charge for isoelectronic elements of the alkali halides. He noted that the size of an ion is determined by the radial distribution of the outermost electron, and is inversely proportional to the effective nuclear charge acting on this electron. Effective nuclear charge Z' is equal to the proton number Z minus the screening constant S: $Z' = Z - S$. The screening constant S expresses the degree to which the inner electrons screen the nucleus of the ionized atom. Isoelectronic ions such as ${}_9F^-$ and ${}_{11}Na^+$ have the electronic structure of ${}_{10}Ne$, which is $1s^2 2s^2 2p^6$ and have an electronic screening constant $S = 4.52$. The interatomic distance in NaF = 2.31 Å. By division of this distance in the inverse ratio of the effective nuclear charge, Pauling derived ionic radii for Na^+ and F^- as follows:

$$\frac{r_{Na^+}}{r_F} = \frac{9 - 4.52}{11 - 4.52} = \frac{4.48}{6.48} = \frac{41\%(2.31)}{59\%(2.31)} = \frac{0.95\text{ Å}}{1.36\text{ Å}}$$

The radii are assumed to be additive, and thus subtraction of the radius of $Na^+ = 0.95$ from the interatomic distance for NaCl = 2.76 yields the value of the radius of 1.81 for Cl^-. Continuation of this substitution process leads to derivation of values for other ions.

In 1952, Ahrens revised Pauling's values by using different values for the alkali ions and by plotting regularities in their radii against ionization potentials. On this basis Ahrens provided the most complete table of ionic radii available, one that has been widely used by geochemists and mineralogists (Table 4.2).

In 1964, Fumi and Tosi derived a new set of ionic or "crystal" radii (Table 4.2) for the alkali halides stated to have a very high degree of accuracy. These radii were based on Born repulsive parameters and electron density plots. The Fumi and Tosi radii are startling, in that they indicate considerably larger values for cation radii and correspondingly smaller values for anion radii than do any of the previous sets. If cations are larger and anions are smaller than previously indicated, then the concept of the radius ratio would appear to be invalid: individual bonds would then be more important than packing considerations in determining site occupancies. Perhaps more work with electron density plots will help resolve this major inconsistency.

In 1969, Shannon and Prewitt derived another set of "effective" ionic radii stated to represent most accurately the values found in crystals (Table 4.2).

Table 4.2. *Ionic radii by different authors for different coordinations*

Ion	E.C.	C.N.	G	A	P	S-P	W-M	F-T
$_{3}Li^{+}$	$1s^2$	IV	-	-	-	.59	.68	.73
		VI	.78	.68	.60	.74	.82	.88
$_{4}Be^{2+}$	$1s^2$	IV	.34	.35	.31	.27	.35	.31
$_{5}B^{3+}$	$1s^2$	III	-	.23	.20	.02	.10	.16
		IV	-	-	-	.12	.20	.26
$_{8}O^{2-}$	$2p^6$	II	-	-	-	1.35	1.26	1.21
		III	-	-	-	1.36	1.28	1.22
		IV	-	-	-	1.38	1.30	1.24
		VI	1.32	1.40	1.40	1.40	1.32	1.26
$_{9}F^{-}$	$2p^6$	II	-	-	-	1.285	1.21	1.145
		III	-	-	-	1.30	1.22	1.16
		IV	-	-	-	1.31	1.23	1.17
		VI	1.33	1.36	1.36	1.33	1.25	1.19
$_{11}Na^{+}$	$2p^6$	VI	.98	.97	.95	1.02	1.10	1.16
		VII	-	-	-	1.13	-	1.27
		VIII	-	-	-	1.16	1.24	1.30
		IX	-	-	-	1.32	-	1.46
$_{12}Mg^{2+}$	$2p^6$	IV	-	-	-	.58	.66	.63
		VI	.78	.66	.65	.72	.80	.86
		VIII	-	-	-	.89	.97	1.03
$_{13}Al^{3+}$	$2p^6$	IV	-	-	-	.39	.47	.53
		V	-	-	-	.48	-	.62
		VI	.57	.51	.50	.53	.61	.67
$_{24}Cr^{3+}$	$3d^3$	VI	.64	.63	.69	.615	.70	.755
$_{25}Mn^{2+}$	$3d^5$	VI	.91	.80	.80	.82	.91	.96
		VIII	-	-	-	.93	1.01	1.07
Mn^{3+}	$3d^4$	VI	.70	.66	.66	.65	-	.79
Mn^{4+}	$3d^3$	VI	.52	.60	.54	.54	.62	.68
$_{26}Fe^{2+}$	$3d^6$	IV	-	-	-	.63	.71	.77
		VI	.83	.74	.76	.77	.86	.91
		VIII	-	-	-	(.91)	-	-
$_{26}Fe^{3+}$	$3d^5$	IV	-	-	-	.49	.57	.63
		VI	.67	.64	.64	.645	.73	.785
$_{27}Co^{2+}$	$3d^7$	VI	.82	.72	.74	.735	.83	.875
Co^{3+}	$3d^6$	VI	-	.63	.63	.61	.69	.65
$_{28}Ni^{2+}$	$3d^8$	VI	.78	.69	.72	.70	.77	.84
Ni^{3+}	$3d^7$	VI	-	-	.62	.60	-	.74
$_{29}Cu^{+}$	$3d^{10}$	IIs-p	-	-	1.18	.46	-	.60
		VI	-	-	.96	-	-	-
Cu^{2+}	$3d^9$	IVsq	-	-	-	.62	-	.76
		V	-	-	-	.65	-	.79
		VI	-	.72	-	.73	-	.87
$_{30}Zn^{2+}$	$3d^{10}$	IV	-	-	-	.60	.68	.74
		VI	.83	.74	.74	.745	.83	.885
$_{31}Ga^{3+}$	$3d^{10}$	IV	-	-	-	.47	-	.61
		V	-	-	-	.55	-	.69

Ion	Config.	CN						
$_{14}Si^{4+}$	$2p^6$	IV	-	-	-	.26	.34	.40
		VI	.39	.42	.41	.40	.48	.54
$_{15}P^{3+}$	$2p^6$	IV	.34	.35	.34	.17	.25	.31
$_{16}S^{2-}$	$3p^6$		1.74	1.84	1.84	-	1.72	-
$_{17}Cl^{-}$	$3p^6$		1.81	1.81	1.81	-	1.72	-
$_{19}K^{+}$	$3p^6$	VI	1.33	1.33	1.33	1.38	1.46	1.52
		VIII	-	-	-	1.51	1.59	1.65
		IX	-	-	-	1.55	-	1.69
		X	-	-	-	1.59	1.67	1.73
		XII	-	-	-	1.60	1.68	1.74
$_{20}Ca^{2+}$	$3p^6$	VI	1.06	.99	.99	1.00	1.08	1.14
		VII	-	-	-	1.07	-	1.21
		VIII	-	-	-	1.12	1.20	1.26
		IX	-	-	-	1.18	-	1.32
		X	-	-	-	1.28	1.36	1.42
		XII	-	-	-	1.35	1.43	1.49
$_{21}Sc^{3+}$	$3p^6$	VI	.83	.81	.81	.73	.83	.87
		VIII	-	-	-	.87	.95	1.01
$_{22}Ti^{3+}$	$3d^1$	VI	.69	.76	-	.67	-	.81
Ti^{4+}	$3p^6$	VI	.64	.64	.68	.605	.69	.745
$_{23}V^{3+}$	$3d^2$	VI	.65	.74	.74	.64	.72	.78
V^{4+}	$3d^1$	VI	.61	.63	.60	.59	.62	.73
V^{5+}	$3p^6$	IV	-	-	-	.355	.44	.495
		V	-	-	-	.46	-	.60
		VI	.59	.59	.59	.54	.62	.68
		VI	.62	.62	.62	.62	-	.76
$_{32}Ge^{4+}$	$3d^{10}$	IV	.41	-	-	.40	-	.54
		VI	-	.53	.53	.54	-	.68
$_{34}Se^{2-}$	$4p^6$	VI	1.91	-	1.98	-	1.88	-
$_{35}Br^{-}$	$4p^5$	VI	1.96	1.95	1.95	-	-	-
$_{37}Rb^{+}$	$4p^6$	VI	1.49	1.47	1.48	1.49	1.57	1.63
		VIII	-	-	-	1.60	1.68	1.74
		XII	-	-	-	1.73	1.74	1.87
$_{38}Sr^{2+}$	$4p^6$	VI	1.27	1.12	1.13	1.16	1.21	1.30
		VIII	-	-	-	1.25	1.33	1.39
		X	-	-	-	1.32	1.40	1.46
		XII	-	-	-	1.44	1.48	1.58
$_{39}Y^{3+}$	$4p^6$	VI	1.06	.92	.93	.89	.98	1.03
		VIII	-	-	-	1.015	1.10	1.155
		IX	-	-	-	1.10	-	1.24
$_{40}Zr^{4+}$	$4p^6$	VI	.87	.79	.80	.72	.80	.86
		VIII	-	-	-	.84	.92	.98
$_{41}Nb^{5+}$	$4p^6$	IV	-	-	-	.32	-	.46
		VI	.69	.69	.70	.64	-	.78
$_{42}Mo^{4+}$	$4d^2$	VI	.68	.70	-	.65	.73	.79
Mo^{6+}	$4p^6$	IV	-	-	-	.42	.50	.56
		VI	-	.62	.62	.60	.68	.74
$_{44}Ru^{4+}$	$4d^4$	VI	.65	.67	-	.62	-	.76
$_{45}Rh^{3+}$	$4d^6$	VI	.68	.68	-	.665	-	.805
Rh^{4+}	$4d^5$	VI	-	-	-	.615	-	.775

Table 4.2. *(cont.)*

Ion	E.C.	C.N.	G	A	P	S-P	W-M	F-T
$_{46}Pd^{2+}$	$4d^8$	VI	-	.80	.86	.86	-	1.00
Pd^{4+}	$4d^6$	VI	-	.65	-	.62	-	.76
$_{47}Ag^{+}$	$4d^{10}$	II	-	-	1.39	.67	-	.81
		VI	1.13	1.26	1.26	1.15	1.23	1.29
		VIII	-	-	-	1.07	1.15	1.21
$_{49}In^{3+}$	$4d^{10}$	VI	.92	.81	.81	.79	-	.93
		VIII	-	-	-	.92	-	1.06
$_{50}Sn^{4+}$	$4d^{10}$	VI	.74	.71	.71	.69	.77	.83
$_{51}Sb^{5+}$	$4d^{10}$	VI	.60	.62	.62	.61	.69	.75
$_{52}Te^{2-}$	$5p^6$	VI	2.11	-	2.21	-	-	-
$_{53}I^{-}$	$5p^6$	VI	2.20	2.16	2.16	-	-	-
$_{55}Cs^{+}$	$5p^6$	VI	1.65	1.67	1.69	1.70	1.78	1.84
		IX	-	-	-	1.78	1.82	1.92
		X	-	-	-	1.81	1.89	1.95
		XII	-	-	-	1.88	1.96	2.02
$_{56}Ba^{2+}$	$5p^6$	VI	1.43	1.34	1.35	1.36	1.44	1.50
		VIII	-	-	-	1.42	1.50	1.56
		IX	-	-	-	1.47	-	1.61
		X	-	-	-	1.52	1.60	1.66
		XII	-	-	-	1.60	1.68	1.74
$_{57}La^{3+}$	$4d^{10}$	VI	1.22	1.14	1.15	1.06	1.13	1.20
		VII	-	-	-	1.10	-	1.24
		VIII	-	-	-	1.18	1.26	1.32
		IX	-	-	-	1.20	-	1.34
		X	-	-	-	1.28	1.36	1.42
		XII	-	-	-	1.32	1.40	1.46

Ion	E.C.	C.N.	G	A	P	S-P	W-M	F-T
$_{74}Ho^{3+}$	$4f^{10}$	VI	-	.91	.97	.89	-	1.03
		VIII	-	-	-	1.02	-	1.16
$_{68}Er^{3+}$	$4f^{11}$	VI	-	.89	.96	.88	-	1.02
		VIII	-	-	-	1.00	-	1.14
$_{69}Tm^{3+}$	$4f^{12}$	VI	-	.87	.95	.87	-	1.01
		VIII	-	-	-	.99	-	1.13
$_{70}Yb^{3+}$	$4f^{13}$	VI	-	.86	.94	.86	-	1.00
		VIII	-	-	-	.98	-	1.12
$_{71}Lu^{3+}$	$4f^{14}$	VI	.99	.85	.93	.85	-	.99
		VIII	-	-	-	.97	-	1.11
$_{72}Hf^{4+}$	$4f^{14}$	VI	.85	.78	.81	.71	-	.85
		VIII	-	-	-	.83	-	.97
$_{73}Ta^{5+}$	$5p^6$	VI	.69	.68	-	.64	.72	.78
$_{74}W^{4+}$	$5d^2$	VI	.68	.70	-	.65	.73	.79
W^{6+}	$5p^6$	IV	-	-	-	.41	.50	.55
		VI	-	.62	-	.58	.68	.72
$_{75}Re^{4+}$	$5d^3$	VI	-	.72	-	.63	-	.77
$_{76}Os^{4+}$	$5d^4$	VI	.67	.69	-	.63	-	.77
$_{77}Ir^{4+}$	$5d^5$	VI	.66	.68	-	.63	-	.77
$_{78}Pt^{4+}$	$5d^6$	VI	-	.65	-	.63	-	.77
$_{79}Au^{+}$	$5d^{10}$	VI	-	1.37	1.37	-	-	-
$_{80}Hg^{2+}$	$5d^{10}$	VI	-	1.10	1.10	1.02	1.10	1.16
		VIII	-	-	-	1.14	1.22	1.28
$_{81}Tl^{+}$	$6s^2$	VI	1.49	1.47	1.40	1.50	-	1.64
Tl^{3+}	$5d^{10}$	VI	1.05	.95	-	.85	-	1.02
		VIII	-	-	-	1.00	-	1.14

$_{58}Ce^{3+}$	**6s**^{1}	VI	1.18	1.07	1.11	1.03	1.09	1.17
		VIII	-	-	-	1.14	1.22	1.29
		XII	-	-	-	1.28	-	1.43
Ce^{4+}	**5p**^{6}	VI	1.02	.94	1.01	.80	.88	.94
		VIII	-	-	-	.97	1.05	1.11
$_{59}Pr^{3+}$	**4f**^{2}	VIII	1.16	-	-	1.14	-	1.28
$_{60}Nd^{3+}$	**4f**^{3}	VIII	1.15	1.04	1.08	.995	-	1.135
$_{61}Pm^{3+}$	**4f**^{4}	VI	-	-	1.06	.98	-	1.12
$_{62}Sm^{3+}$	**4f**^{5}	VI	-	1.00	1.04	.96	-	1.10
		VIII	-	-	-	1.09	-	1.23
$_{63}Eu^{2+}$	**4f**^{7}	VIII	1.25	-	1.12	1.25	-	1.39
Eu^{3+}	**4f**^{6}	VI	-	.98	1.03	.95	-	1.09
		VIII	-	-	-	1.07	-	1.21
$_{64}Gd^{3+}$	**4f**^{7}	VI	-	.97	1.02	.94	-	1.08
		VIII	-	-	-	1.06	-	1.20
$_{65}Tb^{3+}$	**4f**^{8}	VI	-	.93	1.00	.92	-	1.06
		VIII	-	-	-	1.04	-	1.18
$_{66}Dy^{3+}$	**4f**^{9}	VI	-	.92	.99	.91	-	1.05
		VIII	-	-	-	1.03	-	1.17
$_{82}Pb^{2+}$	**6s**^{2}	VI	1.32	1.20	1.20	1.18	1.26	1.32
		VIII	-	-	-	1.29	1.37	1.45
		IX	-	-	-	1.33	-	1.47
		XII	-	-	-	1.49	1.57	1.63
Pb^{4+}	**5d**^{10}	VI	.84	.84	.84	.775	-	.945
		VIII	-	-	-	.94	-	1.08
$_{83}Bi^{3+}$	**6s**^{2}	V	1.1	-	-	.99	-	1.13
		VI	-	.96	-	1.02	-	1.16
		VIII	-	-	-	1.11	-	1.25
Bi^{5+}	**5d**^{10}	VI	-	.74	.74	-	-	-
$_{89}Ac^{3+}$	**6p**^{6}	VI	-	-	1.18	-	-	-
$_{90}Th^{4+}$	**6p**^{6}	VI	1.10	1.02	1.14	1.00	1.08	1.14
		VIII	-	-	-	1.06	1.12	1.20
$_{92}U^{4+}$	**5f**^{2}	VI	1.05	.97	.97	-	-	-
		VIII	-	-	-	1.00	1.08	1.14
		IX	-	-	-	1.05	-	1.19
U^{6+}	**6p**^{6}	II	-	-	-	.45	-	.59
		VI	-	.80	-	.75	.81	.89

Sources:

G: V. M. Goldschmidt (1926). Skrif.Norske Videnskaps-Akademie, Oslo, I, Mat. Naturviske. (1945) *Soil Sci.* 60, No. 1, July.

A: L. H. Ahrens (1952). *Geochim. et Cosmochim. Acta,* 2: 158.

P: Linus Pauling (1960). *The nature of the chemical bond,* 3d ed. Cornell Univ. Press, Ithaca, N.Y.

S–P: R. D. Shannon and C. T. Prewitt (1969). *Acta Crystallogr. Sect. B* 25: 925.

W–M: E. J. W. Whittaker and R. Muntus (1970) *Geochim. et Cosmochim. Acta,* 34: 952–3.

F–T: F. Fumi and M. T. Tosi (1964). *J. Phys. Chem. Solids,* 25: 31.

Their list of radii is extensive and includes values for each cation and anion in different coordination sites. Although Goldschmidt and Pauling were aware of the variation of ionic radius with change in CN, and provided correction factors for increasing or decreasing values of radii with corresponding changes in CN, they did not suggest variations in the radii of the anions with a change in cation CN. Shannon and Prewitt (as do Fumi and Tosi) provide different values for the O^{2-} and F^- anions when these are coordinated by II, III, IV, VI, or VIII cations. Implied, but not stated by these authors, is the presence of oxygen anions of different radius in the same mineral, as in beryl, which contains O^{2-} anions coordinated by both II and III cations. Shannon and Prewitt do give values that appear to be applicable to most minerals, but, because they include the effects of covalency, they also mask them. Normally, departure from ideal ionic sums of radii is used as a measure of covalency in minerals.

Still another set of empirically derived radii was presented by Whittaker and Muntus in 1970. They also provide different radii for cations and anions in different coordinations (see Table 4.2). The values of these investigators appear to lie between those of Shannon and Prewitt and of Fumi and Tosi.

In view of the large differences in ionic radii carefully derived by different investigators, it is surprising that the ionic model has been so successful in evaluation of crystal structure. The reason for such success lies in the fact that ionic and covalent models may yield the same result. If Pauling's concepts of electronegativity differences versus degree of ionic and covalent bonding, and resonance between ionic and covalent structures, are correct, then neither the ionic nor the covalent model per se accurately represents the chemical bonding in most minerals. An example of the two models, ionic and covalent, leading to similar results is afforded by the common mineral, sphalerite, ZnS, for which the interatomic distance is 2.35 Å (see Fig. 3.4). Radii are as follows:

Covalent sp^3 tetrahedral model	(Å)	Ionic model	(Å)
Zn_{sp^3}	1.31	Zn^{2+}	0.60
S_{sp^3}	1.04	S^{2-}	1.84
Sum of radii	2.35		2.44

$$R = r_{Zn^{2+}}/r_{S^{2-}} = 0.326 = \text{CN IV}$$

Thus, both the extreme covalent sp^3 hybrid model and the extreme ionic model place four atoms or ions of S around Zn, and vice versa. According to Pauling's ideas, neither is correct. Pauling's more recent *electroneutrality concept* states that no atom ever carries a positive charge greater than unity; any greater amount of charge becomes neutralized by transfer of anionic electrons to the cation. This is a good place to close our discussion of ionic radii.

Summary

The covalent bond is directional or homopolar. The ionic bond is nondirectional or heteropolar.

The interatomic distance between atoms is governed by the balance of their attractive and repulsive forces.

The radius ratio (cation radius/anion radius) governs the packing of atoms around each other, or coordination.

Electrostatic bond strength is cation charge/cation CN, and is the fractional contribution of each cation to the charge on each anion it surrounds.

Bonds are never entirely ionic; percent ionic bonding may be estimated from differences in electronegativity and ionization potential of atoms in question. Alkali fluorides are the most ionic compounds.

The Si—O bond is 51% ionic and thus resonates as a mixture of ionic and covalent bonding.

The ionic bond requires spherical electron density: thus B-group elements with p valence electrons will form covalent complex anions.

Pauling's five rules codify the behavior of ionic compounds.

Rule 1. The number of anions surrounding a cation is a function of their radius ratio.

Rule 2. The number of coordination polyhedra sharing a common anion corner is limited by the EBS (cation charge/cation CN).

Rule 3. Sharing of corners, edges, and faces of coordination polyhedra leads to their distortion, and decreases stability, because of the electrostatic repulsion of cations.

Rule 4. Polyhedra with small cations of high charge tend not to share anions.

Rule 5. A large number of different kinds of coordination polyhedra in a given mineral tends to decrease its stability. There are many exceptions to this rule.

The closeness of packing of spheres of equal radius is the same for both hexagonal and cubic close-packed arrays. In both HCP and CCP, each anion is the common corner of eight tetrahedra and six octahedra. A maximum of six of these fourteen cation coordination sites is occupied in minerals, with the cations placed in geometric arrays, equidistant from one another. The radius ratio of rigid spheres governs the placement of small spheres in the interstices of large spheres. Minimum radius ratios are tetrahedral (CN IV, $R = 0.225$), octahedral (CN VI, $R = 0.414$), square antiprismatic (CN VIII, $R = 0.645$), cubic (CN VIII, $R = 0.732$), and cubo–octahedral or hexagonal–antiprismatic (CN XII, $R = 1.0$).

Ionic radii are ideal distributions of interatomic distances derived from X-ray analyses. They have been derived by different methods by

Wasastjerna, Goldschmidt, Pauling, Ahrens, Fumi and Tosi, Shannon and Prewitt, and Whittaker and Muntus. All these ionic radii are tabulated and compared.

Bibliography

Ahrens, L. H. (1952). The use of ionization potentials. Part 1. Ionic radii of the elements. *Geochim. et Cosmochim. Acta,* 2: 155–69.

Brown, G. E., and Gibbs, G. V. (1969). Oxygen coordination and the Si—O bond. *Am. Mineral.,* 54: 1528–39.

Fumi, F. G., and Tosi, M. P. (1964). Ionic sizes and Born repulsive parameters in the NaCl-type alkali halides. I. The Huggins–Mayer and Pauling forms. *J. Phys. Chem. Solids,* 25: 31–43.

Goldschmidt, V. M. (1954). *Geochemistry.* Clarendon Press, Oxford.

Pauling, L. (1960). *The nature of the chemical bond,* 3rd ed. Cornell Univ. Press, Ithaca, N.Y., Chap. 13.

(1987). Determination of ionic radii from cation–anion distances in crystal structures: Discussion. *Am. Mineral, 72*: 1016.

Shannon, R. D., and Prewitt, C. T. (1969). Effective ionic radii in oxides and fluorides. *Acta Crystallogr. Sect. B,* 25: 925–45.

Tosi, M. P., and Fumi, F. G. (1964). Ionic sizes and Born repulsive parameters in the NaCl-type alkali halides. II. The generalized Huggins–Mayer form. *J. Phys. Chem. Solids,* 25: 45–52.

Wells, A. F. (1970). *Models in structural inorganic chemistry.* Oxford Univ. Press, New York.

Whittaker, E. J. W., and Muntus, R. (1970). Ionic radii for use in geochemistry. *Geochim. et Cosmochim. Acta,* 34: 952–3.

Wondratschek, Hans (1987. Determination of ionic radii from cation–anion distances in crystal structures. *Am. Mineral.,* 72: 82.

5 Pauling's second rule of electrostatic valency in ionic or coordination compounds

Whereas Pauling's first rule uses ionic radii of cation and anion and their radius ratio R to establish the number of anions that surround each cation (cation CN), it is Pauling's second rule that employs electrostatic valency or bond strength to limit the number of cations that surround each anion (anion CN). Implicit in the second rule is the requirement that electrical charge of anion and cation be neutralized locally by nearest neighbors. In minerals containing one species of anion, such as O^{2-} or F^-, it might be tacitly assumed that the cation array or cation environment around each anion is identical. In many minerals, this is true, and it may be seen that in the mineral rutile, TiO_2, each O^{2-} anion is surrounded by three Ti^{4+}/VI cations that contribute fractional electrostatic bond strengths of 4+/VI, exactly satisfying locally the negative valency of each O^{2-} anion. Similarly, in the mineral fluorite, CaF_2, each F^- anion is surrounded by four Ca^{2+}/VIII cations, each contributing a fractional electrostatic bond strength of 2+/VIII exactly satisfying locally the univalency of each F^- anion. We will follow the practice of using schematic *electrostatic bond strength (EBS) diagrams,* illustrated in Figure 5.1, to express electrostatic valency solutions, and to derive the structural formula for a given mineral.

We will classify rutile and fluorite into the category Pauling Rule 2A group, representing those minerals that contain one unique anion and, hence, one unique EBS solution; all anions are thus equivalent. Other minerals may contain two or more anions with different cation environment, or they may contain two different anions of different valency, such as O^{2-} and F^-. In such minerals, each different anion will require a different electrostatic valency solution. We will classify these in the category Pauling Rule 2B group.

In a third category, the Pauling Rule 2C group, we will include those minerals in which the same anion (e.g., O^{2-}) is surrounded by two or more different cation arrays, none of which satisfies Pauling's second rule, because each anion is either undercharged or overcharged. However, in these minerals electrostatic neutrality is maintained by being spread out over a larger domain. For example, the sum of bond strengths reaching three different O^{2-} will total 6+, although each will individually total more or less than the required 2+. Thus, in the pyroxene mineral group, diopside, $CaMgSi_2O_6$, contains three different oxygen anions; O-1, O-2, and O-3, which are surrounded, respectively, by cation arrays contributing EBS summations of +1.92, +1.58, and +2.50, for a total of 6+. In such minerals it has been observed from X-ray studies that the cations Ca^{2+}, Mg^{2+}, and Si^{4+} do not lie in the centers of their respective coordination polyhedra, because they move away from overcharged anions and closer to undercharged anions. This led the eminent crystallographer Zachariasen (1963) to postulate that EBS is inversely proportional to bond length. We may refer to this idea as Zachariasen's elaboration of Pauling's second rule. It cannot be applied without detailed values of interatomic distances obtainable only from X-ray analysis of crystals. We will now consider several examples of each of these categories of Pauling electrostatic valency types.

Pauling Rule 2A group

We have already cited rutile and fluorite as examples of minerals containing one unique anion. A more complex representative of this category is

Figure 5.1. Electrostatic bond strength diagrams for minerals in which all oxygens have equivalent near neighbor cation arrays: (A) rutile; (B) fluorite; (C) garnet; (D) zircon; (E) xenotime. All oxygen anions in structure are identical.

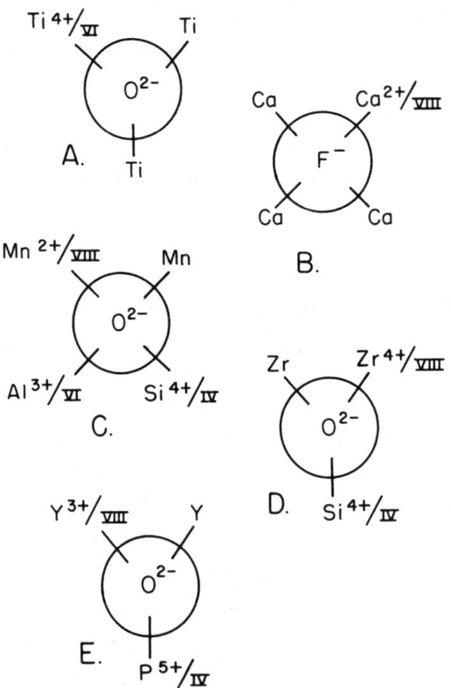

the garnet mineral group. The mineral spessartine is a garnet of the composition $Mn_3Al_2[SiO_4]_3$, in which cation coordination numbers, predicted from radius ratios, place each Mn^{2+} ion in CN VIII, each Al^{3+} ion in CN VI, and each Si^{4+} ion in CN IV. Electrostatic valency considerations, now confirmed by X-ray location of the ions, require that each O^{2-} anion be surrounded by two Mn^{2+} ions contributing 2+/VIII, one Al^{3+} ion contributing 3+/VI, and one Si^{4+} ion contributing 4+/IV, as illustrated in Figure 5.1 (also see Figs. 12.7, 12.8).

At first, it might appear that this arrangement would not be compatible with the garnet formula, but, because the positive charge on each cation is divided equally among all neighboring anions, the garnet formula may be derived from the bond strength diagram as follows:

Mn^{2+}	CN VIII	$\frac{1}{8}$ of each 2 Mn reaches an O^{2-} ion
Al^{3+}	CN VI	$\frac{1}{6}$ of each Al reaches an O^{2-} ion
Si^{4+}	CN IV	$\frac{1}{4}$ of each Si reaches an O^{2-} ion

Thus, it follows that

$$\left.\begin{array}{l} \frac{1}{8}\,Mn^{2+} \\ \frac{1}{8}\,Mn^{2+} \\ \frac{1}{6}\,Al^{3+} \\ \frac{1}{4}\,Si^{4+} \\ 1\,O \end{array}\right\} \times 12 \qquad \begin{array}{l} Mn_3 \\ \\ Al_2 \\ Si_3 \\ O_{12} \end{array}$$

Here, it is important to know that whenever the grouping $[SiO_4]$ appears in a mineral formula, the $[SiO_4]$ tetrahedra are independent of one another and there will be but one unique O^{2-} anion in the structure. Thus, in minerals such as olivine, $(Mg,Fe)_2[SiO_4]$ and zircon, $Zr[SiO_4]$, each O^{2-} anion will be in an identical cation environment. In olivine, each and every O^{2-} anion will be surrounded by one Si, 4+/IV, and three (Mg,Fe), 2+/VI, for a total EBS of 2+. In zircon, each O^{2-} anion will be surrounded by one Si, 4+/IV, and two Zr, 4+/VIII, ions for a total EBS of 2+ (Fig. 5.1). In xenotime, YPO_4, isostructural with zircon, each O^{2-} anion is bonded by one P^{5+}/IV and two Y^{3+}/VIII ions in a zircon-type lattice array (Figs. 5.1, 12.5, 12.6).

Pauling Rule 2B group

Whenever SiO_4 tetrahedra share corners with one another, it is no longer possible for all O^{2-} anions to be equivalent. An oxygen anion common to two tetrahedral corners will be in contact with two Si^{4+} ions, and 4+/IV × 2 will completely saturate the valency of the common O^{2-} anion. An oxygen anion that links one or more corners of two polyhedra is called a *bridging oxygen.* The bridging oxygen will thus have a different cation environment from nonbridging oxygens. A simple example is provided by the mineral thortveitite, $Sc_2[Si_2O_7]$. Here, two SiO_4 tetrahedra are joined by a bridging oxygen to produce the dumbbell-shaped Si_2O_7 group. The bridging oxygen will be in contact with two Si^{4+} ions, and the six remaining oxygens, nonbridging, will each be in contact with only one Si^{4+} ion and two Sc^{3+} ions, the latter in CN VI. The bond strength diagram shown in Figure 5.2 illustrates the presence of two different O^{2-} anions in the structure, implicit in the formula $Sc_2[Si_2O_7]$. (Also see Fig. 12.2.)

Another good example of this category is provided by the mineral beryl, $Al_2Be_3[Si_6O_{18}]$. Here, each SiO_4 tetrahedron shares two of its corner oxygens with adjoining tetrahedra to build hexagonal Si_6O_{18} rings. The six O^{2-} anions forming the center of each ring will be in contact with two Si^{4+} cations, completely neutralizing their charge, and thus may not be in contact with either Al^{3+} or Be^{2+}. The remaining twelve O^{2-} anions will each be in contact with one Si, 4+/IV, one Be, 2+/IV, and one Al, 3+/VI, for a total of 2+. The bond strength diagram for beryl is shown in Figure 5.2. (Also see Fig. 14.2.)

In thortveitite, Si_2O_7 dumbbell groups are linked together by Sc^{3+} ions; in beryl, Si_6O_{18} hexagonal rings are joined together by Al^{3+} and Be^{2+} ions. Each mineral contains nonequivalent O^{2-} ions.

In the sheet silicate or phyllosilicate minerals, each SiO_4 tetrahedron shares three of its four corners with adjoining tetrahedra to form infinite two-dimensional Si_4O_{10} sheets. It will be apparent that the three common tetrahedral corner oxygens will each be in contact with two Si^{4+} ions that will completely neutralize their charge. The fourth, unshared, oxygen will be in contact with only one Si^{4+} ion, and it may contact other cations.

In the mineral talc, $Mg_3[Si_4O_{10}](OH)_2$, the unshared O^{2-} anion will be in contact with three Mg cations, 2+/VI, in addition to one Si ion, 4+/IV, for a total of 2+. The hydroxyl ion, $(OH)^-$, will contact three Mg ions but may not be in contact with an Si ion. Thus, in talc, we have three different anions,

Figure 5.2. Electrostatic bond strength distribution in minerals containing two nonequivalent oxygens, with nonequivalent near neighbor cation arrays. (A) thortveitite; (B) beryl.

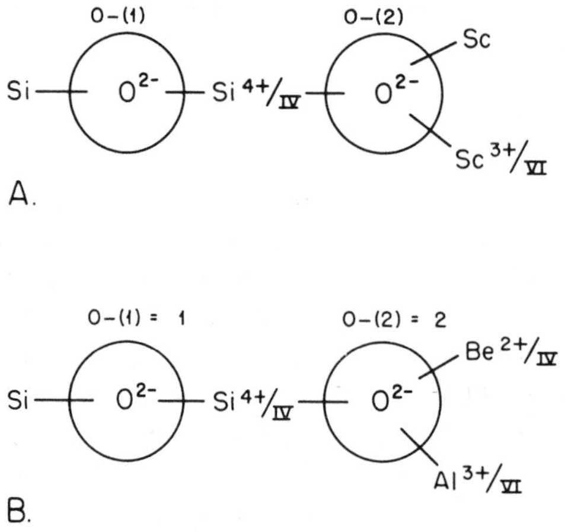

each having its negative valency satisfied in different ways by nearest cation neighbors. We may organize these as a sandwich, with Mg cations lying in between and joining opposed tetrahedral $Si_2O_5(OH)_2$ sheets, as schematically illustrated in Figure 5.3 (also see Fig. 15.2). We have three basal oxygens, O-(1), one apical oxygen, O-(2), in each tetrahedron, and one (OH) ion outside of each tetrahedron. The bond strength diagram of talc is shown in Figure 5.3.

Some minerals from Group 2B incorporate anions of two different valencies into their structure, necessitating two different cation environments. A good example is the rare earth carbonate mineral bastnaesite, $CeF[CO_3]$, containing the anions O^{2-} and F^-. In this mineral, the relatively large cerium ion Ce^{3+} has radius of 1.14 Å for CN VIII and 1.29 Å for CN XII, yielding radius ratios with the O^{2-} anion of 0.838 and 0.948, respectively, which are compatible with either CN VIII or CN XII. The EBS solution, however, suggests that the CN of Ce^{3+} is actually IX rather than VIII or XII. X-ray analysis shows that each Ce^{3+} ion is coplanar with three F^- anions and in contact with these, plus three O^{2-} above and another three O^{2-} below in a geometrically irregular CN IX site. Each F^- anion is surrounded by three Ce^{3+} near neighbors; each O^{2-} anion is in turn surrounded by two Ce^{3+} and one C^{4+} ion of a $[CO_3]$ group. Each carbon atom is actually covalently bonded to a distorted planar triad of O^{2-} anions in a $[CO_3]^{2-}$ anion complex. Thus, each such $[CO_3]^{2-}$ planar anion is surrounded

Figure 5.3. Electrostatic bond strength distribution in minerals containing three different anions, as in talc, O-(1), O-(2), and (OH) = O-(3).

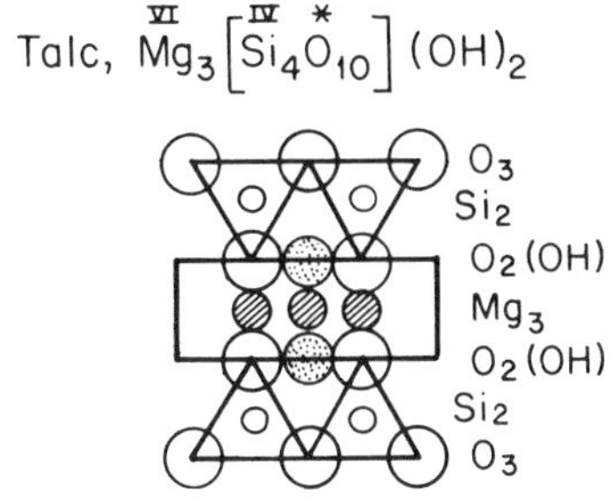

IV II IV III
* $[Si_4\ ①_6\ ②_4](OH)_2$

O-(1)
O^{2-}
Si
Si $\frac{4+}{IV}$
O-(2)
O^{2-}
Mg
O-(3) = (OH)
Mg
(OH)
Mg $\frac{2+}{VI}$
Mg $\frac{2+}{VI}$

by six trivalent cerium ions to yield an electrostatic bond summation of Ce^{3+}/IX = 18/9 = +2, and three such Ce bonds reach each F^- anion for a summation of +1. The EBS diagram is illustrated in Figure 5.4A (also see Figs. 19.7, 19.8).

Still another example is provided by the cubic mineral sulfohalite, $Na_6[SO_4]_2FCl$, which may be alternatively written as $2Na_2[SO_4]\cdot NaF\cdot NaCl$. This mineral has three different anions: divalent oxygen, univalent fluorine, and univalent chlorine. The elegant cubic structure is built of tetrahedra of oxygen anions around each sulfur ion, $[SO_4]^{2-}$, and octahedra of four oxygen anions plus one fluorine anion and one chlorine anion. The four O^{2-} anions around each Na^+ cation are corner oxygens of four different $[SO_4]^{2-}$ tetrahedra, and the F^- and Cl^- anions are centers of octahedral groupings of six Na^+ cations. The coordination numbers of each ion are summarized below, and the EBS diagram is given in Figure 5.4B. The crystal structure is further illustrated in Figure 5.5 to clarify the polymerization of polyhedra.

Sulfohalite $Na_6^{VI}[SO_4^{XII}]_2F^{VI}Cl^{VI}$ or

$2Na_2^{VI}[S^{IV}O_4^{IV}]\cdot Na^{VI}F^{VI}\cdot Na^{VI}Cl^{VI}$

Each S^{6+}	is coordinated by IV O^{2-}
Each Na^+	is coordinated by IV O^{2-}, 1 F^-, 1 Cl^- = CN VI
Each F^-	is coordinated by VI Na
Each Cl^-	is coordinated by VI Na^+

Figure 5.4. Electrostatic bond strength distribution in minerals with different species of anions, O, F, and Cl: (A) bastnaesite; (B) sulfohalite.

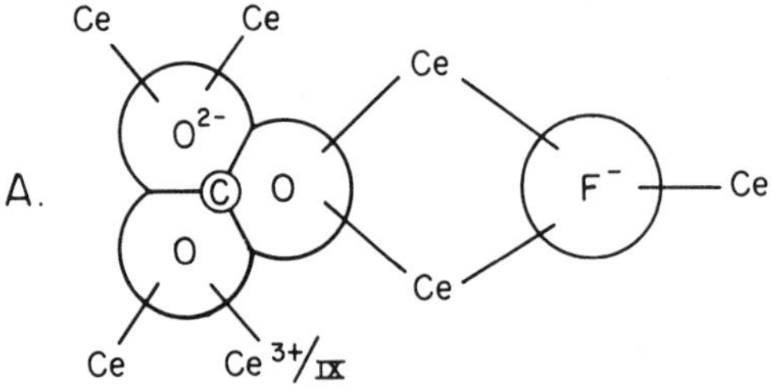

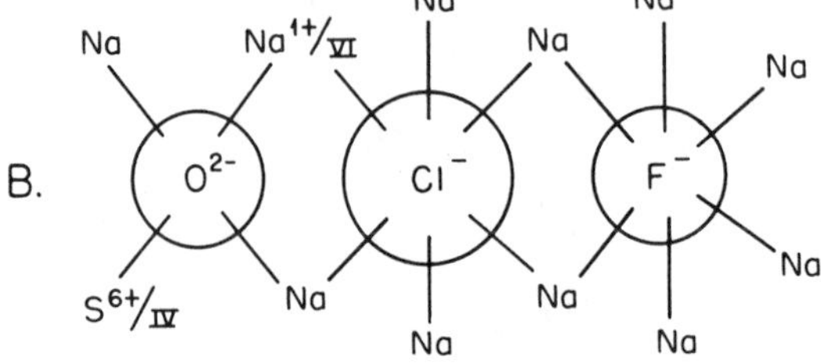

Each O^{2-} is coordinated by III Na^+, 1 S^{6+} = CN IV
Each $[SO_4]^{2-}$ is coordinated by XII Na^+

Pauling Rule 2C group

In many minerals, the nearest neighbor cation arrays around each anion may differ, leading to EBS summations that depart widely from the ideal of +2.00 for oxygen or +1.00 for fluorine anions. Neutrality, then, is not satisfied locally, but must be spread out over a larger domain in the crystal, theoretically resulting in a reduction in stability. Consider the example of diopside, $CaMg[Si_2O_6]$, where three different cation arrays surround oxygens labeled 1, 2, and 3, leading to idealized EBS strength summations of 1.916 around 1, 1.583 around 2, and 2.500 around 3 (Fig. 5.6). In place of the simple formula $Ca^{VIII}Mg^{VI}[Si_2^{IV}O_6]$, we may write $Ca^{VIII}Mg^{VI}[Si_2^{IV}①_2^{IV}②_2^{III}③_2^{IV}]$ to emphasize the presence of the three unequal or different oxygens. From Figure 5.6 and Table 5.1, we see that although none of the three oxygens strictly satisfies Pauling's Rule 2, the sum of the EBS reaching all three oxygens is +6.00, and electric neutrality is assumed to be spread over the three oxygens rather than around each. From the data of Table 5.1, we see that Si, Mg, and Ca

Figure 5.5. Packing model of part of the sulfohalite structure. Large spheres represent O^{2-}, F^-, and Cl^- anions. Corks represent Na^+ ions, and small white spheres, S^6 ions in $[SO_4]^{2-}$ groups.

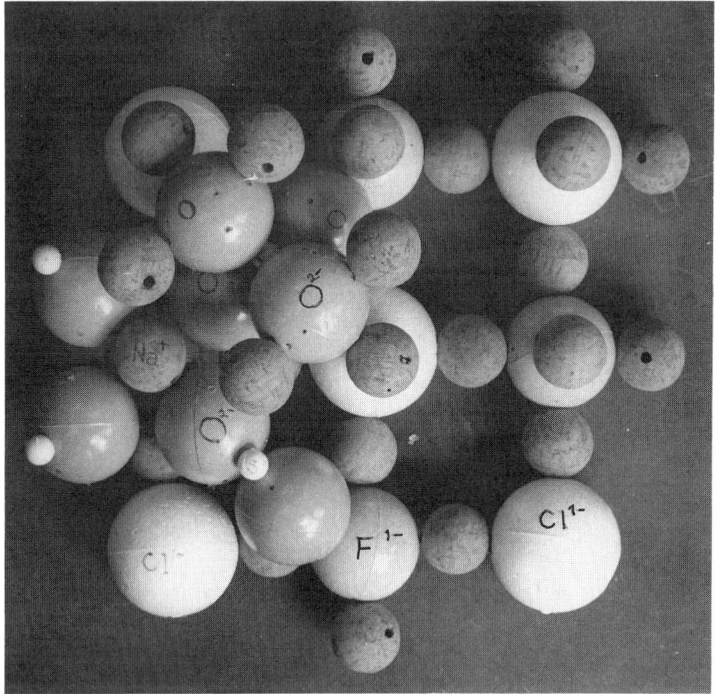

Table 5.1. *Measured and ideal interatomic distances in diopside, and electrostatic bond strength solutions derived from these*

Anion–cation arrays		A—B (Å) obs.	EBS Zachariasen	EBS Pauling	A—X (Å) Ideal
①	Si	1.603	1.12	1.00	1.613
①	Mg	2.175	0.21	0.333	2.080
①	Mg	2.061	0.40	0.333	2.080
①	Ca	2.374	0.28	0.250	2.480
			2.01	1.916	
②	Si	1.591	1.26	1.00	1.613
②	Mg	2.053	0.40	0.333	2.080
②	Ca	2.332	0.34	0.250	2.480
			2.00	1.583	
③	Si	1.675	0.82	1.00	1.613
③	Si	1.690	0.80	1.00	1.613
③	Ca	2.606	0.22	0.250	2.480
③	Ca	2.746	0.16	0.250	2.480
			2.00	2.500	

Note: ① = O-(1) = oxygen designation number.
Source: Clark, Appleman, and Papike (1969).

Figure 5.6. Cation–anion shifts in *C2/c* pyroxene, diopside, resulting from excess electric charge on anions O-(3) and insufficient charge on anions O-(1) and O-(2). See also Table 5.1 for bond strength distribution on all three anions.

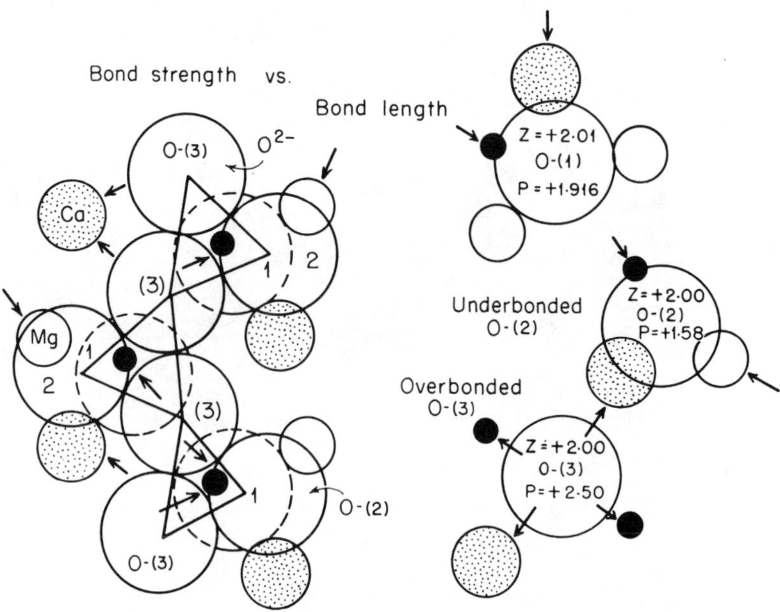

cations are not centrosymmetrically disposed in their oxygen polyhedra, and that such polyhedra (tetrahedron, octahedron, and square antiprism) are considerably distorted from ideal geometry (see Fig. 8.1); we shall return to this subject of polyhedral distortion in Chapter 8.

Zachariasen's elaboration of Pauling's Rule 2

Zachariasen (1963) suggested that EBS is inversely proportional to A—X (Å) bond lengths as accurately measured from single crystal X-ray analyses of borate minerals. Thus an oxygen with a classical Pauling bond strength of less than +2.00 will increase its effective bond strength by drawing closer to neighboring cations than is normal or ideal: bond lengths shorten. Conversely, where the bond strength of an anion exceeds +2.00, bond strength is decreased as bonds to neighboring cations lengthen. Bond strengths empirically assigned on such a basis of departure from ideal A—X distances lead to localized charge balances on each anion, as shown in Figure 5.6 and Table 5.1 for diopside.

Summary

Pauling's second rule employs EBS to limit the number of cations around each anion and the number of polyhedra with a common corner.

Group 2A includes minerals with one species of anion, each with the same environment.

Group 2B includes minerals whose anions have a different environment or different valency.

Group 2C includes minerals each of whose anions is surrounded by a different cation array, the sum of whose bond strengths satisfies electrostatic neutrality, although no one of them is electrostatically neutral.

Each of these groups is illustrated by several mineral examples.

Zachariasen's elaboration of this rule suggests that EBS is inversely proportional to bond length; that is, metal–anion distances will be shorter than ideal for electrically underbonded anions. Here is a prime cause of polyhedral distortion.

Bibliography

Clark, J. R., Appleman, D. E., and Papike, J. J. (1969). Crystal chemical characterization of clinopyroxenes based on eight new structure refinements. *Mineral. Soc. Am. Spec. Pap.*, 2: 31–50.

Pabst, A. (1934). The crystal structure of sulfohalite. *Z. Krist.*, 89: 514.

Pauling, L. (1960). *The nature of the chemical bond,* 3d ed. Cornell Univ. Press, Ithaca, N.Y.

Watanabé, T. (1934). The crystal structure of sulfohalite. *Proc. Imp. Acad. (Tokyo),* 10: 575.

Zachariasen, W. H. (1963). The crystal structure of monoclinic metaboric acid. *Acta Crystallogr., Pt.* 5 16: 385–92.

6 External (nontranslational) and internal (translational) symmetry

Crystals may be regarded as solids enclosing a three-dimensional lattice or array of points that translates its pattern geometrically in space. There are but five two-dimensional lattices from which all minerals are built. These outline points in space as a

1. Square
2. Rectangle
3. Diamond with angles between the edges not 120° or 60°
4. Rhombus with angles of 120° and 60°
5. Parallelogram with angles neither 60°, 90°, nor 120°, and with sides unequal

When these two-dimensional lattices are combined with a second such perpendicularly oriented two-dimensional lattice, a three-dimensional array of points, the *Bravais space lattice,* is built.

It is geometrically possible to construct only fourteen such unique space lattices (Fig. 6.1). Of these fourteen lattices, the seven labeled *P*, for primitive (Fig. 6.1) represent the seven crystal systems into which all minerals and regular solids may be classified and readily recognized (Fig. 6.2).

Although it is possible to outline or enclose, from an extended array of lattice points, more than one such set that repeats itself (translates itself), we choose the smallest coplanar array of points that translates itself in three-dimensional space to define the unit cell. For halite, NaCl, Cl atoms or ions on eight corners of the cell and on all six face centers, and Na atoms or ions on all twelve edges and in the center, outline the unit cube marked D on Figure 6.3. Cubelet A (Fig. 6.3), which is smaller, does not repeat itself in the space array of points. Cubelet B, though smaller than D, repeats itself in our two-dimensional array, but will not repeat itself in vertical translation because Na ions will be positioned directly over the Cl ions shown. Cube C is too large. Hence cube D becomes the only choice for the familiar unit cell of halite (also see Fig. 10.1).

If we closely stack together unit blocks or cells of a mineral, it will be apparent that atoms located on corners are common to eight unit cells, those on edges to four, those on faces to two, and those inside to one. Thus, for halite the unit cube of Figure 6.3D contains:

$$\text{Cl} \longrightarrow 8_{\text{cor}} \times \tfrac{1}{8} = 1 \quad \text{and} \quad 6_{\text{faces}} \times \tfrac{1}{2} = 3; \qquad \text{sum} = 4\ \text{Cl}$$
$$\text{Na} \longrightarrow 12_{\text{edge}} \times \tfrac{1}{4} = 3 \quad \text{and} \quad 1_{\text{ctr}} = 1; \qquad \text{sum} = 4\ \text{Na}$$

Figure 6.1. The fourteen Bravais space lattices. Reprinted with permission from M. J Buerger, *Elementary crystallography,* John Wiley & Sons, 1956.

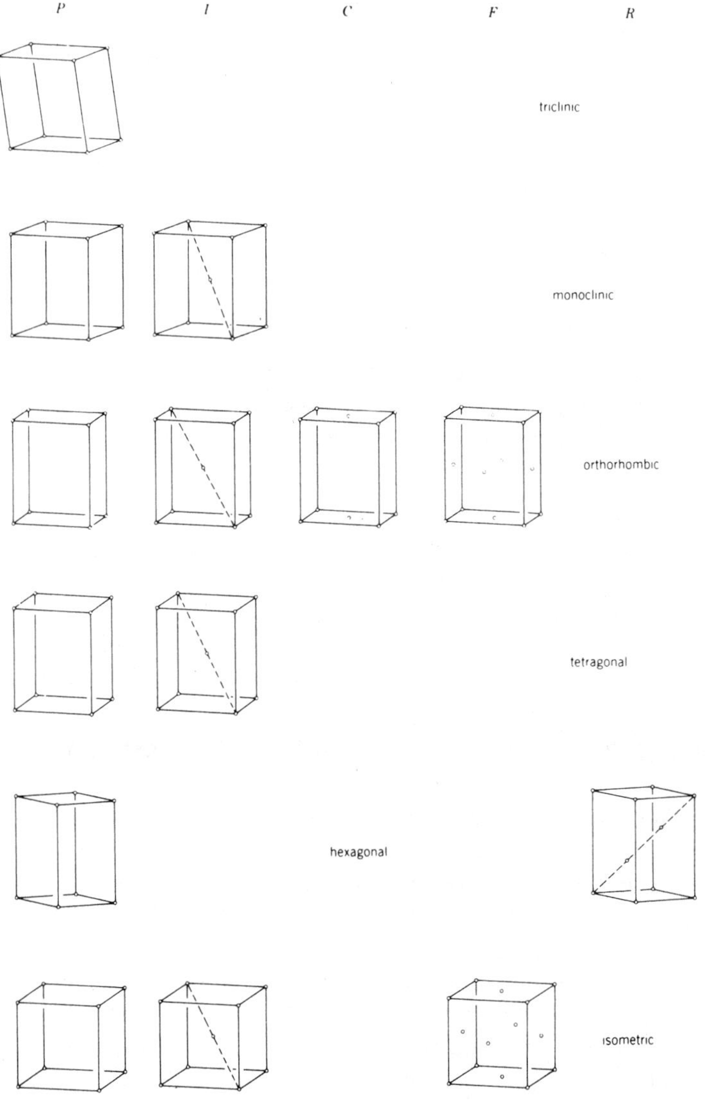

Figure 6.2. Symmetry axes in crystal drawings representing the seven primitive crystal systems. Symbols as in Figure 6.5.

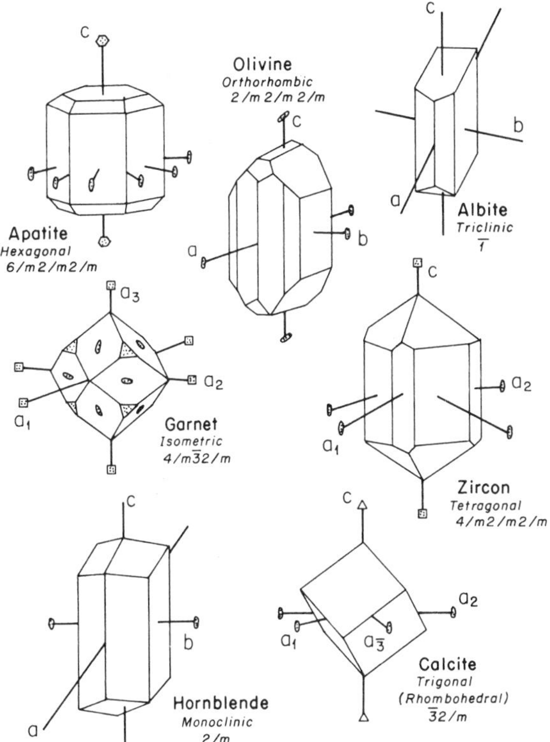

Figure 6.3 Two-dimensional repeat cells in packing model of halite, NaCl. The true unit cell with smallest repeat unit of all atoms is labeled D.

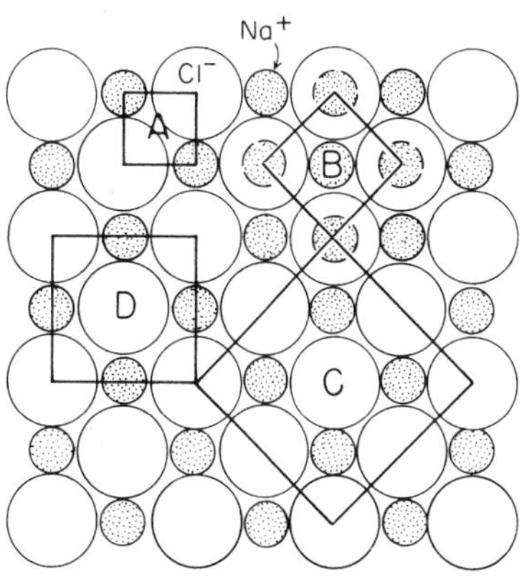

A very small crystal of halite balanced on the tip of one's finger represents a cube made up of many smaller cubelets, the smallest being the unit cube bounded by twelve edges, a (or a_0), of length 5.46 Å (5.46×10^{-8} cm). Without X-ray data, however, our small halite crystal represents a cube that reveals nothing about the internal distribution of the Na and Cl ions of which it is built, and does not allow us to identify the space lattice.

The geometry or external symmetry of the cube (and of all cubes) does display four threefold axes of rotational symmetry passing through cube corners, three fourfold axes passing through the six cube faces, each perpendicular to a mirror plane of symmetry that bisects each cube face parallel to its edges, six twofold axes of symmetry that pass through the opposite twelve edges and each perpendicular to planes of mirror symmetry that pass through each cube diagonal. Because opposite faces, edges, and corners are identical, the crystal has a center of symmetry. Now, thirteen axes + nine mirror planes + the center add up to twenty-three elements of nontranslational (external) symmetry. Thus it is possible to classify the cube into a *crystal class* (point group) of the highest possible symmetry, coded $4/m\ \bar{3}\ 2/m$ of Table 6.1.

There are thirty-two of these crystal classes (Table 6.1), each displaying lower external symmetry in passing from the isometric to the triclinic crystal systems. Class $4/m\ \bar{3}\ 2/m$ is one of five classes that have isometric symmetry; it is also one of thirty-six space groups that have isometric internal symmetry (Table 6.2).

To identify which of the fourteen space lattices and which of the 230 space groups are represented, we must use X-ray analyses to determine internal symmetry. Translational symmetry elements are screw axes and glide planes that translate internal lattice points some regular increment of a unit cell dimension or length when combined with rotation (Figs. 6.4, 6.5; see also Figs. 3.3, 10.1). Atomic positions are located by single crystal X-ray analyses (Weissenberg patterns), whereas different planes of atoms that repeat themselves by three-dimensional translation are identified by X-ray powder diffraction analysis (Debye–Scherrer patterns) of a minute amount of finely powdered material. Usually, X-ray powder diffraction analysis serves to identify the crystal, the space lattice, and the space group. A single crystal X-ray analysis may then be made to locate specific atoms within the lattice, based on the capacity of different atomic species to scatter X-radiation of a particular wavelength. Atoms distant from one another in the periodic table are readily identified. Atoms with similar atomic number, or next to one another, such as ${}_{13}Al$ and ${}_{14}Si$, will scatter X-radiation to an almost identical degree and render their discrimination on this basis impossible. Thus, to determine whether a particular tetrahedron in feldspars is occupied by Al or Si, we must measure the T—O (tetrahedral cation–oxygen) distance and the O—O distance to decide that the larger tetrahedral site contains the Al and the smaller contains the Si.

X-ray powder diffraction patterns are made on film in Debye–Scherrer cameras or on automatic recording charts with X-ray diffractometers. These pat-

Table 6.1. *Symmetry complements of the thirty-two crystal classes (point groups) of the seven crystal systems*

	Axes				Planes	Center			
Crystal system	2	3	4	6	*m*	*c*	Σ	H–M class	
Triclinic							0	1	1
$a,b,c \neq a \wedge b = r$, $b \wedge c = \alpha$, $a \wedge c = \beta \neq 90°$						1	1	$\bar{1}$	2
Monoclinic	1						1	2	3
$a,b,c \neq a \wedge c = \beta \neq 90°$					1		1	m	4
	1				1	1	3	$2/m$	5
Orthorhombic	3						3	2 2 2	6
$a,b,c \neq a \perp b \perp c \perp a$	1				2		3	$m\,m\,2$	7
	3				3	1	7	$2/m\,2/m\,2/m$	8
Trigonal		1					1	3	9
$a_1a_2a_3 = \perp c \neq$		1				1	2	$\bar{3}$	10
$a_1 \wedge a_2 \wedge a_3 = 120°$	3	1					4	3 2	11
		1			3		4	$3\,m$	12
	3	1			3	1	8	$\bar{3}\,2/m$	13
Tetragonal			1				1	4	14
$a_1 = a_2 \perp c \neq$			1				1	$\bar{4}$	15
$a_1 \wedge a_2 \wedge c = 90°$			1		1	1	3	$4/m$	16
	4		1				5	4 2 2	17
			1		4		5	$4\,m\,m$	18
	2		1		2		5	$\bar{4}\,2\,m$	19
	4		1		5	1	11	$4/m\,2/m\,2/m$	20
Hexagonal				1			1	6	21
$a_1a_2a_3 = \perp c \neq$				1	1		2	$\bar{6}$	22
$a_1 \wedge a_2 \wedge a_3 = 120°$				1	1	1	3	$6/m$	23
	6			1			7	6 2 2	24
				1	6		7	$6\,m\,m$	25
	3			1	3		7	$\bar{6}\,m\,2$	26
	6			1	7	1	15	$6/m\,2/m\,2/m$	27
Isometric (cubic)	3	4					7	2 3	28
$a_1 = a_2 = a_3$, all 90°	3	4			3	1	11	$2/m\bar{3}$	29
	6	4	3				13	4 3 2	30
		4	3		6		13	$\bar{4}\,3\,m$	31
	6	4	3		9	1	23	$4/m\,\bar{3}\,2/m$	32

terns show many lines of differing intensity and location or interplanar spacing, which result from the in-phase scattering of X-radiation by each atom in a plane of atoms (although constructive waves enter and leave at the same angle to a crystal plane, and without a change in wavelength, diffraction by atoms in a plane is not directly analogous to optical reflection).

The location of X-ray diffraction lines depends on the wavelength of radiation used (Cu and Fe anodes are the most common), the angle of the collimated X-ray beam with a plane of atoms, and the distance (interplanar spacing, d [Å]) between planes according to Bragg's law of diffraction, $n\lambda = 2d \sin \theta$, where n is an integral number of wavelengths of the X-radiation used, d is the interplanar spacing, and θ is the angle of incidence (Fig. 6.6). The intensity of the particular lines in a pattern is largely a function of the atomic scattering factor (heavy atoms scatter X-radiation more efficiently) and, to some degree, of the angle of incidence and the type of X-ray target (anode) used.

In many cases, space lattice and space group identifications depend as much on the absence of particular Bragg X-ray reflections (diffraction lines) as on the presence of others, because the introduction of a like plane of atoms, halfway between two identical planes of atoms, say at 4.00 Å, will cancel the 4.00 Å spacing and yield a 2.00 Å spacing. Thus, the 5.46 Å spacing of the unit cell edge of halite is canceled by the introduction of the identical plane of face centered atoms at 2.73 Å: a strong line at 2.73 Å thus represents half the distance of the unit cell edge and is given the Bragg diffraction line label d_{200} rather than d_{100}. The absence of the d_{100} reflection and other systematic absences from other planes in the crystal permit the identification of the space lattice as F (face-centered) for halite. Note that if the lattice had been P (primitive) with points only at the corners, both the d_{100} and d_{200} Bragg reflections would have

Table 6.2. *Crystal systems, space lattices, crystal classes, space groups for all minerals and synthetic crystals*

Systems	Lattices & symbols	No. of crystal classes	No. of space groups
Isometric	3, P, I, F	5	36
Tetragonal	2 P(=C), I(=F)	7	68
Hexagonal	1 P(=C)	7	27
Trigonal	1 P(=R)	5	25
Orthorhombic	4 P, C, I, F	3	59
Monoclinic	2 P, C	3	13
Triclinic	1 P	2	2
Total 7	14	32	230

appeared. The student will profit from learning, from published data, the patterns of Bragg reflections that are produced from different types of space lattices.

The description of a space lattice as primitive, *P*, body centered, *I*, or face centered, *F*, from a drawing or projection of atomic locations, will be obvious from Figure 6.1. Note that in the isometric system the centering of a point on

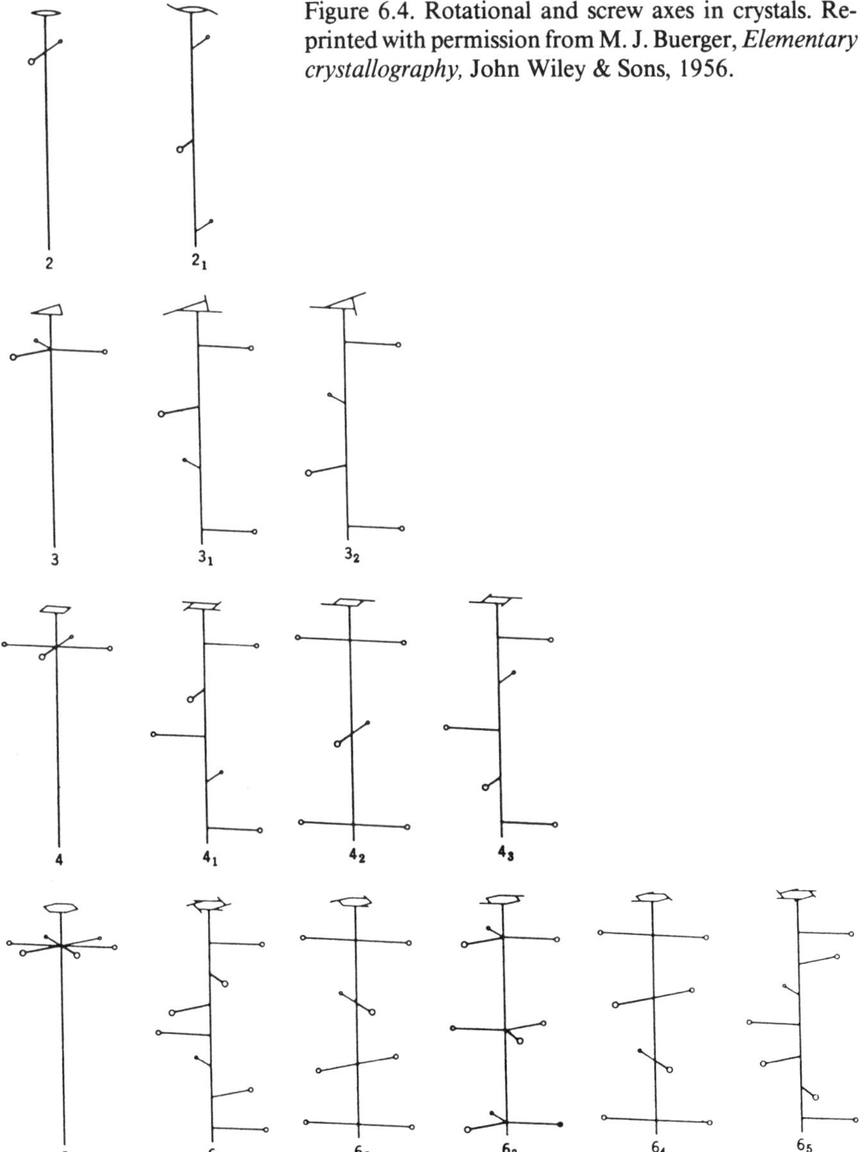

Figure 6.4. Rotational and screw axes in crystals. Reprinted with permission from M. J. Buerger, *Elementary crystallography,* John Wiley & Sons, 1956.

Figure 6.5. Symbols used for space group notations.

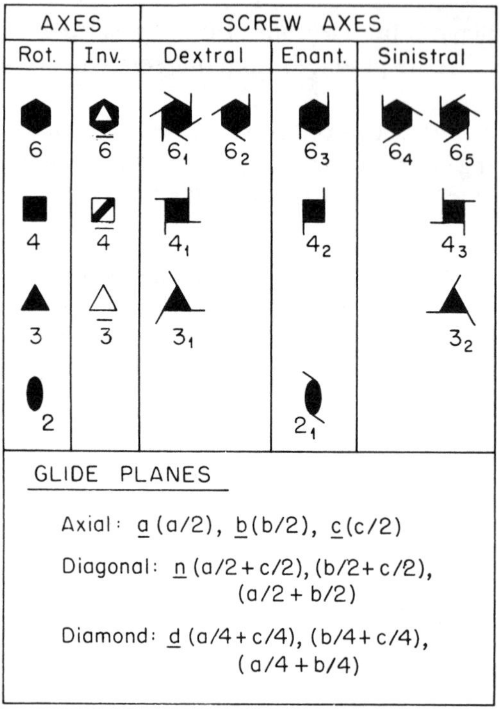

AXES		SCREW AXES				
Rot.	Inv.	Dextral		Enant.	Sinistral	
6	$\bar{6}$	6_1	6_2	6_3	6_4	6_5
4	$\bar{4}$	4_1		4_2		4_3
3	$\bar{3}$	3_1				3_2
2				2_1		

GLIDE PLANES

Axial: a (a/2), b (b/2), c (c/2)

Diagonal: n (a/2 + c/2), (b/2 + c/2), (a/2 + b/2)

Diamond: d (a/4 + c/4), (b/4 + c/4), (a/4 + b/4)

Figure 6.6. Graphic solution of Bragg's law of X-ray diffraction, $n\lambda = 2d \sin \theta$.

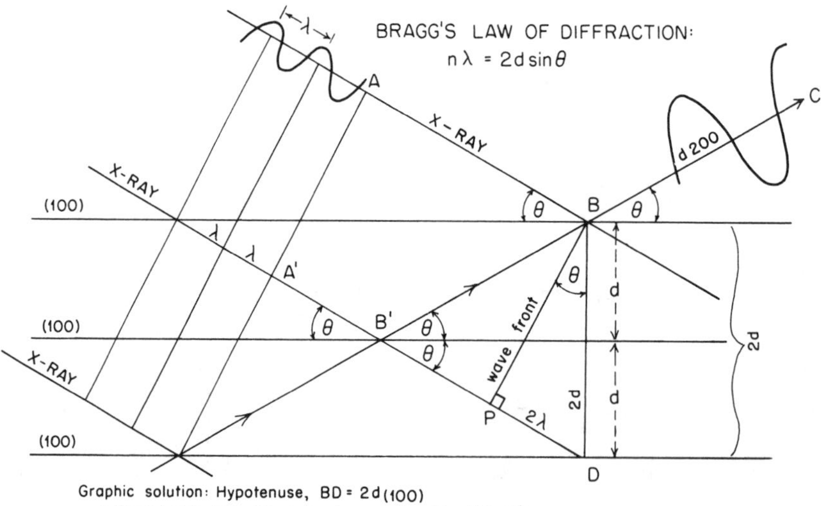

Table 6.3. *Space groups in which common minerals crystallize*

Triclinic Key: no planes of symmetry; only a center and a 1 (1/360°) axis = no symmetry	*P, C,* 1 or $\bar{1}$				
			$C\bar{1}$		Kaolinite Microcline Albite
			$I\bar{1}$		Bytownite
			$P\bar{1}$		Anorthite
Monoclinic Key: one twofold axis + 1 plane of symmetry, or 1 of either	*P, C,* (*A, B*)	*b*/(010)			
			$C2/c$		Augite Diopside Muscovite
			$P2_1/a$		Datolite
			$P2_1/m$		Epidote
			$P2_1/c$		Pigeonite
			$C2/m$		Sanidine Phlogopite
Orthorhombic Key: 3 letters or numbers or their combination after the lattice symbol	*P, C, I, F*	(100)	(010)	(001)	
	P	*b*	*n*	*m*	Diaspore Forsterite Sillimanite
	P	*n*	*n*	*m*	Andalusite
	P	*n*	*m*	*a*	Anthophyllite
	P	*b*	*c*	*a*	Hypersthene
	F	*d*	*d*	*d*	Sulfur
	A	*m*	*m*	*a*	Anhydrite
Tetragonal Key: 4 in first position (*No* 3-axis)	*P* (=*C*), *I* (=*F*)	*c*/(001)	(100)	(110)	
	$P4_2/$	*m*	*n*	*m*	Rutile
	$I4_1/$	*a*	*m*	*d*	Anatase Zircon
	$P\bar{4}2_1$			*m*	Melilite
Hexagonal Key: 6 in first position	*P* (=*C*)	*c*/(0001)	$(11\bar{2}0)$	$(10\bar{1}0)$	
	$P6_3/$	*m*	*m*	*c*	Molybdenite Graphite
	$P6_3/$	*m*			Apatite
	$P\bar{6}$		2	*c*	Bastnaesite Benitoite
Trigonal Key: 3 in first position	*R* (=*P*)	*c*/(0001)	$(11\bar{2}0)$	$(10\bar{1}0)$	
	$R\bar{3}$				Dolomite Ilmenite
	*R*3		*m*		Tourmaline
	$R\bar{3}$		*c*		Calcite Corundum Hematite

Table 6.3. *(cont.)*

Isometric	*P, I, F*	(001)	(111)	(110)	
Key: 3 in second position	*F*	*m*	3	*m*	Fluorite, Galena, Halite
	F	*d*	3	*m*	Spinel, Magnetite, Diamond
	P	*n*	3	*m*	Cuprite
	I	*a*	3	*d*	Analcime, Garnet
	P	*a*	3		Pyrite
	*P*4		3	*n*	Sodalite
	F$\bar{4}$		3	*m*	Sphalerite

Table 6.4. *Minerals of diverse bond type crystallized in space group Pbnm and the related groups Pnma and Pcmn*

Pbnm		
Goethite	FeO(OH)	Covalent–ionic
Forsterite	$Mg_2[SiO_4]$	Ionic–covalent
Ramsdellite	MnO_2	Ionic-metallic
Chrysoberyl	$Al_2[BeO_4]$	Ionic-covalent
Topaz	$Al_2[SiO_4](OH,F)_2$	Ionic-covalent
Danburite	$Ca[B_2Si_2O_8]$	Ionic-covalent
Stibnite	Sb_2S_3	Metallic-covalent
Diaspore	AlO(OH)	Ionic-covalent
Sillimanite	$AlOSi[AlO_4]$	Ionic-covalent
Humite	$Mg(OH,F)_2 \cdot 3Mg_2SiO_4$	Ionic-covalent
Sinhalite	$MgAl[BO_4]$	Ionic-covalent
Pnma		
Anthophyllite	$(Mg,Fe)_7[Si_8O_{22}](OH)_2$	Ionic-covalent
Barite	$Ba[SO_4]$	Ionic-covalent
Chalcostibite	$CuSbS_2$	Metallic-covalent
Pcmn		
Aragonite	$Ca[CO_3]$	Ionic-covalent
Niter	$K[NO_3]$	Ionic-covalent
Cubanite	$CuFe_2S_3$	Metallic-covalent

one face requires that all faces be centered, but in the orthorhombic system, for example, centering the *C*-face does not require centering of the *A*- and *B*-faces, which are not equivalent.

A *space group* is an extended network of reflection planes, glide planes, rotation axes, inversion axes, and screw axes, all based on a system of translation of the space lattice elements. These translational elements relate identical points in space. Any symmetry operation or translation brings all the others into coincidence. Each point in the structure need not be brought back to its original position, but only to a similar position in another unit cell of the lattice.

Familiarity with space group notation and the recognition of various symmetry elements represented on drawings, such as unit cell projections of atomic positions, and on models of crystal structures, are an important part of the study of crystal chemistry. Table 6.3 lists some of the more common space groups in

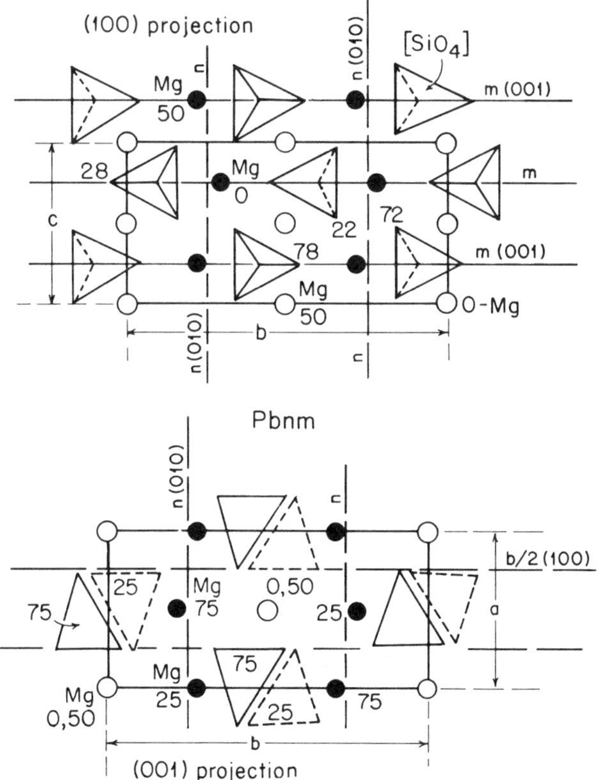

Figure 6.7. Unit cell projections of atomic positions, glide and mirror planes in forsterite, Mg_2SiO_4.

which a large number of the more common minerals crystallize. It illustrates the sequence of presentation of symmetry elements in the seven crystal systems and gives the criteria requisite for assigning a particular space group symbol or label to a particular crystal system.

Several of the 230 space groups are represented by numerous minerals and synthetic compounds, others by one or a few, and still others, though mathematically possible, are not represented by minerals at all. For example, the very common orthorhombic space group *Pbnm* is represented by many minerals embracing a wide variety of chemical composition and chemical bonding, as well (Table 6.4). In the orthorhombic crystal system, where crystal axes *a, b,*

Table 6.5. *The 230 space groups*

Point-group	*Hermann–Mauguin symbols*†		*Schoenflies symbols*
Triclinic			
1	$P1$		C_1^1
$\bar{1}$	$P\bar{1}$		C_i^1, S_2^1
Monoclinic‡	*1st setting (z-axis unique)*	*2nd setting (y-axis unique)*	
2	$P2$ $P2_1$ $B2$. $A2$, $I2$	$P2$ $P2_1$ $C2$. $A2$, $I2$	C_2^1 C_2^2 C_2^3
m	Pm Pb. Pa, Pn Bm. Am, Im Bb. Aa, Ia	Pm Pc. Pa, Pn Cm. Am, Im Cc. Aa, Ia	C_s^1 C_s^2 C_s^3 C_s^4
$2/m$	$P2/m$ $P2_1/m$ $B2/m$. $A2/m$, $I2/m$ $P2/b$. $P2/a$, $P2/n$ $P2_1/b$. $P2_1/a$, $P2_1/n$ $B2/b$. $A2/a$, $I2/a$	$P2/m$ $P2_1/m$ $C2/m$. $A2/m$, $I2/m$ $P2/c$. $P2/a$, $P2/n$ $P2_1/c$. $P2_1/a$, $P2_1/n$ $C2/c$. $A2/a$, $I2/a$	C_{2h}^1 C_{2h}^2 C_{2h}^3 C_{2h}^4 C_{2h}^5 C_{2h}^6
Orthorhombic	*Normal setting*	*Other settings*	
222	$P222$ $P222_1$ $P2_12_12$ $P2_12_12_1$ $C222_1$ $C222$ $F222$ $I222$ $I2_12_12_1$	 $P2_122$, $P22_12$ $P22_12_1$, $P2_122_1$ $A2_122$, $B22_12$ $A222$, $B222$ 	D_2^1, V^1 D_2^2, V^2 D_2^3, V^3 D_2^4, V^4 D_2^5, V^5 D_2^6, V^6 D_2^7, V^7 D_2^8, V^8 D_2^9, V^9

† The Hermann–Mauguin symbols given first are those given in *International Tables* (1952).

and *c* are interchangeable and have in the past been arbitrarily selected, it is possible to transpose these axes and, in so doing, transpose the space group: there are alternatives. Space group *Pbnm* is an alternative to *Pnma* or to *Pcmn:* they are thus equivalent from an overall standpoint of symmetry. Recent crys-

Table 6.5. *(cont.)*

Point-group	Hermann–Mauguin symbols		Schoenflies symbols
	Normal	Other settings	
mm2	$Pmm2$ (Pmm)	$P2mm$, $Pm2m$	C_{2v}^1
	$Pmc2_1$ (Pmc)	$P2_1ma$, $Pb2_1m$, $Pm2_1b$, $Pcm2_1$, $P2_1am$	C_{2v}^2
	$Pcc2$ (Pcc)	$P2aa$, $Pb2b$, $P2aa$	C_{2v}^3
	$Pma2$ (Pma)	$P2mb$, $Pc2m$, $Pm2a$, $Pbm2$, $P2cm$	C_{2v}^4
	$Pca2_1$ (Pca)	$P2_1ab$, $Pc2_1b$, $Pb2_1a$, $Pbc2_1$, $P2_1ca$	C_{2v}^5
	$Pnc2$ (Pnc)	$P2na$, $Pb2n$, $Pn2b$, $Pcn2$, $P2an$	C_{2v}^6
	$Pmn2_1$ (Pmn)	$P2_1mn$, $Pn2_1m$, $Pm2_1n$, $Pnm2_1$, $P2_1nm$	C_{2v}^7
	$Pba2$ (Pba)	$P2cb$, $Pc2a$, $Pba2$, $P2cb$	C_{2v}^8
	$Pna2_1$ (Pna)	$P2_1nb$, $Pc2_1n$, $Pn2_1a$, $Pbn2_1$, $P2_1cn$	C_{2v}^9
	$Pnn2$ (Pnn)	$P2nn$, $Pn2n$	C_{2v}^{10}
	$Cmm2$ (Cmm)	$A2mm$, $Bm2m$	C_{2v}^{11}
	$Cmc2_1$ (Cmc)	$A2_1ma$, $Bb2_1m$, $Bm2_1b$, $Ccm2_1$, $A2_1am$	C_{2v}^{12}
	$Ccc2$ (Ccc)	$A2aa$, $Bb2b$	C_{2v}^{13}
	$Amm2$ (Amm)	$B2mm$, $Cm2m$, $Am2m$, $Bmm2$, $C2mm$	C_{2v}^{14}
	$Abm2$ (Abm)	$B2cm$, $Cm2a$, $Ac2m$, $Bma2$, $C2mb$	C_{2v}^{15}
	$Ama2$ (Ama)	$B2mb$, $Cc2m$, $Am2a$, $Bbm2$, $C2cm$	C_{2v}^{16}
	$Aba2$ (Aba)	$B2cb$, $Cc2a$, $Ac2a$, $Bba2$, $C2cb$	C_{2v}^{17}
	$Fmm2$ (Fmm)	$F2mm$, $Fm2m$	C_{2v}^{18}
	$Fdd2$ (Fdd)	$F2dd$, $Fd2d$	C_{2v}^{19}
	$Imm2$ (Imm)	$I2mm$, $Im2m$	C_{2v}^{20}
	$Iba2$ (Iba)	$I2cb$, $Ic2a$	C_{2v}^{21}
	$Ima2$ (Ima)	$I2mb$, $Ic2m$, $Im2a$, $Ibm2$, $I2cm$	C_{2v}^{22}
mmm	$Pmmm$		D_{2h}^1, V_h^1
	$Pnnn$		D_{2h}^2, V_h^2
	$Pccm$	$Pmma$, $Pbmb$	D_{2h}^3, V_h^3
	$Pban$	$Pncb$, $Pcna$	D_{2h}^4, V_h^4
	$Pmma$	$Pbmm$, $Pmcm$, $Pmam$, $Pmmb$, $Pcmm$	D_{2h}^5, V_h^5
	$Pnna$	$Pbnn$, $Pncn$, $Pnan$, $Pnnb$, $Pcnn$	D_{2h}^6, V_h^6
	$Pmna$	$Pbmn$, $Pncm$, $Pman$, $Pnmb$, $Pcnm$	D_{2h}^7, V_h^7
	$Pcca$	$Pbaa$, $Pbcb$, $Pbab$, $Pccb$	D_{2h}^8, V_h^8
	$Pbam$	$Pmcb$, $Pcma$	D_{2h}^9, V_h^9
	$Pccn$	$Pnaa$, $Pbnb$	D_{2h}^{10}, V_h^{10}
	$Pbcm$	$Pmca$, $Pbma$, $Pcmb$, $Pcam$, $Pmab$	D_{2h}^{11}, V_h^{11}
	$Pnnm$	$Pmnn$, $Pnmn$	D_{2h}^{12}, V_h^{12}
	$Pmmn$	$Pnmm$, $Pmnm$	D_{2h}^{13}, V_h^{13}
	$Pbcn$	$Pnca$, $Pbna$, $Pcnb$, $Pcan$, $Pnab$	D_{2h}^{14}, V_h^{14}
	$Pbca$	$Pcab$	D_{2h}^{15}, V_h^{15}
	$Pnma$	$Pbnm$, $Pmcn$, $Pnam$, $Pmnb$, $Pcmn$	D_{2h}^{16}, V_h^{16}
	$Cmcm$	$Amma$, $Bbmm$, $Bmmb$, $Ccmm$, $Amam$	D_{2h}^{17}, V_h^{17}
	$Cmca$	$Abma$, $Bbcm$, $Bmab$, $Ccmb$, $Acam$	D_{2h}^{18}, V_h^{18}
	$Cmmm$	$Ammm$, $Bmmm$	D_{2h}^{19}, V_h^{19}
	$Cccm$	$Amaa$, $Bbmb$	D_{2h}^{20}, V_h^{20}
	$Cmma$	$Abmm$, $Bmcm$, $Bmam$, $Cmmb$, $Acmm$	D_{2h}^{21}, V_h^{21}
	$Ccca$	$Abaa$, $Bbcb$, $Bbab$, $Cccb$, $Acaa$	D_{2h}^{22}, V_h^{22}
	$Fmmm$		D_{2h}^{23}, V_h^{23}
	$Fddd$		D_{2h}^{24}, V_h^{24}
	$Immm$		D_{2h}^{25}, V_h^{25}

tallographic convention calls for making the longest axis *b* and the shortest *c*, but this is by no means widely accepted.

One of the many common minerals that crystallize with *Pbnm* symmetry is forsterite, the Mg end member of the olivine series. Unit cell projections of the

Table 6.5. *(cont.)*

Point-group	*Hermann–Mauguin symbols*		*Schoenflies symbols*
	Normal	*Other settings*	
mmm (*cont.*)	$Ibam$	$Imcb$, $Icma$	D_{2h}^{26}, V_h^{26}
	$Ibca$	$Icab$	D_{2h}^{27}, V_h^{27}
	$Imma$	$Ibmm$, $Imcm$, $Imam$, $Immb$, $Icmm$	D_{2h}^{28}, V_h^{28}
Tetragonal	*Normal*	*Larger cell*	
4	$P4$	$C4$	C_4^1
	$P4_1$	$C4_1$	C_4^2
	$P4_2$	$C4_2$	C_4^3
	$P4_3$	$C4_3$	C_4^4
	$I4$	$F4$	C_4^5
	$I4_1$	$F4_1$	C_4^6
$\bar{4}$	$P\bar{4}$	$C\bar{4}$	S_4^1
	$I\bar{4}$	$F\bar{4}$	S_4^2
$4/m$	$P4/m$	$C4/m$	C_{4h}^1
	$P4_2/m$	$C4_2/m$	C_{4h}^2
	$P4/n$	$C4/n$	C_{4h}^3
	$P4_2/n$	$C4_2/n$	C_{4h}^4
	$I4/m$	$F4/m$	C_{4h}^5
	$I4_1/a$	$F4_1/a$	C_{4h}^6
422	$P422$	$C422$	D_4^1
	$P42_12$	$C422_1$	D_4^2
	$P4_122$	$C4_122$	D_4^3
	$P4_12_12$	$C4_122_1$	D_4^4
	$P4_222$	$C4_222$	D_4^5
	$P4_22_12$	$C4_222_1$	D_4^6
	$P4_322$	$C4_322$	D_4^7
	$P4_32_12$	$C4_322_1$	D_4^8
	$I422$	$F422$	D_4^9
	$I4_122$	$F4_122$	D_4^{10}
$4mm$	$P4mm$	$C4mm$	C_{4v}^1
	$P4bm$	$C4mb$	C_{4v}^2
	$P4_2cm$ ($P4cm$)	$C4_2mc$ ($C4mc$)	C_{4v}^3
	$P4_2nm$ ($P4nm$)	$C4_2mn$ ($C4mn$)	C_{4v}^4
	$P4cc$	$C4cc$	C_{4v}^5
	$P4nc$	$C4cn$	C_{4v}^6
	$P4_2mc$ ($P4mc$)	$C4_2cm$ ($C4cm$)	C_{4v}^7
	$P4_2bc$ ($P4bc$)	$C4_2cb$ ($C4cb$)	C_{4v}^8
	$I4mm$	$F4mm$	C_{4v}^9
	$I4cm$	$F4mc$	C_{4v}^{10}
	$I4_1md$ ($I4md$)	$F4_1dm$ ($I4dm$)	C_{4v}^{11}
	$I4_1cd$ ($I4cd$)	$F4_1dc$ ($I4dc$)	C_{4v}^{12}
$\bar{4}2m$	$P\bar{4}2m$	$C\bar{4}2m$	D_{2d}^1, V_d^1
	$P\bar{4}2c$	$C\bar{4}c2$	D_{2d}^2, V_d^2
	$P\bar{4}2_1m$	$C\bar{4}m2_1$	D_{2d}^3, V_d^3
	$P\bar{4}2_1c$	$C\bar{4}c2_1$	D_{2d}^4, V_d^4

atoms parallel to (100) and to (001) planes are illustrated in Figure 6.7. They show the locations of *b* glides, *b*/2 (100); *n* glides, *a*/2 + *c*/2 (010); and mirror planes, *m* (001). In addition to learning how to locate the symmetry planes on such unit cell projection diagrams, readers should count all of the atoms in the

Table 6.5. *(cont.)*

Point-group	*Hermann–Mauguin symbols*		*Schoenflies symbols*
	Normal	*Larger cell*	
$\bar{4}2m$ (*cont.*)	$P\bar{4}m2$	$C\bar{4}2m$	D_{2d}^5, V_d^5
	$P\bar{4}c2$	$C\bar{4}2c$	D_{2d}^6, V_d^6
	$P\bar{4}b2$	$C\bar{4}2b$	D_{2d}^7, V_d^7
	$P\bar{4}n2$	$C\bar{4}2n$	D_{2d}^8, V_d^8
	$I\bar{4}m2$	$F\bar{4}2m$	D_{2d}^9, V_d^9
	$I\bar{4}c2$	$F\bar{4}2c$	D_{2d}^{10}, V_d^{10}
	$I\bar{4}2m$	$F\bar{4}m2$	D_{2d}^{11}, V_d^{11}
	$I\bar{4}2d$	$F\bar{4}d2$	D_{2d}^{12}, V_d^{12}
$4/mmm$	$P4/mmm$	$C4/mmm$	D_{4h}^1
	$P4/mcc$	$C4/mcc$	D_{4h}^2
	$P4/nbm$	$C4/amb$	D_{4h}^3
	$P4/nnc$	$C4/acn$	D_{4h}^4
	$P4/mbm$	$C4/mmb$	D_{4h}^5
	$P4/mnc$	$C4/mcn$	D_{4h}^6
	$P4/nmm$	$C4/amm$	D_{4h}^7
	$P4/ncc$	$C4/acc$	D_{4h}^8
	$P4_2/mmc$ ($P4/mmc$)	$C4_2/mcm$ ($C4mcm$)	D_{4h}^9
	$P4_2/mcm$ ($P4/mcm$)	$C4_2/mmc$ ($C4/mmc$)	D_{4h}^{10}
	$P4_2/nbc$ ($P4/nbc$)	$C4_2/acb$	D_{4h}^{11}
	$P4_2/nnm$ ($P4/nnm$)	$C4_2/amn$	D_{4h}^{12}
	$P4_2/mbc$ ($P4/mbc$)	$C4_2/mcb$ ($C4/mcb$)	D_{4h}^{13}
	$P4_2/mnm$ ($P4/mnm$)	$C4_2/mmn$ ($C4/mmn$)	D_{4h}^{14}
	$P4_2/nmc$ ($P4/nmc$)	$C4_2/acm$	D_{4h}^{15}
	$P4_2/ncm$ ($P4/ncm$)	$C4_2/amc$	D_{4h}^{16}
	$I4/mmm$	$F4/mmm$	D_{4h}^{17}
	$I4/mcm$	$F4/mmc$	D_{4h}^{18}
	$I4_1/amd$ ($I4/amd$)	$F4_1/ddm$	D_{4h}^{19}
	$I4_1/acd$ ($I4/acd$)	$F4_1/ddc$	D_{4h}^{20}
Trigonal and Hexagonal			
3	$P3$ ($C3$)	($H3$)	C_3^1
	$P3_1$ ($C3_1$)	($H3_1$)	C_3^2
	$P3_2$ ($C3_2$)	($H3_2$)	C_3^3
	$R3$		C_3^4
$\bar{3}$	$P\bar{3}$ ($C\bar{3}$)	($H\bar{3}$)	C_{3i}^1, S_6^1
	$R\bar{3}$		C_{3i}^2, S_6^2
32	$P312$ ($C312$)	($H32$)	D_3^1
	$P321$ ($C32$)	($H312$)	D_3^2
	$P3_112$ ($C3_112$)	($H3_12$)	D_3^3
	$P3_121$ ($C3_12$)	($H3_112$)	D_3^4
	$P3_212$ ($C3_212$)	($H3_22$)	D_3^5
	$P3_221$ ($C3_22$)	($H3_212$)	D_3^6
	$R32$		D_3^7
$3m$	$P3m1$ ($C3m$)	($H31m$)	C_{3v}^1
	$P31m$ ($C31m$)	($H3m$)	C_{3v}^2

unit cell to get the unit cell formula, to connect appropriate atom centers in order to locate octahedral and tetrahedral groups and their linkages, to derive EBS distributions, and to look at the disposition and number of shared edges between polyhedra. Ball or packing models can then be built from these atomic maps, giving a better idea of atom or ion packing. Crystal chemistry then comes alive.

A tabulation of the 230 space groups is reproduced in Table 6.5.

Table 6.5. *(cont.)*

Point-group	*Hermann–Mauguin symbols*		*Schoenflies symbols*
	Normal	*Larger cell*	
$3m$	$P3c1$ ($C3c$)	($H31c$)	C_{3v}^{3}
(*cont.*)	$P31c$ ($C31c$)	($H3c$)	C_{3v}^{4}
	$R3m$		C_{3v}^{5}
	$R3c$		C_{3v}^{6}
$\bar{3}m$	$P\bar{3}1m$ ($C\bar{3}1m$)	($H\bar{3}m$)	D_{3d}^{1}
	$P\bar{3}1c$ ($C\bar{3}1c$)	($H\bar{3}c$)	D_{3d}^{2}
	$P\bar{3}m1$ ($C\bar{3}m$)	($H\bar{3}1m$)	D_{3d}^{3}
	$P\bar{3}c1$ ($C\bar{3}c$)	($H\bar{3}1c$)	D_{3d}^{4}
	$R\bar{3}m$		D_{3d}^{5}
	$R\bar{3}c$		D_{3d}^{6}
6	$P6$ ($C6$)	($H6$)	C_{6}^{1}
	$P6_1$ ($C6_1$)	($H6_1$)	C_{6}^{2}
	$P6_5$ ($C6_5$)	($H6_5$)	C_{6}^{3}
	$P6_2$ ($C6_2$)	($H6_2$)	C_{6}^{4}
	$P6_4$ ($C6_4$)	($H6_4$)	C_{6}^{5}
	$P6_3$ ($C6_3$)	($H6_3$)	C_{6}^{6}
$\bar{6}$	$P\bar{6}$ ($C\bar{6}$)	($H\bar{6}$)	C_{3h}^{1}
$6/m$	$P6/m$ ($C6/m$)	($H6/m$)	C_{6h}^{1}
	$P6_3/m$ ($C6_3/m$)	($H6_3/m$)	C_{6h}^{2}
622	$P622$ ($C62$)	($H62$)	D_{6}^{1}
	$P6_122$ ($C6_12$)	($H6_12$)	D_{6}^{2}
	$P6_522$ ($C6_52$)	($H6_52$)	D_{6}^{3}
	$P6_222$ ($C6_22$)	($H6_22$)	D_{6}^{4}
	$P6_422$ ($C6_42$)	($H6_42$)	D_{6}^{5}
	$P6_322$ ($C6_32$)	($H6_32$)	D_{6}^{6}
$6mm$	$P6mm$ ($C6mm$)	($H6mm$)	C_{6v}^{1}
	$P6cc$ ($C6cc$)	($H6cc$)	C_{6v}^{2}
	$P6_3cm$ ($C6_3cm$)	($H6_3mc$)	C_{6v}^{3}
	$P6_3mc$ ($C6_3mc$)	($H6_3cm$)	C_{6v}^{4}
$\bar{6}m2$	$P\bar{6}m2$ ($C\bar{6}m$)	($H\bar{6}2m$)	D_{3h}^{1}
	$P\bar{6}c2$ ($C\bar{6}c$)	($H\bar{6}2c$)	D_{3h}^{2}
	$P\bar{6}2m$ ($C\bar{6}2m$)	($H\bar{6}m$)	D_{3h}^{3}
	$P\bar{6}2c$ ($C\bar{6}2c$)	($H\bar{6}c$)	D_{3h}^{4}
$6/mmm$	$P6/mmm$ ($C6/mmm$)	($H6/mmm$)	D_{6h}^{1}
	$P6/mcc$ ($C6/mcc$)	($H6/mcc$)	D_{6h}^{2}
	$P6_3/mcm$ ($C6/mcm$)	($H6/mmc$)	D_{6h}^{3}
	$P6_3/mmc$ ($C6/mmc$)	($H6/mcm$)	D_{6h}^{4}

Summary

Crystals are solids enclosing three-dimensional lattices of points. These lattices are built from five two-dimensional lattices combined perpendicularly to form fourteen three-dimensional arrays of points, the Bravais space lattices.

Seven primitive lattices represent the seven crystal systems.

In minerals, the smallest coplanar array of points that translate in three-dimensional space defines the unit cell. Thus, all minerals crystallize in one each of the 7 crystal systems, 32 crystal classes, and 230 space groups.

Table 6.5. *(cont.)*

Point-group	*Hermann–Mauguin symbol*	*Schoenflies symbol*
Cubic		
23	$P23$	T^1
	$F23$	T^2
	$I23$	T^3
	$P2_13$	T^4
	$I2_13$	T^5
$m3$	$Pm3$	T_h^1
	$Pn3$	T_h^2
	$Fm3$	T_h^3
	$Fd3$	T_h^4
	$Im3$	T_h^5
	$Pa3$	T_h^6
	$Ia3$	T_h^7
432	$P432$ ($P43$)	O^1
	$P4_232$ ($P4_23$)	O^2
	$F432$ ($F43$)	O^3
	$F4_132$ ($F4_13$)	O^4
	$I432$ ($I43$)	O^5
	$P4_332$ ($P4_33$)	O^6
	$P4_132$ ($P4_13$)	O^7
	$I4_132$ ($I4_13$)	O^8
$\bar{4}3m$	$P\bar{4}3m$	T_d^1
	$F\bar{4}3m$	T_d^2
	$I\bar{4}3m$	T_d^3
	$P\bar{4}3n$	T_d^4
	$F\bar{4}3c$	T_d^5
	$I\bar{4}3d$	T_d^6
$m3m$	$Pm3m$	O_h^1
	$Pn3n$	O_h^2
	$Pm3n$	O_h^3
	$Pn3m$	O_h^4
	$Fm3m$	O_h^5
	$Fm3c$	O_h^6
	$Fd3m$	O_h^7
	$Fd3c$	O_h^8
	$Im3m$	O_h^9
	$Ia3d$	O_h^{10}

The external symmetry of a crystal allows us to determine its crystal system and class, but X-ray analysis of its internal symmetry is necessary to find its space lattice and space group. This is done by single crystal analysis, which can locate specific atoms in the structure, and by powder diffraction, which measures the distances between like planes of atoms.

Bibliography

Azaroff, L. V., and Buerger, M. J. (1958). *The powder method.* McGraw-Hill, New York.

Bloss, F. D. (1971). *Crystallography and crystal chemistry.* Holt, Rinehart & Winston, New York.

Bragg, W. L. (1933). *The crystalline state.* Vol. 1. *A general survey.* G. Bell, London.

Bragg, W. L., Claringbull, G. F., and Taylor, W. H. (1965). *Crystal structures of minerals. The crystalline state,* Vol. 4. Cornell University Press, Ithaca, N.Y.

Buerger, M. J. (1963). *Elementary crystallography.* Wiley, New York.

(1971). *Introduction to crystal geometry.* McGraw-Hill, New York.

International tables for X-ray crystallography, Vol. I (1952). Kynoch Press, Birmingham, England.

Klein, C., and Hurlbut, C. S., Jr. (1985), *Manual of mineralogy* (after J. D. Dana), 20th ed. Wiley, New York.

Klug, H. P., and Alexander, L. E. (1954). *X-ray diffraction procedures.* Wiley, New York.

Lipson, H. S. (1970). *Crystals and X-rays.* Wykeham, London.

7 Crystal field theory

Crystal field theory describes the net change in crystal energy resulting from the orientation of *d* orbitals of a transition metal cation inside a coordinating group of anions, also called ligands. Although considered to be largely electrostatic in origin, there is always a covalent contribution. Because it is simpler to treat the theory using an electrostatic model, we do so here.

Any free transition metal ion is classed as fivefold degenerate, because all five *d* orbitals are equal in energy. However, when the ion is placed in an octahedral field of six O^{2-} anions (the ligands), the degeneracy is lifted by a splitting of the crystal field into low and high energy levels. Consider an octahedral array of O^{2-} anions surrounding a transition metal cation. Oxygen anions will be positioned on the six corners of a regular octahedron on opposite ends of reference axes *x*, *y*, and *z*, analogous to the *a*, *b*, and *c* axes of crystals. Transition metal ion orbitals d_{xy}, d_{xz}, and d_{yz} (Fig. 7.1) each orient their four lobes of high electron density between octahedral axes *x*, *y*, and *z* and thus do not point directly at any of the six O^{2-} oxygen anions; *d* orbital electrons are not repelled, and all six anions can approach more closely to the cation, reducing the A—X (M–O) distance and contracting the octahedron. This contraction results in a lowering of crystal energy and an increase in stability. Each electron in orbitals d_{xy}, d_{xz}, and d_{yz} is lowered in energy by an amount $\Delta_0 = \frac{2}{5}$ (after Burns 1970a); it is two-fifths more stable than in the free or degenerate ion. These three low energy orbitals, d_{xy}, d_{xz}, and d_{yz}, constitute the t_{2g} level of crystal field theory.

Conversely, each electron located in the $d_{x^2-y^2}$ or d_{z^2} orbitals will be repelled, because their lobes of high electron density point directly at oxygen anions located on the octahedral corners. Accordingly, each such electron is raised in energy by an amount $\Delta_0 = \frac{3}{5}$ above that in a free degenerate ion. This higher energy state consisting of orbitals $d_{x^2-y^2}$ and d_{z^2} constitutes the e_g level of crystal field theory. Some find it easier to think of the oxygen anions as being repelled by the *d* orbital electrons, rather than the reverse, but the result is the same.

Now consider the electron occupancy of *d* orbitals of first-period transition metal ions (high spin configuration) from ${}_{21}Sc^{3+}$ to ${}_{30}Zn^{2+}$ (Table 7.1). Note that ions Sc^{3+} and Ti^{4+} have lost their *d* electrons in the ionization process, so $\Delta_0 = 0$. Ions Mn^{2+} and Fe^{3+} place one electron in each of the five 3*d* orbitals, resulting in an algebraic cancellation of increased and decreased energies, to a net $\Delta_0 = 0$, or 0/5; Zn^{2+} has two electrons paired in each of the five 3*d* orbitals, also effecting a cancellation of CFSE (crystal field stabilization energy) to $\Delta_0 = 0$. In all other cases, the transition metal cations of Table 7.1 will split the octahedral crystal field into low (t_{2g}) and high (e_g) energy levels and gain CFSE; each will have an octahedral site preference energy commensurate with its value of Δ_0.

It is important to see that Cr^{3+}, with three electrons in t_{2g}, will permit all six coordinating octahedral O^{2-} anions to approach closely to form a small or contracted, but regular, octahedron. In contrast, the addition of a fourth electron into $d_{x^2-y^2}$ or d_{z^2}, as in Mn^{3+} or Cr^{2+}, will induce negative charge repulsion along either the *x*-*y* octahedral axes or along the *z* axis, depending on which orbital holds the fourth *d* electron. Accordingly, the repulsion of negative charge will cause the octahedron to become distorted (Fig. 7.2); $d_{x^2-y^2}$ occu-

Figure 7.1. When a transition metal cation lies within an octahedral crystal field formed of six O^{2-} anions or ligands, cation orbitals, $d_{x^2-y^2}$ and d_{z^2} orient along octahedral axes, *x*, *y*, and *z*, in high energy level, *eg*; they are repelled by the six O^{2-} ligands. Cation orbitals, d_{xy}, d_{yz}, and d_{xz} orient between octahedral axes, lie in low energy level, t_{2g}, and are attracted by, drawing closer to, the cation.

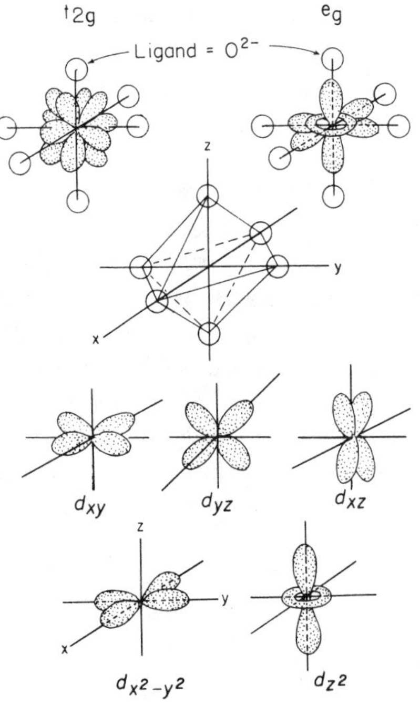

pancy will induce a tetragonal compressed or oblate octahedral distortion, whereas d_{z^2} occupancy will introduce a tetragonal elongate or prolate octahedral distortion. These distortions are classed as Jahn–Teller distortions in crystal field theory. Note that the addition of a fifth *d* electron, as in Mn^{2+} or Fe^{3+}, will cancel both the CFSE and the octahedral distortion.

Because crystal field theory has been successful in explaining cation ordering in the spinel mineral group, it will prove instructive to examine several examples. All members of this group have sixteen octahedral cations and eight tetrahedral cations along with thirty-two oxygen anions in one unit cell containing eight formula units ($Z = 8$). Spinel formulae confuse students because some mineralogists write $Mg^{IV}Al_2^{VI}O_4$ and others write $Al_2^{VI}[Mg^{IV}O_4]$: the latter is chosen here because it provides more crystal chemical information. The entry $[MgO_4]$ clearly informs us that the Mg cation is four-coordinated by O^{2-}, a fact often unknown in this chemically very complex spinel group.

Although many natural and synthetic compounds crystallize with the spinel structure, the most appropriate example to consider first is chromite, $Cr_2^{VI}[Fe^{IV}O_4]$. From Table 7.1 we see that Cr^{3+}, with three unpaired *d* electrons,

Table 7.1. *d orbital occupancy and octahedral crystal field stabilization energies for first series transition metal ions in high spin configuration*

Z	Ion	d els.	d-orbital occupancy: t_{2g} $(d_{xy}$	d_{xz}	$d_{yz})$	e_g $(d_{x^2y^2}$	$d_{z^2})$	C.F.S.E. Δ_o
21	Sc^{3+}	0	○	○	○	○	○	0/5
22	Ti^{4+}	0	○	○	○	○	○	0/5
22	Ti^{3+}	1	(↑)	○	○	○	○	2/5
23	V^{3+}	2	(↑)	(↑)	○	○	○	4/5
24	Cr^{3+}	3	(↑)	(↑)	(↑)	○	○	6/5
24	Cr^{2+}	4	(↑)	(↑)	(↑)	(↑)	○	3/5
25	Mn^{3+}	4	(↑)	(↑)	(↑)	○	(↑)	3/5
25	Mn^{2+}	5	(↑)	(↑)	(↑)	(↑)	(↑)	0/5
26	Fe^{3+}	5	(↑)	(↑)	(↑)	(↑)	(↑)	0/5
26	Fe^{2+}	6	(↑↓)	(↑)	(↑)	(↑)	(↑)	2/5
27	Co^{2+}	7	(↑↓)	(↑↓)	(↑)	(↑)	(↑)	4/5
28	Ni^{2+}	8	(↑↓)	(↑↓)	(↑↓)	(↑)	(↑)	6/5
29	Cu^{2+}	9	(↑↓)	(↑↓)	(↑↓)	(↑↓)	(↑)	3/5
30	Zn^{2+}	10	(↑↓)	(↑↓)	(↑↓)	(↑↓)	(↑↓)	0/5

has CFSE $\Delta_0 = \frac{6}{5}$, whereas Fe^{2+} has only four unpaired *d* electrons and $\Delta_0 = \frac{2}{5}$: thus Cr^{3+} enters both octahedral sites, causing the ion with the larger ionic radius, Fe^{2+}, to enter the tetrahedral site. Crystal field theory states that Cr^{3+}, $\Delta_0 = \frac{6}{5}$, gains more stabilization energy than any other competing cation through occupancy of the octahedral, rather than the tetrahedral, site.

The formula for magnetite is often written Fe_3O_4 in order to avoid more specific location of the Fe^{2+} and Fe^{3+} ions in the structure. The data of Table 7.1 show that Fe^{2+}, $\Delta_0 = \frac{2}{5}$, gains octahedral site preference over Fe^{3+}, $\Delta_0 = 0$, but complications arise because $Fe^{2+}:Fe^{3+} = 1:2$, and CN VI sites:CN IV sites = 2:1. Crystal field relations place the one Fe^{2+} ion in the octahedron, leaving the two Fe^{3+} ions to occupy a tetrahedral site and an octahedral site. The formula is now correctly written as $Fe^{2+VI}Fe^{3+VI}[Fe^{3+IV}O_4^{IV}]$. Magnetite is classed as an inverse spinel compared to chromite, which is a normal spinel.

Another example of the application of crystal field theory to spinels is provided by hausmannite, Mn_3O_4, which is distorted from cubic (isometric) to

Figure 7.2. Comparison of regular, prolate-distorted, and oblate-distorted octahedra. Distortions are caused by unbalanced distribution of electrons in high and low energy *d* orbitals of a transition metal ion.

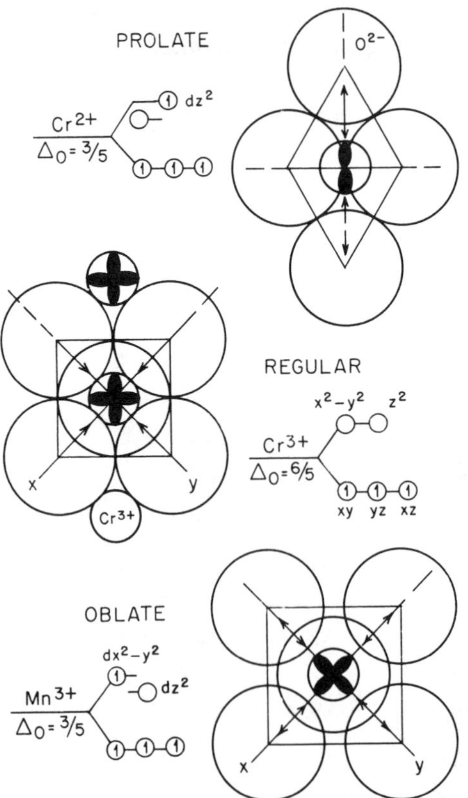

tetragonal symmetry by Jahn–Teller distortions of octahedral sites occupied by the Mn^{3+} ions. Note that Mn^{3+} has $\Delta_0 = \frac{3}{5}$ and Mn^{2+} has $\Delta_0 = 0$; thus the smaller radius Mn^{3+} is placed in the octahedral sites, and the much larger Mn^{2+} ion is placed in the tetrahedral site. Because the length of the *c* axis is greater than that of the *a* axis, the distortion is assumed to be prolate, with the fourth electron in Mn^{3+} entering d_{z^2}.

The mineral spinel for which the group is named (see Chapter 18) has the formula $Al_2^{VI}[Mg^{IV}O_4^{IV}]$, and neither Al^{3+} nor Mg^{2+} possesses any *d* orbital electrons or any crystal field stabilization energy. Why, then, does the smaller radius Al^{3+} ion occupy both octahedral sites, and the larger Mg^{2+} ion the tetrahedral site? It is a general truth of crystal chemistry that reduced interatomic distances, especially for ions with high formal charge, lead to increased stability; so Al^{3+} is favored in the octahedral sites and is said to have higher lattice energy. Such lattice energy must be very comparable to crystal field energy in magnitude because the synthetic Al_2NiO_4 spinel is said by McClure (1957) to be very disordered, with roughly equal amounts of Al^{3+} and Ni^{2+}, $\Delta_0 = \frac{6}{5}$, in both octahedral and tetrahedral sites. All other Al-rich spinels show the Al^{3+} ion occupying both octahedral sites. Burns (1970b) has estimated that $\Delta_0 = 1/(\text{M–O distance})^5$, or that CFSE is inversely proportional to interatomic distance, so the smaller octahedral site will always have a greatly increased CFSE.

Location of a transition metal inside a coordination polyhedron other than an octahedron can also confer CFSE on a crystal, but *d* orbital lobes that point to corners of an octahedron do not point toward the corners of either a cube or a tetrahedron. Again, however, energy is lowered in these CN VIII and CN IV polyhedra when *d* orbital lobes point away from anions, and raised when they point at or near to the anions disposed on corners of these polyhedra. The disposition of these lobes of electron density with respect to the geometric location of oxygen anions confers lower CFSE values for cubic coordination than for octahedral coordination, and still lower values for tetrahedral coordination. Because of differing geometries, the low t_{2g} and high e_g energy levels of the octahedron are reversed in both the cube and the tetrahedron. Try to draw a cube or a tetrahedron around the lobes of the five *d* orbitals, and it will be apparent that CFSE energies will be lower because of the difficulty in orienting the lobes as efficiently as can be done in the octahedron.

According to Hund's classical selection rules (see Chapter 1), transition metal electrons should occupy *d* orbitals singly before doubly, favoring the high spin over the low spin state. This appears to be the rule in most oxide and silicate minerals. In some cases, however, spin pairing may lead to a lower crystal energy than would a high spin configuration. In many sulfides, Fe^{2+} and Co^{2+} ions adopt the low spin configuration. Spin pairing of *d* orbital electrons leads to increased π bond formation (see Chapter 3), and CFSE, which in turn reduces interatomic distances (M–S) and strengthens metal–sulfur bonds. This once again reinforces our suggestion that a covalent component is always to be expected in so-called ionic compounds. Where π bonds form between transition metal and sulfur, a molecular orbital extends throughout the crystal, endowing it with metallic luster and electrical conductivity.

Summary

Crystal field theory describes the net change in crystal energy resulting from the orientation of *d* orbitals of a transition metal cation inside a coordinating group of anions (ligands).

Inside an octahedron of oxygen anions, a high spin transition metal ion (fivefold degenerate in the free state, because its five *d* orbitals are equal in energy) has its valence electrons split into three low energy orbitals, d_{xy}, d_{xz}, d_{yz}, and two high energy orbitals, $d_{x^2-y^2}$, d_{z^2}. Ions Sc^{3+} and Ti^{4+} lose their *d* electrons in ionizing and have $\Delta_0 = 0$. Mn^{2+} and Fe^{3+}, with one electron in each of five 3*d* orbitals, cancel high and low energy levels so that $\Delta_0 = 0$, as does Zn^{2+} with two electrons paired in each of the five *d* orbitals. All other transition metal cations gain crystal field stabilization energy, and those with d^4 or d^9 configurations (e.g., Mn^{3+} and Cu^{2+}) cause Jahn–Teller distortions of the octahedron.

Crystal field relations cause various distortions in minerals crystallizing with the spinel structure and may be a dominant factor in cation ordering. In polyhedra other than octahedra, crystal field stabilization energy of *d* orbitals is lowered when the orbital lobes point away from anions and raised when they point toward anions.

CFSE values are lower in magnitude in cubic and tetrahedral coordination than in octahedral cation coordination.

In sulfide minerals, spin pairing of *d* orbital electrons leads to lower crystal energy through increase in both π bond formation and CFSE (Fe^{2+} in low-spin configuration will have $\Delta_0 = \frac{12}{5}$ because all six *d* electrons are paired in the three low energy orbitals).

Bibliography

Ballhausen, C. J. (1962). *Introduction to ligand field theory.* McGraw-Hill, New York.

Burns, R. G. (1970a). *Mineralogical applications of crystal field theory.* Cambridge Univ. Press, London.

(1970b). Crystal field spectra and evidence of cation ordering in olivine minerals. *Am. Mineral.*, 55:1608–32.

Cotton, F. A., and Wilkinson, G. (1966). *Advanced inorganic chemistry,* 2d ed. Interscience, New York.

Dunitz, J.D., and Orgel, L. E. (1957). Electronic properties of transition metal oxides. II. Cation distribution amongst octahedral and tetrahedral sites. *J. Phys. Chem. Solids,* 3:318–33.

McClure, D. S. (1957). The distribution of transition metal cations in spinels. *J. Phys. Chem. Solids,* 3:311–17.

8 Polyhedral distortion

The tetrahedron, octahedron, and cube are geometric solids that possess the basic external symmetry elements required by the isometric crystal system, specifically four threefold axes of rotational symmetry, whereas planar triangles and squares will possess one threefold and one fourfold rotation axis, respectively. Close examination of the geometry of these polyhedral or planar forms by single crystal X-ray analyses of crystals shows that these requirements are rarely satisfied. Atomic groupings in most minerals are considerably distorted, and, strictly speaking, isometric tetrahedra, octahedra, and cubes rarely occur. Regular, symmetrically disposed groupings of atoms, however, are to be expected in those crystals whose atoms are polymerized by the use of covalent hybrid bond orbitals such as sp^2 (120°) planar, equilateral triangular as in graphite; sp^3 (109°28′) regular tetrahedral as in diamond; dsp^2 (90°) regular, square planar (or rectangular) as in cooperite (PtS); d^2sp^3 (90°) regular octahedral orbitals of Fe combined with S—S sp^3 orbitals as in pyrite.

Those crystals in which ionic bonding predominates will have their symmetry distorted by local imbalance of electrical charge, by cation repulsion across a shared polyhedral edge or bridge, or by Jahn–Teller distortions of crystal field theory. In the pyroxene diopside (see Fig. 5.6, Table 5.1), it was noted that unequal EBS distribution over three different oxygens was associated with unequal distances in $Si—O_4$ tetrahedra, $Mg—O_6$ octahedra, and in $Ca—O_8$ square antiprisms; the oxygen polyhedra are distorted as shown (Fig. 8.1), and the cations are not centrosymmetrically oriented in them. Such phenomena are also found in the magnesian olivine forsterite (Fig. 8.2).

In the idealized structure of olivine, the oxygen anions form a perfect hexagonal close-packed array, with Mg ions in regular octahedra and Si ions in regular tetrahedra. The actual structure is not at all close packed but is built of kinked Mg—O octahedral chains that run parallel to the *c* axis in a zigzag pattern. The chains are built of two types of polyhedra: elongated $Mg—O_6$ octahedra designated M-1 that run zigzag parallel to *c*, and alternately attached extra links representing a larger octahedron designated M-2 (Fig. 8.2). Each oxygen in the structure links two M-1 octahedra to one M-2 octahedron and one tetrahedron. Both the drawings and the interatomic distances for O—O, Si—O, and Mg—O reveal large distortions from isometric coordination polyhedra. Here, the cause of these distortions lies in the repulsion of positive charges on Mg^{2+} and Si^{4+} (ideal charges) cations that approach too closely on opposite sides of O—O bridges. Note that each tetrahedron shares three of its six edges with octahedra, each M-1 octahedron shares six of its twelve edges with M-2 octahedra and tetrahedra, and each M-2 octahedron shares three of its six edges with the other polyhedra.

Shared edges, O—O, shorten from an ideal 2.80-Å distance to 2.56 Å, and unshared edges lengthen to 3.39 Å, resulting in the considerable distortion of all polyhedra. Regular and distorted octahedra are compared in packing models (Fig. 8.3), where Mg—O_6 octahedra in periclase are regular because all twelve edges are shared, eliminating the possibility of electrostatically induced distortion; Al—O_6 "octahedra" in corundum are distorted to trigonal antiprisms because Al—Al repulsion occurs across one shared face and not the others (also see Figs. 18.2, 18.3); and M-1 and M-2 octahedra in forsterite show the distortions described above. Thus, in periclase, the oxygen anions are in cubic close packing, where all octahedral edges are shared; whereas, in both forsterite and corundum the oxygens only approximate a hexagonal close-packed array. Cur-

Figure 8.1. Comparison of regular with distorted square antiprismatic, octahedral, and tetrahedral polyhedra in diopside.

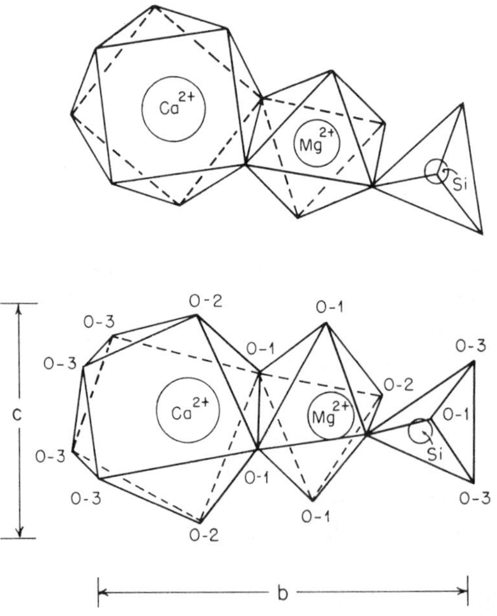

iously, the electrostatic phenomenon of cation–cation repulsion that induces the shared-edge shortening results in O—O overlap, a covalent phenomenon.

Electrostatically induced distortion of polyhedra also occurs where crystal field splitting of octahedral sites places a fourth or ninth *d* electron in either one of the axially directed orbitals $d_{x^2-y^2}$ or d_{z^2} (see Chapter 7) and causes the prolate octahedral distortion observed in hausmannite, $Mn_2^{3+}[MnO_4]$, which then assumes a tetragonal symmetry rather than the isometric symmetry of a normal spinel.

Many minerals in which Cu^{2+}, $3d^9$, is bonded to oxygen show characteristic square planar arrays of O around Cu, with fifth and sixth oxygens at corners of very distorted, prolate or elongated, "octahedra." Here, as elsewhere in crystal chemistry, two interpretations are possible. The crystal field argument would be based on the unbalanced electron distribution that places paired electrons in $d_{x^2-y^2}$ and a single electron in d_{z^2}; an alternative would be the formation of a dsp^2 square planar hybrid (see Chapter 3) on Cu^{2+} in which the four orbitals of the hybrid point toward corners of a square; the ninth electron, not clearly accounted for, may be involved in π bonding.

The same arguments may be offered for the compound CrF_2, where Cr^{2+} has a $3d^4$ configuration; the prolate octahedral distortion observed could result from placement of the fourth *d* electron in d_{z^2} (Jahn–Teller distortion) or by the formation of a square planar hybrid bond orbital on Cr^{2+}. In each example, the problem might be solved by determining paramagnetic susceptibilities: high magnetic response would indicate unpaired electrons and favor the crystal field

Figure 8.2. Distortion of octahedra and tetrahedra in forsterite, Mg_2SiO_4, induced by shared-edge shortening.

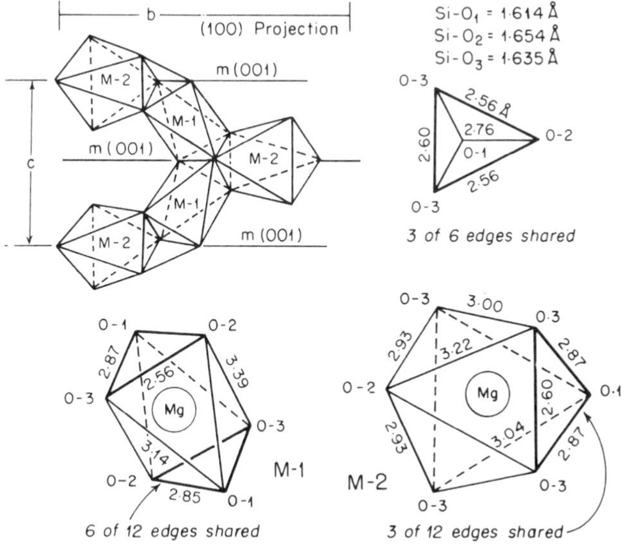

argument; low magnetic response would indicate electron pairing and favor the hybrid bond orbital argument. Pyrite, FeS_2, and hauerite, MnS_2, though they are isostructural, show totally different magnetic properties. Pyrite is essentially diamagnetic, and hauerite is strongly paramagnetic; *d* electrons of Fe^{2+} in pyrite must be paired in low spin configuration, whereas *d* electrons of Mn^{2+} in hauerite must be unpaired or in high spin configuration (see Chapter 3).

Some minerals have their oxygen atoms linked to other atoms by both covalent and ionic bonds. In barite, $Ba[SO_4]$ (Colville and Staudhammer 1967), sp^3 hybrid covalent bonds link S to O in $[SO_4]^{2-}$ groups, and these groups in turn are held together by Ba—O bonds of predominantly ionic character. In barite, the large-radius Ba^{2+} ion ($r_{Ba} = 1.60$ Å) is coordinated by twelve oxygen atoms or anions that form parts of seven different $[SO_4]^{2-}$ tetrahedra. Cation repulsion across the large number of O—O edges shared between covalent $[SO_4]$ and ionic $Ba—O_{12}$ polyhedra can perturb and distort the otherwise more regular $[SO_4]$ groups. Pauling (1960) has noted that the $[SO_4]$ group, although built predominantly of sp^3 hybrid covalent bonds, also has a small ionic component, and the Ba—O bonds must, in turn, have a covalent component of 17%, calculated from electronegativity differences. Repeated here, for emphasis, is Pauling's warning that, although most atoms are charged, they carry a far lower charge than is indicated by formal oxidation states.

To sum up, it appears that polyhedral distortion in crystals results from a variety of crystal chemical phenomena that are essentially electrostatic in ori-

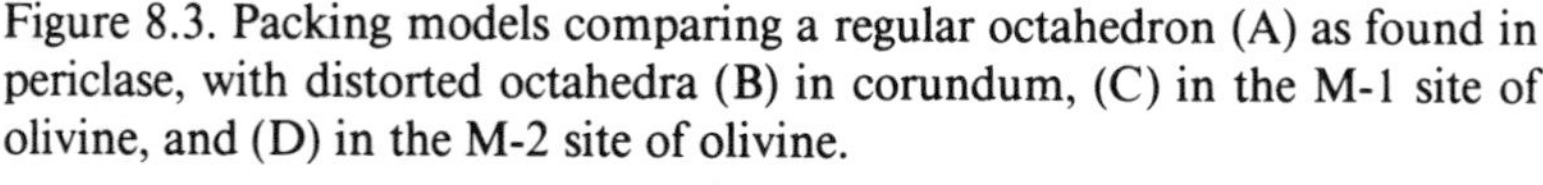
Figure 8.3. Packing models comparing a regular octahedron (A) as found in periclase, with distorted octahedra (B) in corundum, (C) in the M-1 site of olivine, and (D) in the M-2 site of olivine.

gin. These phenomena include ionization, cation repulsion across shared anion bridges such as O—O, local imbalance of electrostatic bond strength (uneven charge distribution), and Jahn–Teller distortions associated with unbalanced occupancy of axially directed d_{z^2} and $d_{x^2-y^2}$ orbitals in octahedral fields.

Polyhedral regularity of geometry is to be expected in crystals polymerized essentially by directed hybrid bond covalent orbitals, sp^2, sp^3, dsp^2, and d^2sp^3.

Summary

Strictly isometric coordination polyhedra rarely occur in crystals in which ionic bonding predominates, though they may be found in crystals polymerized by covalent hybrid bond orbitals. Crystals with predominantly ionic bonding, local electrostatic imbalance, cation repulsion across a shared edge or bridge, or Jahn–Teller effects contain distorted polyhedra. Examples of each of these cases are discussed.

Bibliography

Birle, J. D., Gibbs, G. V., Moore, P. B., and Smith, J. V. (1968). Crystal structures of natural olivines. *Am. Mineral.,* 53:807–24.

Belov, N. V., Belova, E. N., Andrianova, N. N., and Smirova, R. F. (1951). Parameters of olivine. *Dokl. Akad. Nauk SSSR,* 81:399.

Bragg, L., Claringbull, G. F., and Taylor, W. H. (1965). *Crystal structures of minerals. The crystalline state,* Vol. IV. Cornell Univ. Press, Ithaca, N.Y.

Burns, R. G. (1970). *Mineralogical applications of crystal field theory.* Cambridge University Press, London.

Brown, G. E., Jr. (1982). Olivines and silicate spinels. *Mineral. Soc. Am. Rev. Mineral.,* 5:275–381.

Colville, A. A., and Staudhammer, K. (1967). A refinement of the structure of barite. *Am. Mineral.,* 52:1877–80.

Hazen, R. M., and Finger, L. W. (1985). Crystals at high pressure. *Sci. Am.,* 252:110–17.

Pauling, L. (1960). *The nature of the chemical bond,* 3d ed. Cornell Univ. Press, Ithaca, N.Y.

9 Diadochy and isostructural crystals

The proliferation of terms used to describe aspects of substitution of ions or atoms in crystals includes diadochy, isomorphism, isostructuralism, isotypism, limited substitution, and solid solution. Because these terms have been inconsistently used, they are defined here in terms of their etymology.

Diadochy (from Greek, *διαδεχομαι*, to receive from one another, to succeed, to substitute for) is obviously the most appropriate term to use for the substitution of atoms in crystals. As used here, it covers the limited substitution of ions in a particular mineral species, as well as complete solid solution of two or more ions in related mineral species.

Isomorphism (from Greek, *ἰσομορφος*, equal in form) is intended to relate a group of minerals of similar or identical morphology and analogous formula that contain different ions in a given coordination site; for example, calcite, $Ca^{VI}[CO_3]$, magnesite, $Mg^{VI}[CO_3]$, rhodochrosite, $Mn^{VI}[CO_3]$, and siderite, $Fe^{VI}[CO_3]$. The terms *isomorphous* or *isomorphism* have been incorrectly used to imply that the group shows solid solution relations. Ca and Mg of calcite and magnesite, respectively, for example, show only very limited solid solution relations under natural conditions, whereas the others show extensive to complete solid solution.

Isostructural (Greek *ἰσο* + Latin *structura*, meaning equal in structure). As stringently defined here, isostructural crystals must exhibit equivalence of coordination numbers of all cations and anions, space group, formula, and the number of formulae *(Z)* or atoms in one unit cell. It clearly does not imply solid solution relations, and frequently none exist.

The terms *limited solid solution* and *complete solid solution* are treated here as aspects of diadochy, and isomorphism is abandoned in favor of isostructuralism, although only nuances separate them.

Diadochy and isostructuralism are subdivided in Figure 9.1, and a discussion of each type follows.

Ion–ion and coupled substitutions

The great pioneer ("father") of geochemistry and crystal chemistry was V. M. Goldschmidt. On the basis of the very limited experimental data available in 1920–40, he foresaw and formulated many principles of crystal chemistry that are still valid today. He taught that ionic radius and electric charge are principal factors controlling diadochy in minerals crystallizing from magmas. Three of his substitution categories included *camouflage, capture,* and *admission.*

Camouflage represents the substitution of one element, often a trace element, for a major element in a host mineral when both elements have similar radii (± 15%) and identical electric charge; for example, Hf^{4+VIII} ($r = 0.83$ Å) replaces Zr^{4+VIII} ($r = 0.84$ Å) in the distorted square antiprismatic coordination site of zircon, $Zr[SiO_4]$ (Figs. 12.5, 12.6), both elements tending to concentrate in very late stages of magmatic crystallization.

Capture refers to the substitution of one minor element having similar ionic radius to, but higher electric charge than, the host element it replaces. The excess charge must be balanced by another substitution that reduces charge, and a coupled substitution often results; for example, $Th^{4+}Si^{4+} \leftrightarrows Ce^{3+}P^{5+}$ in monazite, (Ce,La) $[PO_4]$; $Nb^{5+}Fe^{3+} \leftrightarrows 2Ti^{4+}$ in sphene (titanite), CaTi (O,F) $[SiO_4]$; or $Ba^{2+}Al^{3+} \leftrightarrows K^{+}Si^{4+}$ in orthoclase, $K[AlSi_3O_8]$.

Admission refers to the substitution of an element of similar ionic radius, but lower charge, for a host element; in the three examples used, the admission process is coupled to the capture process to balance electric charge.

Figure 9.1. Classification scheme for diadochic and isostructural compounds.

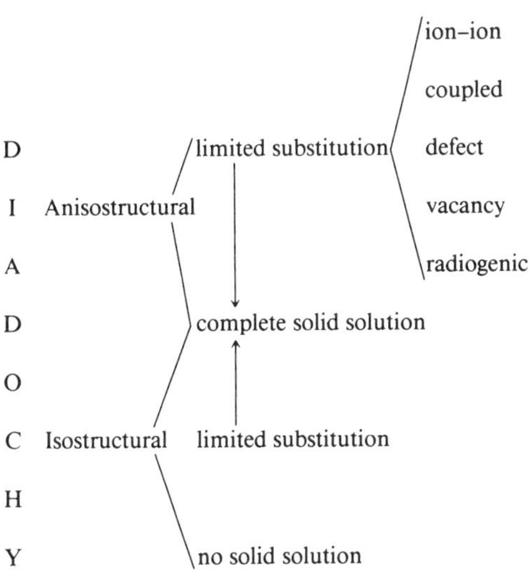

Defect or omission substitution

Defect or omission substitution may occur when a crystal structure possesses void sites that may only be partially occupied by other ions. The classic example is pyrrhotite, $Fe_{1-x}S$ or $Fe_{11}S_{12}$, in a lattice that does not have an equal amount of Fe and S atoms. This mineral is thus Fe deficient, and the lattice is said to be defective. Note that the Fe and S are covalently bonded, and electric charge need not be balanced in the same manner as in ionic structures. The common mineral magnetite, $Fe^{3+}Fe^{2+}[Fe^{3+}O_4]$, may alter by oxidation processes to maghemite (γ-Fe_2O_3), which is really an ion-deficient and iron-deficient spinel structure in which the combined three octahedral and tetrahedral sites are only partially occupied, all by trivalent Fe; the formula may be written as $[Fe_{1.78}\square_{0.22}]$ $[Fe_{0.88}\square_{0.12}O_4]$.

Another fine example of defect substitution occurs in the complex oxide mineral pyrochlore, $Na,Ca^{VIII}Nb_2{}^{VI}O_6{}^{IV}F^{IV}$, which was once considered rare but is now the principal ore mineral of niobium. When alloyed with iron, niobium produces high-quality steel. The chemical formula cited above is ideal and is seldom met with in nature. Because pyrochlore is built of an octahedral framework in which each NbO_6 octahedron shares all of its six corner oxygens with other octahedra, as much as 75% of the Na, Ca, and F ions can be removed, leaving gaping holes in a defect lattice of formula $\square^{VIII}_{1.5}(Ca,Ba)_{0.5}Nb^{VI}_2O^{IV}_4(OH)^{IV}_3$. Neutrality is maintained by a compensating substitution of $(OH)^-$ for O^{2-}, according to the scheme $NaCaO_3 \leftrightharpoons \square\square(OH)_3$.

In other varieties of pyrochlore, where the mineral is concentrated in a chemical weathering process under tropical humid conditions, Na and Ca are removed and Ba is partially substituted, leaving voids in the structure. Here we are obviously dealing with weak one-electron ionic bonds Na—O, Na—F, Ca—O, Ca—F that can be broken much more readily than the stronger, five-electron bonds Nb^{5+}—O that are undoubtedly in part covalent (see Figs. 18.18, 18.19).

Vacancy substitution

One of the most interesting types of diadochy results from the substitution or entry of large stray ions or even molecules into structural voids that exist in virtually perfect crystals as tunnels. For example, in beryl, $Al_2Be_3[Si_6O_{18}]$, the columns of this hexagonal lattice normally have empty tunnels that pass through $[Si_6O_{18}]$ groups that lie parallel to the *c* axis of the crystal. These tunnels are large enough in diameter to accept stray ions of Cs^+ ($r = 1.88$ Å), the largest of all cations, as well as water molecules ($r = 1.38$ Å). Still other minerals carry residual surface charges on cleavage or other faces that may attract oppositely charged ions. Another fine example of vacancy substitution occurs in amphiboles that contain a distorted tenfold coordination site that is commonly empty, for example, tremolite, $\square Ca_2Mg_5[Si_8O_{22}](OH)_2$. Here, a coupled substitution also involving these vacant sites takes place according to the scheme, $Na^{+X}Al^{3+IV} \leftrightharpoons \square^{X}Si^{4+IV}$.

Crystal field substitution

It should be evident that crystal field stabilization energy (see Chapter 7) may play a significant role in diadochy. A study made by Curtis (1964) showed that effective concentration of Cr^{3+} ($\Delta_0 = \frac{6}{5}$) and Ni^{2+}($\Delta_0 = \frac{6}{5}$) in place of the Mg^{2+} in octahedral sites of early formed olivine and pyroxene of the Skaergaard Complex, Greenland, could readily by explained by CFSE processes. Conversely, rejection of Mn^{3+} and Cu^{2+} from these same octahedral sites was related to their Jahn–Teller distortions and probable destabilization of octahedral sites, resulting in residual concentration of these elements in late magmatic liquids.

Radiogenic substitution

Radiogenic substitution is a complex process that may introduce a large ion into a small site because of radioactive decay and, in so doing, may result in the ultimate destruction of the ordered crystal lattice of the host mineral. For example, zircon, alluded to earlier, always carries small amounts of uranium and thorium, both of which decay, largely by α-decay processes, to lead and helium. Entry of U^{4+} ($r = 1.00$ Å) and Th^{4+} ($r = 1.04$ Å) into the CN VIII sites of Zr ($r = 0.84$ Å) results in a small lattice expansion, depending on the amount of each that is introduced. Radioactive decay of the U^{4+} and Th^{4+} to Pb^{2+} ($r = 1.32$ Å) expands the CN VIII sites and the lattice considerably. Depending on radioactive dosage (concentration + time), the zircon lattice may become disordered (Holland and Gottfried 1955), bit by bit, in a process called *metamictization,* in which water enters into radiation-induced fractures and eventually alters and destroys the lattice: the zircon becomes amorphous. Such changes in the degree of crystallinity are accompanied by a lowering of the indices of refraction that can be correlated with an increase of radiogenic lead in zircon (Gottfried, Jaffe, and Senftle 1959).

Isostructural compounds and applied crystal chemistry

To this point, we have considered diadochy largely from an ionic viewpoint, although we know from electronegativity differences that most chemical bonds in minerals carry a considerable covalent component. A covalent component, however, does not negate the ionic component. According to Pauling (1960), it merely transfers and reduces formal charges on cations so that Si^{4+} may really be Si^{1+}, but ions remain electrically charged. This reduction and transfer of charge to a resultant charge near unity embodies Pauling's electronegativity principle. How important, then, are electronegativity, ionization potential, and the chemical bond type in diadochy? Some of the answer must come from the large number of minerals and synthetic compounds that are isostructural but do not show limited substitution, nor do they form any solid solution series.

A list of $A^{VI}X^{VI}$ structures, all isostructural with halite, is given in Table 9.1, along with a comparison of their electronegativity values, percentages of ionic bonding calculated from these, and some physical and chemical properties. Eight of these occur as natural minerals, and nine are synthetic compounds. These data reveal a wide spectrum of chemical bond types, with index of refraction n and luster especially revealing. There is little evidence that any of these form solid solution series, although all have the identical crystal structure of $Na^{VI}Cl^{VI}$. Why, then, do such chemically heterogeneous compounds grow with the identical crystal structure? The important principle here is that atoms in space are disordered and can lower their energy by ordering themselves into the lattice of a growing crystal. Atoms must come into tangential ionic contact or be overlapped in covalent or metallic union, and the VI–VI coordination space lattice becomes one of cubic close packing of the larger atom, with the smaller partner filling the octahedral voids. Thus twelve large atoms surround one another, while six of these atoms surround the smaller atom, and vice versa. Our model predicts that atom pairs in these crystals fill space very economically, but it does not tell us whether the bonds are ionic, covalent, metallic, or a combination of them. We can deduce the bond type from physical properties or from electronegativity predictions. Close-packed atomic arrays form low free energy, high stability assemblages, whether the principal bond be ionic, cova-

Table 9.1. *Compounds isostructural with halite* $4/Na^{VI}Cl^{VI}Fm3m$

Compound (all VI–VI)	Name	n	% Ionic bonding	Luster	Solubility
NaF	Villiaumite	1.328	91	Vitreous	H_2O
LiF	Synthetic	1.392	70	Vitreous	H_2O
CsF	Synthetic	1.478	93	Vitreous	H_2O
KCl	Sylvite	1.490	70	Vitreous	H_2O
NaCl	Halite	1.544	67	Vitreous	H_2O
RbBr	Synthetic	1.553	63	?	H_2O
RbI	Synthetic	1.647	51	?	H_2O
LiBr	Synthetic	1.784	55	?	H_2O
LiI	Synthetic	1.955	43	?	H_2O
AgCl	Cerargyrite	2.06	26	Adamantine	$(NH_4)(OH)$
MnO	Manganosite	2.16	63	Adamantine	HCl
NiO	Bunsenite	2.23	51	Adamantine	HCl
MgS	Synthetic	2.27	34	Adamantine	PCl_3
MnS	Alabandite	Opaque	22	Submetallic	HCl
PbS	Galena	Opaque	12	Metallic	HNO_3
TiC	Synthetic	Opaque	22	Metallic	Aqua regia

lent, or metallic. Closeness of packing, irrespective of bond type, is the hallmark of stability.

Solid-solution series

Solid solution series are too well known to mineralogists and chemists to need much elaboration here. The olivine series $Mg_2SiO_4 \leftrightharpoons Fe_2SiO_4$, the orthopyroxene series $Mg_2Si_2O_6 \leftrightharpoons Fe_2^{VI}Si_2^{IV}O_6$ complete at high pressures, the plagioclase series $NaAlSi_3O_8 \leftrightharpoons CaAl_2Si_2O_8$ at high temperature, the garnet series

$$(Mg \leftrightharpoons Fe \leftrightharpoons Mn \leftrightharpoons Ca)_3(Al \leftrightharpoons Fe^{3+} \leftrightharpoons Cr^{3+})_2[SiO_4]_3$$

and many others could be cited. Here we deal with complete solid solution series and isostructural compounds as well, although some of these require high pressure or high temperature to stabilize particular members of a series.

That element concentration in magmas can restrict solid solution series that might otherwise form is exemplified by the yttrium-bearing manganese garnet spessartine. Jaffe (1951) suggested that yttrium could enter the garnet lattice by use of the coupled substitution $Y^{3+VIII}Al^{3+IV} \leftrightharpoons Mn^{2+VIII}Si^{4+IV}$, and proposed the formula $(Mn_{3-x}Y_x)Al_2[Si_{3-x}Al_xO_4]_3$. He found a maximum of about 2.50% Y_2O_3 in garnet, in which $x = 0.1$ in a formula ratio of $(Mn_{2.9}Y_{0.1})$. Yoder and Keith (1951) later were able to synthesize the end member $Y_3Al_2[AlO_4]_3$, using Jaffe's substitution scheme and 1 g of Y_2O_3 purchased for ten dollars by Jaffe for the study. This modest investment paid dividends and continues to do so, since Yoder and Keith synthesized this yttrium garnet that has become known as YAG.

Introduction of Fe^{3+} in place of Al^{3+} then followed, with the growth, in the laboratory, of YIG, yttrium iron garnet. These studies, made some thirty-five years ago, showed clearly that the same element could occupy two different coordination sites in the same mineral. Al^{3+} occupies octahedral and tetrahedral sites in YAG, in sillimanite, $AlOSi[AlO_4]$, and in muscovite, $KAl_2[Si_3AlO_{10}](OH)_2$. In andalusite, Al^{3+} occupies a distorted fivefold coordination site in addition to the tetrahedral site, giving the formula $AlOSi[AlO_4]$. Synthesis of YIG provided a rare example of occupancy of both octahedral and tetrahedral sites by Fe^{3+} in the same mineral (also see Chapter 12). This also occurs in the defect lattice of maghemite described in this chapter.

Synthetic rare earth garnets, such as the prototypes YAG and YIG, have had a wide variety of uses over the years. YIG was used successfully in radar devices, and YAG has been used as a gemstone and most recently as the crystal source of the laser beam in apparatus used in eye surgery. These examples, as well as numerous synthetic gems (star sapphire, alexandrite, and diamond), synthetic zeolites used for molecular sieves in petroleum refining processes, and even a synthetic rock developed to store radioactive waste material, all point to increasing commercial application of crystal chemical principles.

Summary

Diadochy is the limited substitution of ions in a particular mineral species, as well as complete solid solution of two or more ions in related mineral species.

Isomorphism relates a group of minerals of similar morphology and formula that contain different ions in a given coordination site. It does not necessarily imply solid solution within the group.

Isostructural minerals must be identical in all coordination numbers, space group, and number of formulae per unit cell. This does not necessarily imply solid solution.

Limited substitution and complete solid solution are treated as aspects of diadochy, and the term *isomorphism* is abandoned in favor of the term *isostructuralism.*

V. M. Goldschmidt explained substitution by camouflage (i.e., substitution of one element for another of identical charge and similar radius), capture (i.e., substitution of an element with similar radius but higher electric charge), and admission (i.e., substitution of an element with lower charge but similar radius). Admission plus capture balances the electrical charge and results in a coupled substitution.

Where a crystal lattice contains voids that are only partially occupied, defect substitution may result.

Vacancy substitution results when large stray ions or molecules occupy large structural voids.

Crystal field stabilization energy plays a significant role in diadochy.

Substitution of radioactive ions often leads to destruction of the lattice of the host mineral by metamictization, where the daughter product of radioactive decay is much too large to occupy the volume of the site of the original radioactive ion.

Many complete solid solution series are known. Some are limited by requirement of high temperature or high pressure for stability; others are limited by availability of particular elements (e.g., yttrium garnet).

Bibliography

Curtis, C. D. (1964). Application of the crystal field theory to the inclusion of trace transition elements in minerals during magmatic differentiation. *Geochim. et Cosmochim. Acta,* 28: 389–403.

Goldschmidt, V. M. (1937). The principles of distribution of chemical elements in minerals and rocks. *J. Chem. Soc. for 1937, Pt. 1* 655–73.

(1954). *Geochemistry,* ed. Alex Muir. Oxford at the Clarendon Press, London, pp. 80–125.

Gottfried, D., Jaffe, H. W., and Senftle, F. E. (1959). Evaluation of the lead-alpha (Larsen) method for determining the ages of igneous rocks. U.S. Dept. of Interior, Geol. Survey, Bull. 1097-A, pp. 30–32.

Holland, H. D., and Gottfried, D. (1955). The effect of nuclear radiation on the structure of zircon. *Acta Crystallogr.*, 8: *Pt. 6* 291–300.

Jaffe, H. W. (1951). The role of yttrium and other minor elements in the garnet group. *Am. Mineral.*, 36: 133–55.

(1955). Precambrian monazite and zircon from the Mountain Pass rare earth district, San Bernardino County, California. *Geol. Soc. Am. Bull.*, 66: 1250.

Yoder, H. S., and Keith, M. L. (1951). Complete substitution of aluminum for silicon: The system $3MnO \cdot Al_2O_3 \cdot 3SiO_2 - 3Y_2O_3 \cdot 5Al_2O_3$. *Am. Mineral.*, 36: 519–33.

10 Density, volume, unit cells, and packing

Density is the concentration of matter measured in mass per unit volume. It is recorded in increments of g/cm^3, and is commonly symbolized by the Greek letter ρ. The density of a solid can be obtained by one of the following three methods:

1. $\rho_{(obs)}$ can be determined by direct measurement involving weighing the solid in air and then in a liquid of low density such as toluene, an organic liquid for which density is precisely known and may be corrected for any temperature.
2. $\rho_{(calc)}$ can be determined by calculation from the formula or chemical composition and from the volume of the unit cell measured from X-ray data, using the formula

$$\rho_{(calc)} = \frac{\text{Formula weight} \times Z \times 1.6603 \times 10^{-24}}{\text{Unit cell volume } (\text{Å}^3) \times 10^{-24}} \quad \text{g/cm}^3$$

   where formula weight is the sum of the mass of its constituent atoms, Z is the number of formulae in one unit cell, and 1.6603×10^{-24} is the reciprocal of Avogadro's number (6.0228×10^{23}). This effectively converts formula weight to gram atom weight, and unit cell volume in Å^3 is multiplied by 10^{-24} to convert the volume to cm^3.
3. $\rho_{(opt)}$ can be determined by calculation from the optically measured average index of refraction n_{av}, divided by the specific refractivity K of the compound, according to the formula of Gladstone and Dale, where $(n - 1)/K = \rho$ (for details see Chapter 11).

Measured and calculated values of density should agree to within ± 0.1 g/cm^3, but they often deviate by larger increments because of experimental error, voids or cavities present in the solid, or fallacious assumptions regarding the chemical composition or purity of the sample. When values are measured and calculated from X-ray data for pure synthetic end members of a solid solution series – for example, pure diopside, $CaMgSi_2O_6$ and hedenbergite, $CaFeSi_2O_6$ – the results should show very close agreement. Densities obtained on chemically analyzed natural mineral solid solutions may be subject to greater error because of the greater chemical complexity of the mineral; for example, small amounts of Na, Al, Ti, Fe^{3+}, or Cr may have been overlooked.

Accurate densities of solids are usually obtained by carefully selecting under the binocular microscope a 25–50-mg pure fragment, then weighing it in air and in toluene, using a sensitive Berman density microbalance. The ratio of these weights, multiplied or corrected by the density of the toluene at the temperature of measurement, gives the density in g/cm^3. Toluene is a good liquid to use because of its low surface tension (27 dyn/cm at 20°C, compared with 73 dyn/cm at 18° for water) and because of its low density compared to that of minerals ($\rho = 0.8867$ at 20°C). Other methods, which involve weighing larger fragments or single crystals on various microbalances or, conversely, weighing fine powders in a pycnometer or suspended in heavy liquids of known density, are all subject to greater experimental error for the reasons already given.

The mineralogist should have some knowledge of the densities of common minerals for use as a yardstick in such calculations. Easy to remember are quartz, 2.65 g/cm^3; magnetite, pyrite, and hematite, near 5.0; native terrestrial or meteoritic Ni-bearing iron, 8.0; and others given in Table 10.1.

Volume

The volume V of a crystal or of any solid is usually obtained from X-ray measurements that yield the dimensions of the unit cell and are most commonly reported in Å^3 (less used is the SI unit or nanometer; 1 nm = 10 Å,

Table 10.1. *Range in density of minerals*

Mineral	Range in density (g/cm^3)
Ice	0.91
Hydrated borates	1.4–2.5
Hydrated zeolites	2.0–2.5
Quartz	2.65
Feldspars	2.56–2.78
Ferromagnesian silicates	3.0–4.4
Pyrite, magnetite, hematite	5.0
Other oxides	3.7–7.0
Native elements (nonmetal)	2.0–3.5
Native elements (metal)	7.9–21.5

which results in unit cell dimensions being reported in increments of less than unity).

Unit cell volumes are calculated for minerals crystallizing in the seven crystal systems according to the following formulae (V is in Å^3):

Isometric:

$$V = a^3$$

Tetragonal:

$$V = a^2c$$

Hexagonal:

$$V = a^2c \sin 60° \ (\sin 60° = 0.866)$$

Trigonal (rhombohedral):

$$V = (a^3 \sin \alpha)\sqrt{\frac{\cos \alpha - \cos 2\alpha}{1 + \cos \alpha}}$$

Orthorhombic:

$$V = abc$$

Monoclinic:

$$V = abc \sin \wedge \beta$$

Triclinic:

$$V = (abc)\sqrt{1 - \cos^2 \alpha - \cos^2 \beta - \cos^2 \gamma + 2 \cos \alpha \cos \beta \cos \gamma}$$

The reader should understand that the unit cell of a mineral contains one or many formulae of the chemical composition that characterizes the species. Although one unit cell of a simple structure like halite contains only eight atoms (4 Na + 4 Cl), those of more complex structures contain many formulae and atoms; for example, native sulfur contains sixteen formulae of S_8 rings for a total of 128 atoms in one unit cell, and tourmaline contains 162 atoms in the complex unit cell represented by

$$3/NaMg_3Al_6[Si_6O_{18}](BO_3)_3(OH)_4$$

where 3/ means $Z = 3$.

In crystal chemical studies, it is often necessary to distinguish between the number of formulae per unit cell and the number of atoms per unit cell, particularly the number of larger atoms or anions that occupy most of the volume of the cell. Thus the number of oxygen (or O^{2-}) anions per unit cell (Table 10.2) is often cited in studies made of the packing of atoms in response to crystallization under different conditions of temperature and pressure. Because the formula may be cited differently in different texts, the value ascribed to Z may vary, but the number of oxygens or other atoms per unit cell is fixed.

The formula for talc is cited in many texts as $Mg_3Si_4O_{10}(OH)_2$; in other widely used texts it is $Mg_6Si_8O_{20}(OH)_4$ because the doubled formula is more directly comparable to those of amphiboles: talc and mica have $O_{20}(OH)_4$ whereas amphiboles have $O_{22}(OH)_2$, providing a quick recognition of these mineral families from their formulae. Enstatite, given as $MgSiO_3$ in one text and $Mg_2Si_2O_6$ in another, will have $Z = 16$ or $Z = 8$, depending on the formula used, but the total number of oxygens in one unit cell is fixed at forty-eight. An added complication may arise for minerals crystallizing in the trigonal crystal system. Here the unit cell may be delineated by a unit rhombohedron with three equal axes a and an angle between edges α, or by a unit hexagon, each giving different volumes and Z numbers. For example, hematite, Fe_2O_3, has a unit cell described either by $a_{rh} = 5.427$ Å, $\alpha = 55°18'$, with $Z = 2$, or by $a_{hex} = 5.035$ Å and $c_{hex} = 13.749$ Å, with $Z = 6$. The hexagonal unit cell will always have three times the volume of its rhombohedral analogue, and it is useful to know that such minerals tend to translate their unit cells in multiples of three. This can provide valuable information when calculating the density of a mineral. Where Z is not known and the volume and the formula are known, it is usually possible to calculate the density by estimating the value of Z if one has some idea of the density ranges of the common minerals (Table 10.1). To calculate the density of corundum from its formula Al_2O_3 of mass (M) 101.94, the hexagonal unit cell dimensions $a_{hex} = 4.76$ Å and $c_{hex} = 12.996$ Å, and $1/N = 1.6603 \times 10^{-24}$, we use

$$\rho_{(calc)} = \frac{101.94 \times 1.6603 \times 10^{-24}}{V = a^2c \times 0.866 \times 10^{-24}} = 0.6638 \times Z \text{ g/cm}^3$$

If the mineralogist is aware that corundum should be relatively dense, because its lattice is built on a nearly hexagonally close-packed array of O^{2-} anions with four of every six octahedral sites occupied with Al^{3+} ions (from Pauling's Rule 2; see Chapter 5), and that Z is likely to be a multiple of three, then from $\rho_{(calc)} = 0.6683 \times Z$, we would guess that $Z = 3$, 6, or 9, giving $\rho = 1.99$, 3.98, or 5.97 g/cm^3. It should be obvious that 1.99 is too low, 5.97 is too high, and that 3.98 is the correct value. If the indices of refraction are known or can be

Table 10.2. *Formula units, Z, and oxygen atoms per unit cell*

Mineral	Z	O^{2-} unit cell	Mineral	Z	O^{2-} unit cell
Olivine	4	16	Orthoclase	4	32
Augite	4	24	Albite	4	32
Hypersthene	8 or 16	48	Anorthite	8	64
Tremolite	2 or 4	48	Garnet	8	96
Anthophyllite	4 or 8	96	Leucite	8	96
Biotite	1 or 2	24	Sillimanite	4	20
Kaolinite	1 or 2	18	Andalusite	4	20
Talc	2 or 4	48	Kyanite	4	20
Muscovite	2 or 4	48	Tourmaline	1 or 3	31_{rh} 93_{hex}
Staurolite	2	48	Beryl	2	36
Analcite	16	112	Hematite	2 or 6	6_{rh} 18_{hex}
Cordierite	4	72	Melilite	2	14

measured, $\omega = 1.768$ and $\epsilon = 1.760$, then $n_{av} = (2\omega + \epsilon)/3 = 1.7653$, and for various estimates of the specific refractivity k for Al_2O_3 from 0.186 to 0.193, then, from the law of Gladstone and Dale, $(n - 1)/K = \rho$, corundum will have $\rho_{(opt)} = 3.96$–4.11, verifying the choice of $Z = 6$ for the hexagonal unit cell.

Now compare the simple isometric unit cells of kamacite, native iron of meteorites; native gold; diamond; and halite (Fig. 10.1). Note that kamacite has body-centered disposition of its Fe atoms, and, when properly counted as $8 \times (\frac{1}{8})_{cor} = 1$, $1 \times 1_{ctr} = 1$, its nine Fe atoms reduce to two per unit cell. Drawn to scale, the structure would show eight Fe atoms in cubic array around every other Fe atom, CN VIII, with all six faces of each unit cube shared. In this orientation, each Fe atom at a cube center would have its $4s^2$ valence electrons overlapped with those of the eight Fe atoms that surround it, resulting in a delocalized metallic orbital that runs through the crystal and explains many of

Figure 10.1. Unit cell projections on (100) locating all atoms in the cubic minerals: native iron or kamacite, native gold, diamond, and halite. Atomic locations and their relationships to mirror planes (m), glide planes (a, n, and d), and screw axes (4_1, 4_2, and 4_3) determine the number of atoms per unit cell, their coordination numbers, the formula, and the space group assignment.

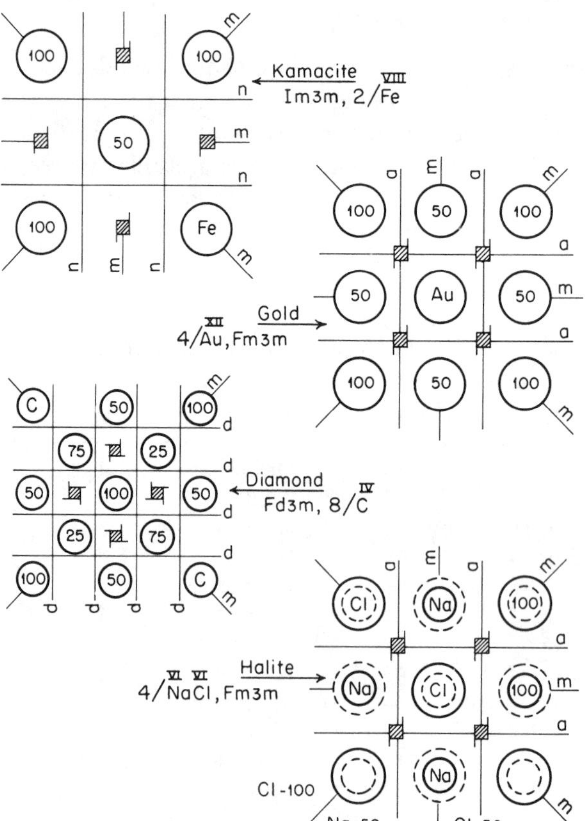

its physical properties. The unit cell formula and the space group are thus written as $2/Fe^{VIII}$ and *Im*3*m*. Note also that screw axes 4_2 perpendicular to *n* axial glides accomplish translation of all atoms in a manner identical to or equivalent to that caused by body centering and need not be noted in the space group code; the body centering is, however, a function of the operation of screw axes and glide planes.

Now consider native gold (Fig. 10.1), in which the lattice is face centered with Au atoms located on eight corners and six face centers of the unit cube. When properly counted (see Chapter 6) the fourteen Au atoms reduce to four per unit cell (Fig. 10.1). The array of Au atoms will be in cubic close packing (CCP) in which each Au atom is surrounded by twelve other Au atoms in CN XII. This array is very conducive to the overlap of the spherical $6s^1$ orbitals on each Au atom, and an extended molecular orbital will carry electrons throughout the crystal, in bands, explaining the metallic luster with its high reflectivity, electrical conductivity, and malleability, all characteristic of native gold. The unit cell formula and space group are thus $4/Au^{XII}$ and *Fm*3*m*.

Also illustrated in Figure 10.1 is the unit cell projection of diamond, in which there are shown eighteen carbon atoms disposed on eight corners, six face centers, and in four of eight cubelets located tetrahedrally at the 75-level (NW and SE) and at the 25-level (NE and SW) inside the unit cell. When the eighteen atoms are properly counted, as explained in Chapters 3 and 6, only eight C atoms belong to the unit cell illustrated. This arrangement and number of atoms results because of disposition of the sp^3 hybrid covalent bonds between all C atoms (see Fig. 3.2).

Note that the three fourfold rotational axes found in kamacite and in native gold are absent from diamond. They are replaced by twelve dextral and sinistral 4_1 and 4_3 screw axes that alternate on and outcrop on each cube face; each such 4_1 and 4_3 axis is perpendicular to diamond glides (*d*), also four per cubic face (Fig. 10.1). In Figure 3.3, each carbon atom labeled 4, 3, 2, and 1 represents the 100-, 75-, 50-, and 25-elevation levels of the unit cell above the 0 base level ($4 = 100$ and 0); the clockwise (4_1) and counterclockwise (4_3) rotations translate each C atom a net of $\frac{1}{4}$ the distance along the body diagonals of the unit cube. The diamond glides accomplish the same translation of atoms across *d* glides located at $\frac{1}{8}$, $\frac{3}{8}$, $\frac{5}{8}$, and $\frac{7}{8}$ of the way along each cube face or pair of faces. Reflection across a plane at $\frac{1}{8}$ is accompanied by translation of each C atom, from 100-75-50-25-0 levels (4-3-2-1-0). The unit cell formula and space group are $8/C^{IV}$ and *Fd*3*m*. The twelve *d* glides replace the six mirror planes *m* of kamacite and gold.

The fourth unit cell (see Fig. 10.1) of the familiar halite structure places fourteen chlorine (or sodium) ions on corners and face centers as in native gold, and then adds another thirteen ions of sodium (or chlorine) in the spaces or voids between one another. Both Cl^- and Na^+ are in interchangeable positions with the larger ion, Cl^-, occupying positions of cubic close packing as in native gold. The voids between atomic or ionic populations are all octahedral sites, and all are filled, so six Cl^- ions surround each Na^+ ion, and vice versa. At the same time twelve Cl^- ions surround one another. In halite, all octahedral voids in the CCP array of Cl^- ions are occupied, and all tetrahedral voids are empty. In native gold all such interstitial sites are empty; in kamacite, all voids are tetrahedral, and all are empty and traversed by the 4_2 screw axes. The unit cell of halite contains twice the number of atoms as that of gold, but the symmetry is identical: for halite, $4/Na^{VI}Cl^{VI}$ and $Fm3m$.

All four examples of Figure 10.1 bring atoms close together but use different principal bond types: kamacite and native gold use metallic orbital overlap; diamond, sp^3 covalent orbital overlap; and halite, approximately tangential contact with minimal orbital overlap. The electronegativities of Na ($\Delta\chi = 0.9$) and of Cl ($\Delta\chi = 3.0$) give $\Delta\chi = 2.1$, equivalent to only 66% ionic bonding on Pauling's scale. Thus halite, readily soluble in water and with a low index of refraction, $n = 1.544$, should not be considered a totally ionic compound.

Packing of atoms

The packing index (PI) refers to the degree of efficiency of the occupancy of space by atoms, according to the equation

$$\mathrm{PI} \times 10 = \frac{\text{Vol. of ions in 1 unit cell (Å}^3\text{)}}{\text{Vol. of 1 unit cell (Å}^3\text{)}}$$

An "oxygen anion packing index" intended to express the volume of 1 oxygen anion in cm^3 is calculated as follows:

$$\mathrm{cm^3/O^{2-}} = \frac{\text{Vol. of 1 mole } V_m(\text{cm}^3) \text{ of compound}}{\text{No. of O}^{2-} \text{ anions in 1 mole compound}}$$

In reciprocal relationship, it has been expressed as an "oxygen packing index" or the number of oxygens per cm^3:

$$\mathrm{O^{2-}/cm^3} = \frac{\text{No. of oxygens in 1 mole of compound}}{\text{Vol. of 1 mole, } V_m \text{ (cm}^3\text{) of compound}}$$

Packing indices calculated from these three equations for several selected minerals are compared in Table 10.3. Note that for equal mass of 60.09, the density ρ of stishovite is 4.28, about 1.6 times greater than that of quartz, 2.655 g/cm^3. On this basis, on a scale of 100, 54.9% of space is occupied by O and Si atoms in quartz, and 91.8% in stishovite. Alternatively, the volume of oxygen in 1 mole of quartz is 11.33 cm^3, compared with 7.02 cm^3 in stishovite.

The PI equation requires that atoms be spherical, with a volume $r^3 \times 4\pi/3$, and that orbitals not be overlapped in covalent bonding. It was shown in Chapter 4 that atoms that have their valence electrons in *s* orbitals form spherical ions, whereas those atoms that place valence electrons in *p* or *d* orbitals are spherically symmetrical only if each orbital is singly or doubly occupied (half filled or filled). Thus Fe^{2+}, with six *d* electrons in five orbitals, and many other

Table 10.3. *Comparison of packing indices calculated for minerals*

Mineral	Mass	/ ρ	= V_m (cm^3)	nO_m	V_m (Å^3)	× Z	= V_{uc} (Å^3)	PI	cm^3/O	O/cm^3
Anorthite	278.13	2.76	100.77	8	167.31	8	1338.48	5.41[a]	12.60[b]	.0794[b]
Cordierite	585.01	2.505	233.54	18	387.75	4	1551.02	4.94	12.97	.0771
Åkermanite	272.66	2.94	92.74	7	153.98	2	307.96	5.84	13.24	.0755
Calcite	100.09	2.715	36.865	3	61.21	6	367.26	5.83	12.29	.0814
Quartz	60.09	2.655	22.632	2	37.58	3	112.74	5.49	11.32	.0883
Epidote	483.25	3.43	140.89	13	233.93	2	467.85	6.51	10.83	.0923
Forsterite	140.73	3.22	43.70	4	72.56	4	290.26	6.51	10.92	.0916
Fayalite	203.79	4.39	46.42	4	77.08	4	308.30	5.39	11.60	.0862
Diopside	216.58	3.22	67.26	6	111.68	4	446.71	6.54	11.21	.0892
Enstatite	100.41	3.209	31.29	3	51.95	16	831.24	6.97	10.43	.0959
Spinel	142.28	3.55	40.08	4	66.55	8	532.03	6.76	10.02	.0998
Andradite	508.21	3.859	131.69	12	218.66	8	1749.28	6.97	10.97	.0912
Grossular	450.47	3.594	125.33	12	208.11	8	1664.87	7.21	10.44	.0958
Andalusite	162.05	3.145	51.33	5	85.55	4	342.21	6.33	10.31	.0970
Sillimanite	162.05	3.247	49.91	5	82.86	4	331.46	6.55	9.98	.1002
Kyanite	162.05	3.60	45.01	5	74.74	4	298.96	7.28	9.00	.1111
Aragonite	100.09	2.94	34.04	3	56.53	4	226.12	7.06	11.35	.0881
Corundum	101.96	4.00	25.49	3	42.32	6	253.94	7.74	8.50	.1176
Anatase	79.90	3.90	20.49	2	33.97	4	135.88	6.47	10.23	.0978
Brookite	79.90	4.14	19.30	2	32.04	8	256.35	6.87	9.64	.1037
Rutile	79.90	4.29	18.64	2	30.94	2	61.88	7.11	9.39	.1065
Perovskite	135.98	4.03	33.74	3	56.02	1	56.02	8.15	11.25	.0889
Stishovite	60.09	4.28	14.04	2	23.21	2	46.62	9.18	7.02	.1424
Gold	196.97	19.3	10.205	—	16.945	4	67.78	7.41	—	—
Diamond	12.00	3.51	3.419	—	5.676	8	45.41	3.36	—	—
Sphalerite	97.43	4.1	23.76	—	39.46	4	157.82	3.55	—	—

[a] Packing index (PI = V ions$_{uc}$ (Å^3)/V_{uc} (Å^3) × 10.

[b] Oxygen PI cm^3/O = V_m (cm^3)/O_m and O/cm^3 = 1/(cm^3/O).

ions are not spherically symmetrical, and concepts of packing based on spherical ionic volumes are inherently flawed. Indeed, it is the asymmetry of the Fe^{2+} ion that endows it with a preference for distorted rather than symmetrical octahedral sites when both occur in the same mineral. This can be a major factor in cation ordering of asymmetric Fe^{2+} and symmetrical Mg^{2+} ions competing for occupancy of two such sites as in pyroxenes and olivines. Implicit in this concept is the probability that asymmetry of electron density enhances formation of covalent bonds. The closeness of packing of a crate of lemons cannot be equated with that of a crate of oranges.

The other two PI formulae, cm^3/O^{2-} or O^{2-}/cm^3, are based on the assumption that the molar volume (mass/density) of a mineral is a function of the closeness of packing of the oxygen anions alone, and ignores any contribution of large-radius cations. Compare the value obtained for the open-structured framework of quartz, SiO_2, and the densely packed mineral perovskite, $CaTiO_3$:

	PI × 10	cm^3/O	O/cm^3
Quartz	5.49	11.32	0.0884
Perovskite	8.15	11.25	0.0889

The packing index does represent ideal perovskite as being much more densely packed than quartz, whereas the oxygen packing indices are misleading, in that they represent these two minerals as being identical in packing. In ideal perovskite, large voids in the oxygen-packed structure ($r_{O^{2-}} = 1.40$ Å) are almost exactly filled by large-radius calcium ($r_{Ca^{2+}} = 1.35$ Å). Perovskite represents a cubic close-packed assemblage of large ions, $O^{2-} + Ca^{2+}$, whereas it is an open-packed structure of O^{2-} anions taken alone. Thus, cm^3/O overestimates the volume occupied by one oxygen in a mole of perovskite.

Summary

Density, or concentration of matter measured in mass/unit volume (g/cm^3) is symbolized by ρ. It may be measured directly, calculated from the chemical composition (formula) and X-ray measurement of the unit cell, or calculated from the mean index of refraction n divided by the specific refractivity, K of the compound: $(n-1)/K = \rho$. Density is measured by weighing a pure fragment of the mineral in air and in toluene.

Volume is obtained by X-ray measurement of the dimensions of the unit cell.

The unit cell may contain one or more formulae of the chemical composition of the mineral being measured. For a particular mineral, the number of atoms per unit cell is fixed, whereas the number of formulae per unit cell may vary with the mode of presentation of the formula. In the trigonal system, the true unit cell may be designated by a unit rhombohedron or by a hexagonal cell with a volume three times as large.

Unit cells of four simple isometric minerals are compared and the atoms in them are counted. Kamacite (2 Fe per unit cell) and native gold (4 Au per unit cell) are bonded by overlap of metallic orbitals; diamond (8 C per unit cell) is bonded by sp^3 covalent bonds; and halite (4 Na, 4 Cl per unit cell) is bonded by predominantly ionic bonds.

The packing index, or degree of efficiency of occupancy of space by atoms, may be calculated in three different ways. None of these calculations takes into account all the complexities of crystal chemistry: each method of calculation has a different inherent flaw. Packing indices are calculated by three different methods and tabulated.

Bibliography

Fairbairn, H. W. (1943). Packing in ionic minerals. *Geol. Soc. Am. Bull.,* 54: 1305–74.

Pauling, L. (1960). *The nature of the chemical bond,* 3d ed. Cornell Univ Press, Ithaca, N.Y., chap. 11.

11 Refractivity and polarizability

Optical mineralogy

Optical mineralogy is the study that deals with the identification of minerals from their properties, measured under the polarizing microscope, on crushed grains mounted in liquid reference media. For transparent or nonopaque minerals, with which we will concern ourselves here, the technique basically requires the matching of the index of refraction n of crushed fragments of a mineral with that of a liquid standard whose index of refraction is precisely known to ± 0.002, and can be verified by mounting a drop of the liquid in a refractometer.

All values of n are calibrated for Na light, although measurements are usually made under blue-filtered white light under a microscope fitted with two polarizing prisms (calcite or polaroid) mounted at 90°, one below and one above the rotating object stage of a petrographic microscope. The technique involves the preparation and study of a number of fragment mounts in liquids of different refractive index, until one, two, or all three of the indices of refraction of a particular mineral are determined by matching with the reference liquids. These data, along with a determination of the parallelism or nonparallelism of vibration directions in the crystal with those of the polarizer (lower prism) and analyzer (upper prism) are usually sufficient to identify any transparent mineral. This last statement implies that light vibrates in particular directions in minerals, and, in the general case, it vibrates faster in some directions than in others.

A mineral may have three different indices of refraction – high (γ), intermediate (β), and low (α) – parallel to three mutually perpendicular vibration directions or vibration velocity vectors: **Z** (slow), **Y** (intermediate), and **X** (fast), respectively, if the mineral crystallizes in the triclinic, monoclinic, or orthorhombic crystal systems.

A mineral may have only two vibration velocities – slow, with ϵ parallel to the c axis, and fast, with ω perpendicular to the c axis – for optically positive uniaxial minerals, with these directional velocities reversed for optically negative uniaxial minerals. All uniaxial minerals, whether positive or negative, crystallize in either the tetragonal, hexagonal, or trigonal crystal systems. A mineral will have only one index of refraction n and thus only one vibration velocity constant in all directions if it crystallizes in the isometric crystal system or if it is an amorphous compound.

Now, a bit of data about light, which may be regarded as an electrical disturbance that is part of a spectrum of electromagnetic radiation ranging from short wavelength radiation, X-rays (λ = 1–10 Å), through ultraviolet radiation, visible radiation (λ = 3900–7900 Å), to infrared, microwave, and radio-wave radiation of very long wavelength. The velocity of light for this entire electromagnetic spectrum is the same, 3×10^{10} cm/s in a vacuum. An electric vector and a magnetic vector of any such radiation are oriented perpendicular to one another and *vibrate in directions* that are perpendicular, that is, *at right angles to, the direction of propagation of the radiation.*

We will be concerned here with the electric vector, specifically with its vibration direction and velocity in the prism of our microscope and in the minerals we study. Sodium light, our reference standard, vibrates in a vacuum with a wavelength $\lambda = 5893$ Å at the stated velocity $c = 3 \times 10^{10}$ cm/s with a frequency $\nu = 5.09 \times 10^{14}$ cps. These properties in a vacuum are close enough to those in air so that we may treat them as the same. The ratio of these two velocities is close to unity and equal to the index of refraction of air ($n_{air} = c_{vac}/c_{air} = 1.000$). The basic law of the physics of light is $c = \lambda\nu$.

Although light is refracted at most angles of incidence on a crystal plate, at vertical incidence it is retarded but not refracted. When Na light enters the mineral, it is retarded and undergoes a reduction in its wavelength, but its frequency remains unchanged.

If we consider diamond, which has the very high index of refraction $n_{Na} = 2.418$, we can relate the velocity, wavelength, and frequency of Na light in diamond to that in air.

$$\lambda_{Na,\,dmnd} = \frac{\lambda_{Na,\,air}}{n_{Na,\,dmnd}} = \frac{5893\ \text{Å}}{2.418} = 2437\ \text{Å}$$

$$c_{Na,\,dmnd} = \frac{c_{Na,\,air}}{n_{Na,\,dmnd}} = \frac{3 \times 10^{10}\ \text{cm/s}}{2.418} = 1.241 \times 10^{10}\ \text{cm/s}$$

$$\nu_{Na,\,air} = \frac{c_{Na,\,air}}{\lambda_{Na,\,air}} = \frac{3 \times 10^{10}\ \text{cm/s}}{5893\ \text{Å}} = 5.09 \times 10^{14}\ \text{cps}$$

$$\nu_{Na,\,dmnd} = \frac{c_{Na,\,dmnd}}{\lambda_{Na,\,dmnd}} = \frac{1.241 \times 10^{10}\ \text{cm/s}}{2437\ \text{Å}} = 5.09 \times 10^{14}\ \text{cps}$$

To reiterate, then, Na light entering a mineral from air will have its velocity and wavelength reduced and its frequency unchanged.

All minerals of the seven crystal systems fall into one of five optical classes or categories:

1. *Isotropic* or *omniaxial,* in which light vibrates with equal velocity in all directions (isometric minerals)
2. *Anisotropic, uniaxial positive*
3. *Anisotropic, uniaxial negative,* in both of which light vibrates with equal velocity in all directions in one plane perpendicular to the *c* axis, and in which light vibrates with changing velocity in planes parallel to the *c* axis (tetragonal, hexagonal, and trigonal minerals)
4. *Anisotropic biaxial positive*
5. *Anisotropic biaxial negative,* in both of which light vibrates with equal velocity in two planes whose location or orientation is controlled by the magnitude of the differences of the three vibration velocities and resultant indices of refraction (orthorhombic, monoclinic, and triclinic minerals)

A plane in which light vibrates with equal velocity in a 360° rotation is defined optically as a *circular section,* and the direction normal to it is the *optic axis.* From the foregoing discussion, it is evident that isometric minerals have an infinite number of circular sections and an infinite number of optic axes, hence the class, omniaxial.

Tetragonal, hexagonal, and trigonal minerals have but one circular section and, thus, one optic axis: uniaxial positive when the fast velocity and low index of refraction lie parallel to the circular section, and uniaxial negative when the slow velocity and high index of refraction lie parallel to the circular section.

Orthorhombic, monoclinic, and triclinic minerals have two circular sections, and the two optic axes perpendicular to these lie in the plane containing the fast (**X**) and slow (**Z**) vibration directions. This **X** – **Z** plane is called the *optic plane* (OP) and is usually referred to a prominent crystallographic plane, for example, OP = (010).

Thus, minerals crystallizing in systems based on three unequal crystallographic axes will have three unequal vibration velocities and three indices of refraction, α, β, and γ. Such minerals, containing two optic axes perpendicular to two circular sections, are biaxial positive and biaxial negative. In these minerals, the intermediate velocity **Y** and intermediate index of refraction β lie parallel to both circular sections. We can model all of these five optical classes by constructing indicatrix models or surfaces that show how the velocity of light changes with direction. By setting each radius of such an indicatrix equal to the one, two, or three indices of refraction of a mineral, we obtain the five indicatrix models shown in Figure 11.1; with $n = r_{\text{indic}}$, all planes become elliptical in the general case, and circular in the special case.

Maximum ellipticity is associated with three indices of refraction that are separated by equal intervals (e.g., $\alpha = 1.600$, $\beta = 1.640$, and $\gamma = 1.680$) and in which circular sections and optic axes both lie at 90° within a biaxial indicatrix.

When the intermediate index of refraction β lies nearer to the low index of refraction α, the ellipticity of the indicatrix is reduced and the mineral is optically biaxial positive (e.g., $\alpha = 1.600$, $\beta = 1.620$, and $\gamma = 1.680$). For an analogous biaxial negative mineral, representative indices are $\alpha = 1.600$, $\beta = 1.660$, and $\gamma = 1.680$. As the β index of refraction moves closer to the γ index of refraction or to the α index of refraction, pairs of biaxial circular sections and the optic axes perpendicular to them approach one another and eventually fuse to produce the uniaxial indicatrix models of Figure 11.1.

In a biaxial positive mineral, it is the α and β indices of refraction that become equal to ω, the low index of refraction and radius of the circular section of a prolate indicatrix (Fig. 11.1); typical indices are $\omega = 1.544$ and $\epsilon = 1.553$, as in quartz. In a biaxial negative mineral these relations are reversed, and β and γ of the biaxial indicatrix become equal to ω, the high index of refraction and radius of an oblate indicatrix (Fig. 11.1); typical indices are $\omega = 1.658$ and $\epsilon = 1.486$, as in calcite.

When a circular section of any mineral lies parallel to the stage of the microscope, the grain will appear dark gray when the polars are crossed (polarizer below and analyzer above the mineral grain), no matter how the stage is rotated. Engaging the convergent lens (located immediately below the stage and the grain) and removing the ocular of the microscope will reveal interference patterns or figures as black maltese crosses, unchanged by stage rotation, for uniaxial minerals and black maltese crosses that split into black crescentic bars

Figure 11.1. Indicatrix models and interference patterns or figures for the five optical classes of crystals using a polarizing microscope with a 50× objective and a superposed 550 μm gypsum retardation plate.

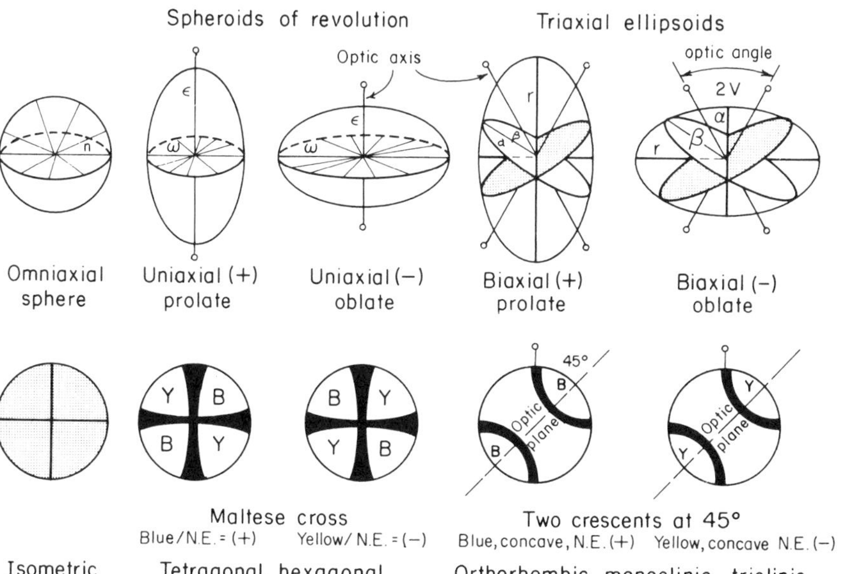

(isogyres) and recombine to form crosses, with stage rotation, in biaxial minerals. Superposition of an accessory gypsum plate into a slot in the microscope tube will add a 550-μm retardation to that produced by the mineral. A net addition or subtraction of wavelength will occur, introducing a blue color into the NE and SW quadrants of the interference figure for positive minerals, or a yellow color into the same quadrants for a negative mineral (Fig. 11.1). Thus, location of the circular section(s) of a mineral, measurement of the associated index of refraction, and observation of the interference figure as indicated become the prime goals of the optical mineralogist. This technique remains the most rapid, most widely used, and best method for the identification of transparent minerals and synthetic compounds.

Refractivity

Over 100 years ago, in 1864, Gladstone and Dale reported that every liquid has a specific refractive energy (specific refractivity) composed of the sum of the specific refractivities of its component elements, modified by the manner of combination, and unaffected by changes in temperature. This specific refractivity accompanies a given liquid when it is mixed with other liquids. The product of this specific refractive energy and its density is, when added to unity, the index of refraction. When the law of Gladstone and Dale, usually expressed in the form $(n - 1)/\rho = K$, is applied to crystalline materials, n is the average or mean index of refraction, ρ is the density, and K is the specific refractive energy of the solid or mineral. Following Larsen and Berman (1921, 1934), k is the specific refractivity of a constituent oxide, p is the weight percent of the constituent oxide, and the sum of the products $k \cdot p = K$ is the specific refractive energy of the mineral; the average index of refraction n_{av} is given as n for isometric minerals or amorphous materials, $(2\omega + \epsilon)/3$ for tetragonal, hexagonal, and trigonal minerals, and $(\alpha + \beta + \gamma)/3$ for orthorhombic, monoclinic, or triclinic minerals, or for omniaxial, uniaxial, and biaxial materials, respectively. Note that an amorphous material such as glass may be regarded as a viscous liquid, turning the full circle between noncrystalline and crystalline materials. These equations used herein for the average index of refraction represent reasonable solutions for the indicatrix models of Figure 11.1 and are more commonly used than $\sqrt[3]{\omega^2\epsilon}$ for uniaxial minerals and $\sqrt[3]{\alpha\beta\gamma}$ for biaxial minerals.

Larsen and Berman (1921, 1934), Tilley (1922), Jaffe (1956), Jaffe, Meyrowitz, and Evans (1953), Jaffe and Molinski (1962), and Mandarino (1976, 1978, 1979, 1981) have applied the law to minerals with considerable success (see also Allen 1956), and requests for reprints of Jaffe's 1956 paper rapidly exceeded his supply. Empirically derived specific refractivity constants k used by Larsen and Berman were modified by Jaffe, then by Mandarino, and again in this work by Jaffe. The new k constants reported herein (Tables 11.1, 11.2) were derived by the author from many mineral and synthetic compounds for which analytical, density, and optical data were available and judged to be reliable. (For comparison, Table 11.3 lists k constants determined by Mandarino 1976). Two sets of specific refractivities are presented:

1. Those derived from minerals of a wide spectrum of chemical composition and recommended for general application (Table 11.1)
2. Those derived specifically from rock-forming silicate minerals using the data from the first four volumes of Deer, Howie, and Zussman (1962, 1963 and their abridged 1966 volume) as well as additional data from the recent literature.

Figure 11.2 plots the measured density versus that calculated using the new k constants of Table 11.2 for many different silicate mineral types, using chemical composition and optical data published in volumes 1–4 (1962–1963) of Deer et al. Why are two sets necessary? Recall the phrase "modified by the manner of combination" in the original statement of the law by Gladstone and

Figure 11.2. Plot comparing measured density with that calculated from chemical and optical data of Deer, Howie, and Zussman, vols. 1, 2, 3, 4, (1962, 1963).

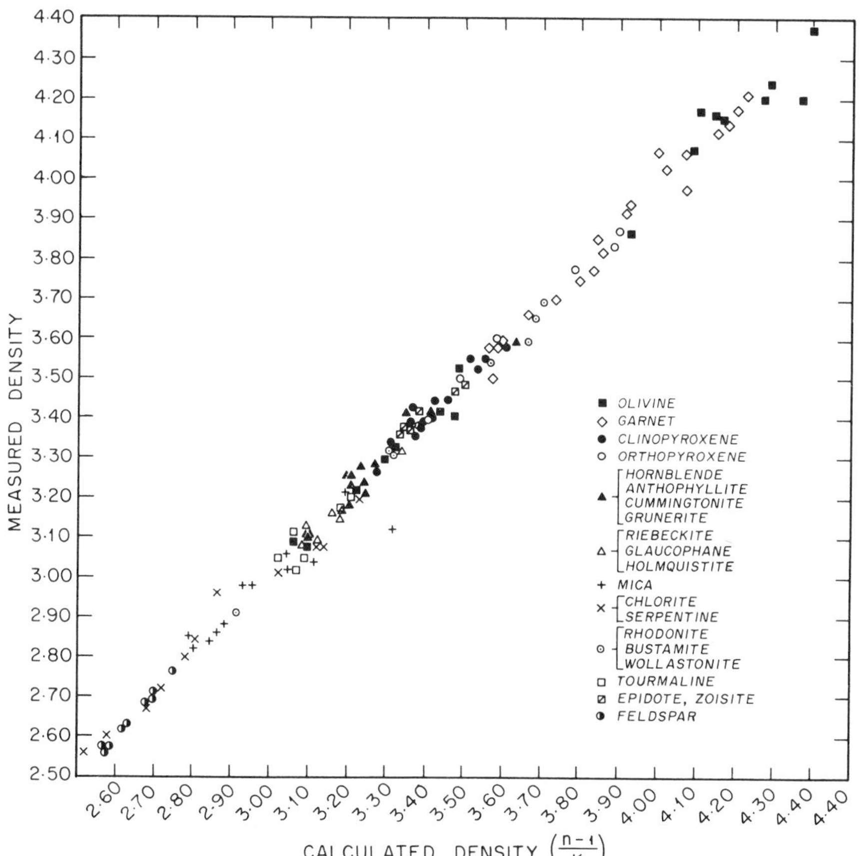

Dale (1864), made before the discovery of X-rays and the advent of modern crystallography. Specific refractive energy values should indeed be expected to vary with the manner of combination of metallic elements with the principal nonmetallic constituent of minerals, oxygen. Here, for the first time, an attempt

Table 11.1. *Relation of specific refractivity k of oxides to valence electron complement and coordination number of cation for mineral components; K values for general use*

Z	V.el.	Comp.	*k*	C.N.	Z	V.el.	Comp.	*k*	C.N.
1	$1s^1$	$(NH_4)_2O$	.482	IX	20	$4s^2$	CaO	.226	VI
		H_2O	.340	--				.217	VII
3	$2s^1$	Li_2O	.298	VI				.215	VIII
4	$2s^2$	BeO	.252	IV				.212	IX
5	$2s^22p^1$	B_2O_3	.240	III				.209	XII
			.220	III+IV	21	$4s^23d^1$	Sc_2O_3	.248	VI
			.214	IV	22	$4s^23d^2$	TiO_2	.398	VI
6	$2s^22p^2$	CO_2	.217	III	23	$4s^23d^3$	V_2O_5	.366	IV
7	$2s^22p^3$	N_2O_5	.245	III		$4s^23d^2$	V_2O_4	.277	VI
8	$2s^22p^4$	O	.203			$4s^23d^1$	V_2O_3	.247	VI
9	$2s^22p^5$	F	.043		24	$4s^13d^5$	CrO_3	.330	IV
11	$3s^1$	Na_2O	.178	VIII		$4s^13d^2$	Cr_2O_3	.229	VI
12	$3s^2$	MgO	.241	IV	25	$4s^23d^5$	Mn_2O_7	.335	IV
			.200	VI		$4s^23d^2$	MnO_2	?	opaque
			.201	VIII		$4s^23d^1$	Mn_2O_3	.259	VI
13	$3s^23p^1$	Al_2O_3	.218	IV		$4s^2$	MnO	.288	IV
			.204	IV+VI				.190	VI
			.199	V+VI				.181	VIII
			.187	VI	26	$4s^23d^1$	Fe_2O_3	.332	IV
14	$3s^23p^2$	SiO_2	.206	IV				.280	VI
			.187	VI			(hem.)	.404	VI
15	$3s^23p^3$	P_2O_5	.190	IV		$4s^2$	FeO	.205	IV
16	$3s^23p^4$	SO_3	.177	IV				.185	VI
		S_8	.502	II				.180	VIII
17	$3s^23p^5$	Cl	.303		27	$4s^2$	CoO	.174	VI
19	$4s^1$	K_2O	.186	XII	28	$4s^2$	NiO	.169	VI

is made to correlate specific refractivity with valency, chemical bond type, and coordination number. It appears that, as the coordination of a cation increases, its radius increases (see Table 4.1), the chemical bond becomes more ionic, the electronegativity decreases, the electron density decreases, and the specific refractivity decreases. Conversely, increase in the oxidation state and valency, coupled with low coordination numbers characteristic of covalent bonds, re-

Table 11.1. *(cont.)*

Z	V.el.	Comp.	*k*	C.N.	Z	V.el.	Comp.	*k*	C.N.
29	$4s^14d^1$	CuO	.172	VI	58	$6s^24f^1$	Ce_2O_3	.144	IX
	$4s^1$	Cu_2O	.285	II	59	$6s^24f^1$	Pr_2O_3	.142	IX
30	$4s^2$	ZnO	.158	IV	60	$6s^24f^1$	Nd_2O_3	.138	IX
			.151	VI	62	$6s^24f^1$	Sm_2O_3	.132	IX
31	$4s^24p^1$	Ga_2O_3	.159	VI	63	$6s^24f^1$	Eu_2O_3	.126	IX
32	$4s^24p^2$	GeO_2	.165	IV	64	$6s^25d^1$	Gd_2O_3	.120	VIII
33	$4s^24p^3$	As_2O_5	.169	IV	65	$6s^24f^1$	Tb_2O_3	.116	VIII
	$4s^24p^1$	As_2O_3	.199	L&B	66	$6s^24f^1$	Dy_2O_3	.111	VIII
34	$4s^24p^2$	SeO_2	.147	L&B	67	$6s^24f^1$	Ho_2O_3	.106	VIII
35	$4s^24p^5$	Br	.214	L&B	68	$6s^24f^1$	Er_2O_3	.103	VIII
37	$5s^1$	Rb_2O	.127	XII	69	$6s^24f^1$	Tm_2O_3	.101	VIII
38	$5s^2$	SrO	.139	IX	70	$6s^24f^1$	Yb^2O_3	.098	VIII
39	$5s^24d^1$	Y_2O_3	.168	VIII	71	$6s^25d^1$	Lu_2O_3	.095	VIII
40	$5s^24d^2$	ZrO_2	.216	VIII	72	$6s^25d^2$	HfO_2	.135	VIII
41	$5s^14d^4$	Nb_2O_5	.248	VI	73	$6s^25d^3$	Ta_2O_5	.138	VI
42	$5s^14d^5$	MoO_3	.240	IV	74	$6s^25d^4$	WO_3	.140	IV
43-46	Tc - Pd		No data		75	$6s^25d^5$	Re_2O_7	.123	?
47	$5s^1$	Ag_2O	.140	VI	76 - 79	Os -Au		No data	
48	$5s^2$	CdO	.121	VI	80	$6s^2$	HgO	.136	VIII
49	$5s^25p^1$	In_2O_3	.133	VI	81	$6s^26p^1$	Tl_2O	.120	XII
50	$5s^25p^2$	SnO_2	.141	VI	82	$6s^2$	PbO	.137	IX
51	$5s^25p^3$	Sb_2O_5	.198	?	83	$6s^26p^1$	Bi_2O_3	.154	?
52	$5s^25p^2$	TeO_2	.199	VI	84 - 89	Po - Ac		No data	
53	$5s^25p^5$	I_2O_5	.177	L&B	90	$7s^26d^2$	ThO_2	.108	VIII
55	$6s^1$	Cs_2O	.120	XII	91	Pa	No data		
56	$6s^2$	BaO	.125	XII	92	$7s^25f^4$	UO_3	.134	II
57	$6s^25d^1$	La_2O_3	.149	XII	93	Np	No data		

Table 11.2. *Specific refractivities k for rock-forming silicates*

Oxide	Olivine	Humite	Zircon	Sphene	Garnet	Topaz	Sill.	Andal.	Kyan.	Staur.
SiO_2	.206	.206	.206	.206	.206	.206	.206	.206	.206	.206
B_2O_3	-	-	-	-	-	-	-	-	-	-
Al_2O_3	.186	.193	-	.200	.186	.195	.204	.199	.192	.186
V_2O_3	-	-	-	-	.247	-	-	-	-	-
Cr_2O_3	.250	-	-	-	.245	-	-	-	-	-
Fe_2O_3	.280	.280	-	.275	.280	-	.280	.280	.280	.280
Mn_2O_3	-	-	-	-	-	-	-	-	-	-
TiO_2	.400	.400	-	.392	.450	-	-	-	-	-
ZrO_2	-	-	.200	-	-	-	-	-	-	-
ThO_2	-	-	.112	.110	-	-	-	-	-	-
MgO	.200	.199	-	.200	.201	-	-	-	-	.204
FeO	.191	.191	-	.183	.180	-	-	-	-	.193
MnO	.188	.191	-	.180	.181	-	-	-	-	.192
CaO	.220	.210	-	.212	.216	-	-	-	-	-
NiO	-	-	-	-	-	-	-	-	-	.169
CoO	-	-	-	-	-	-	-	-	-	.172
ZnO	-	-	-	-	-	-	-	-	-	.150
BeO	-	-	-	-	-	-	-	-	-	-
SrO	-	-	-	-	-	-	-	-	-	-
BaO	-	-	-	-	-	-	-	-	-	-
Li_2O	-	-	-	-	-	-	-	-	-	-
Na_2O	-	-	-	.181	.181	-	-	-	-	-
K_2O	-	-	-	.189	.189	-	-	-	-	-
Rb_2O	-	-	-	-	-	-	-	-	-	-
Cs_2O	-	-	-	-	-	-	-	-	-	-
$(NH_4)_2O$	-	-	-	-	-	-	-	-	-	-
Y_2O_3	-	-	.170	.170	.164	-	-	-	-	-
Ce_2O_3	-	-	-	.144	-	-	-	-	-	-
H_2O	.340	.340	.340	.340	.340	.340	.340	.340	.340	.340
F	-	.043	-	-	-	.043	-	-	-	-
O	-	.203	-	-	-	.203	-	-	-	-
Cl	-	-	-	-	-	-	-	-	-	-
CO_2	-	-	-	-	-	-	-	-	-	-

Abbreviations: sill. = sillimanite, andal. = andalusite, kyan. = kyanite, staur. = staurolite, ctoid. = chloritoid, datol. = datolite, epid. = epidote, melil. = melilite, cord. =

Table 11.2. *(cont.)*

Ctoid.	Datol.	Epid.	Melil.	Beryl	Cord.	Tourm.	Axin.	Opx	Cpx	Amph.	Alk. amph.
.206	.206	.206	.206	.206	.206	.206	.206	.206	.206	.206	.206
-	.213	-	-	-	-	.223	.220	-	-	-	-
									.186jd.		
.186	.193	.190	.229	.193	.213	.193	.198	.200	.200	.208	.186
-	-	-	-	-	-	-	-	-	.230	-	-
-	-	-	-	-	-	-	-	.250	.225	-	-
.280	.280	.280	.280	.280	.280	.280	.280	.265	.265	.280	.270
-	-	.259	-	-	-	-	-	-	-	-	-
.398	-	.400	-	-	-	-	-	.350	.350	.400	.398
-	-	-	-	-	-	-	-	-	-	-	-
-	-	.110	-	-	-	-	-	-	-	-	-
.200	-	.200	.239	.200	.200	.200	.200	.203	.204	.196	.196
.180	-	.185	.215	.180	.180	.185	.185	.183	.205	.182	.180
.187	-	.188	.215	.180	.180	.180	.180	.185	.198	.195	.175
-	.210	.228	.216	.210	.210	.212	.225	.210	.210	.210	.210
-	-	-	-	-	-	-	-	.169	.169	-	-
-	-	-	-	-	-	-	-	-	-	-	-
-	-	-	.156	-	-	-	-	-	-	-	-
-	-	-	-	.260	-	-	-	-	-	-	-
-	-	-	-	-	-	-	-	-	-	-	-
-	-	-	-	-	-	-	-	-	-	-	-
-	-	-	-	.350	-	.350	-	-	.313	.300	.300
-	-	-	.181	.181	.181	.181	-	.181	.181	.181	.168
-	-	-	-	-	-	-	-	.189	.189	.189	.180
-	-	-	-	-	-	-	-	-	-	-	-
-	-	-	-	.122	-	-	-	-	-	-	-
-	-	-	-	-	-	-	-	-	-	-	-
-	-	-	-	-	-	-	-	-	-	-	-
-	-	.144	-	-	-	-	-	-	-	-	-
-	-	.340	.340	.340	.340	.340	.340	-	-	.340	.340
-	-	-	-	-	-	.043	-	-	-	.043	.043
-	-	-	-	-	-	.203	-	-	-	.203	.203
-	-	-	-	-	-	-	-	-	-	.303	.303
-	-	-	-	-	-	-	-	-	-	-	-

cordierite, tourm. = tourmaline, axin. = axinite, opx. = orthopyroxene, cpx. = clinopyroxene, jd. = jadeite, amph. = Ca, Mg, Fe amphiboles, alk. amph. = alkalic amphiboles, . . .

Table 11.2. *(cont.)*

Oxide	Pxoid.	Mica	Chlor.	Talc	Felds.	Neph.	Leuc.	Scap.	Analcime	Natrolite
SiO_2	.206	.206	.206	.206	.206	.206	.206	.206	.206	.206
B_2O_3	-	-	-	-	.227	-	-	-	-	-
Al_2O_3	.200	.194	.200	.200	.215	.217	.217	.217	.217	.217
V_2O_3	-	.247	-	-	-	-	-	-	-	-
Cr_2O_3	-	.250	.250	-	-	-	-	-	-	-
Fe_2O_3	.280	.315	.280	.280	.280	-	-	-	-	-
Mn_2O_3	-	-	-	-	-	-	-	-	-	-
TiO_2	.400	.400	.400	-	-	-	-	-	-	-
ZrO_2	-	-	-	-	-	-	-	-	-	-
ThO_2	-	-	-	-	-	-	-	-	-	-
MgO	.204	.204	.196	.204	-	-	-	-	-	-
FeO	.205	.175	.175	.173	.180	-	-	-	-	-
MnO	.190	.171	.187	-	-	-	-	-	-	-
CaO	.223	.210	.210	-	.212	.212	-	.200	-	-
NiO	-	-	-	-	-	-	-	-	-	-
CoO	-	-	-	-	-	-	-	-	-	-
ZnO	-	-	-	-	-	-	-	-	-	-
BeO	-	-	-	-	-	-	-	-	-	-
SrO	-	-	-	-	.135	-	-	-	-	-
BaO	-	.125	-	-	.126	-	-	-	-	-
Li_2O	-	.230	-	-	-	-	-	-	-	-
Na_2O	.181	.181	.181	-	.163	.181	.181	.165	.176	.169
K_2O	.189	.189	-	-	.178	.185	.194	.178	.190	-
Rb_2O	-	.129	-	-	-	-	-	-	-	-
Cs_2O	-	.122	-	-	-	-	-	-	-	-
$(NH_4)_2O$	-	-	-	-	.482	-	-	-	-	-
Y_2O_3	-	-	-	-	-	-	-	-	-	-
Ce_2O_3	-	-	-	-	-	-	-	-	-	-
H_2O	-	.340	.340	.340	-	-	-	.340	.340	.340
F	-	.043	-	-	-	-	-	.043	-	-
O	-	.203	-	-	-	-	-	.203	-	-
Cl	-	.303	-	-	-	-	-	.303	-	-
CO_2	-	-	-	-	-	-	-	.217	-	-
SO_3			-	-	-	-	-	.177	-	-

Abbreviations (cont.): . . . pxoid. = pyroxenoid, chlor. = chlorite, felds. = feldspars, neph. = nepheline, leuc. = leucite, scap. = scapolite.

Table 11.3. *Specific refractivities k derived by Mandarino for general use*

Component	Atomic Number	Molecular Weight	*k*	Remarks	Reliability Indicator
H_2O	1	18.02	0.340		H
Li_2O	3	29.88	0.307		H
$(NH_4)_2O$	—	52.08	0.483		H
Na_2O	11	61.98	0.190		H
K_2O	19	94.20	0.196		H
Cu_2O	29	143.09	0.234		M
Rb_2O	37	186.94	0.128		H
Ag_2O	47	231.74	0.168		M
Cs_2O	55	281.81	0.119		H
Au_2O	79	409.94	(0.152)		?
Hg_2O	80	417.18	0.144		L
			(0.134)		?
Tl_2O	81	424.74	0.115		M - H
Fr_2O	87	462	(0.115)		?
BeO	4	25.01	0.240		H
MgO	12	40.30	0.200		H
			0.225	sulfates & selenates	H
SO	16	48.06	0.335		M
CaO	20	56.08	0.210		H
VO	23	66.94	(0.207)		H
CrO	24	68.00	(0.202)		H
MnO	25	70.94	0.197		H
FeO	26	71.85	0.188		H
CoO	27	74.93	0.179		H
NiO	28	74.71	0.176		H
CuO	29	79.55	0.170		H
ZnO	30	81.37	0.158		H
SrO	38	103.62	0.145		H
PdO	46	122.40	0.190		L
CdO	48	128.40	0.130		H
SnO	50	134.69	(0.140)		?
BaO	56	153.34	0.128		H
PtO	78	211.09	0.118		L
HgO	80	216.59	0.123		M
PbO	82	223.19	0.133		M
RaO	88	242.00	(0.120)		?
B_2O_3	5	69.62	0.215		H
C_2O_3	6	72.02	0.267		H
N_2O_3	7	76.01	(0.325)		?
Al_2O_3	13	101.96	0.207		H
			0.242	sulfates & selenates	H
			0.176	nesosilicates & inosilicates	H
P_2O_3	15	109.95	(0.315)		?
Sc_2O_3	21	137.91	0.257		H

Component	Atomic Number	Molecular Weight	*k*	Remarks	Reliability Indicator
Ti_2O_3	22	143.80	(0.267)		H
			(0.227)	nesosilicates & inosilicates	H
V_2O_3	23	149.88	(0.279)		H
			(0.237)	nesosilicates & inosilicates	H
Cr_2O_3	24	151.99	(0.290)		H
			(0.247)	nesosilicates & inosilicates	H
Mn_2O_3	25	157.87	0.301		H
			0.256	nesosilicates & inosilicates	H
Fe_2O_3	26	159.69	0.315		H
			0.268	silicates	H
Co_2O_3	27	165.86	(0.329)		H
Ni_2O_3	28	165.42	(0.339)		H
Ga_2O_3	31	187.44	0.170		H
As_2O_3	33	197.84	0.235		M
Y_2O_3	39	225.81	0.170		M
			(0.195)		?
In_2O_3	49	277.64	0.130		H
Sb_2O_3	51	291.50	0.203		M
La_2O_3	57	325.82	0.148		H
Ce_2O_3	58	328.24	0.144		H
Pr_2O_3	59	329.81	0.141		H
Nd_2O_3	60	336.48	0.137		H
Pm_2O_3	61	342	(0.133)		?
Sm_2O_3	62	348.70	0.130		H
Eu_2O_3	63	351.92	0.126		H
Gd_2O_3	64	362.50	0.123		H
Tb_2O_3	65	365.85	0.119		H
Dy_2O_3	66	373.00	0.115		H
Ho_2O_3	67	377.86	0.112		H
Er_2O_3	68	382.52	0.108		H
Tm_2O_3	69	385.87	0.104		H
Yb_2O_3	70	394.08	0.101		H
Lu_2O_3	71	397.94	0.097		H
Tl_2O_3	81	456.74	0.053		M
Bi_2O_3	83	465.96	0.153		M

Constants in brackets were derived by extrapolation or interpolation. Reliability indicators: L (low), M (medium), H (high).

Component	Atomic Number	Molecular Weight	*k*	Remarks	Reliability Indicator
CO_2	6	44.01	0.211		H
SiO_2	14	60.08	0.208		H
PO_2	15	62.97	0.236		M
SO_2	16	64.06	0.262		M
TiO_2	22	79.90	0.393		H
VO_2	23	82.94	0.393		H
CrO_2	24	83.99	(0.394)		H
MnO_2	25	86.94	0.394		H
GeO_2	32	104.59	0.167		H
SeO_2	34	110.96	0.195		H
ZrO_2	40	123.22	0.211		H
SnO_2	50	150.69	0.143		H
TeO_2	52	159.60	0.201		M
CeO_2	58	172.12	(0.205)		?
HfO_2	72	210.49	0.115		M
PtO_2	78	227.09	0.151		L
PbO_2	82	239.19	0.105		M
ThO_2	90	264.04	0.167		M
UO_2	92	270.03	(0.100)		?
N_2O_5	7	108.01	0.242		H
P_2O_5	15	141.94	0.183		H
Cl_2O_5	17	150.90	0.220		H
V_2O_5	23	181.88	0.340		H
As_2O_5	33	229.84	0.162		H
Br_2O_5	35	239.81	0.180		H
Nb_2O_5	41	265.81	0.268		M
Sb_2O_5	51	323.50	(0.153)		?
I_2O_5	53	333.81	0.195		H
Ta_2O_5	73	441.89	0.151		M
Bi_2O_5	83	497.96	(0.139)		?
SO_3	16	80.06	0.177		H
CrO_3	24	99.99	0.335		M - H
SeO_3	34	126.96	0.164		H
MoO_3	42	143.94	0.237		M
TeO_3	52	175.60	0.172		M
WO_3	74	231.85	0.171		M
UO_3	92	286.03	0.118		H
S_2O_7	16	176.12	0.133		M
Cl_2O_7	17	182.90	0.182		M
Mn_2O_7	25	221.87	0.348		M
Br_2O_7	35	271.80	(0.156)		?
I_2O_7	53	365.80	0.168		M
Re_2O_7	75	484.40	0.130		L
F^-	9	19.00	0.047		M
Cl^-	17	35.45	0.318		M
Br^-	35	79.90	0.217		M
I^-	53	126.90	0.227		M
O^{2-}	8	16.00	0.203		H
S^{2-}	16	32.06	0.628	sulfur-bearing silicates	H

Source: Reprinted with permission from J. A. Mandarino, The Gladstone and Dale relationship, *Can. Mineral.*, 14: 498–502.

sults in a concentration of electron density (overlapped orbitals) and a resulting increase in specific refractivity.

Values for the k constants are derived for oxide constituents of minerals rather than for individual elements or ions because the refractivity of cation and anion or metal and nonmetal are interdependent and specific only for a given coordination. If k values were derived for cations or metals, the values obtained for the nonmetals, principally oxygen anions, would vary all over the map; there is nothing constant about the refractivity of oxygen by itself in minerals. It must be emphasized that the specific refractivity K does not measure the index of refraction; rather, it measures the rates of retardation of the electric vector **E** of Na light with the mass–volume ratio or density (Fig. 11.3). Values presented here, as well as by others, are averaged from data that show a wide range for some constituents and a narrow range for others, consistent with the factors cited earlier. Note that k values for constituents such as MgO and Al_2O_3 show little variation in different silicate and oxide minerals, whereas those for Fe_2O_3, CuO, and others vary widely between silicate and oxide phases. Evidently this reflects the increased effects of covalent bonding Fe^{3+}—O^{2-} (formal valencies, not ionic charge), and the value for k ($Fe_2^{VI}O_3$) is 0.280 in silicates and a whopping 0.404 in hematite (Table 11.1). The covalency and resulting molecular orbitals permit the refractivity to remain large and undiluted in the oxide, whereas this effect is diluted in the more ionic silicate species.

It has been emphasized that refractivity values, like electronegativity values, are empirically derived and may be expected to vary. Values of K obtained from data on polymorphs of $Si^{IV}O_2^{II}$ are quartz, $K = 0.206$; tridymite, $K = 0.206$; cristobalite, $K = 0.211$; and coesite, $K = 0.203$. The average is the recommended value of 0.206 cited in Table 11.1. Note that for the high pressure polymorph stishovite ($Si^{VI}O_2^{III}$) the increased coordination number causes a decrease of K to 0.187.

A variety of minerals containing tetrahedrally coordinated aluminum give values of $k(Al_2O_3) = 0.215, 0.217, 0.215, 0.220, 0.213, 0.229, 0.227, 0.211$ and an average of 0.218. With coordination number increased to VI, $k(Al_2O_3)$ becomes 0.187 (Table 11.1). The effect of a change in valency or a change in oxidation state on refractivity is very well shown by the oxides of manganese, in which k increases dramatically with valence and is accompanied by a decrease in coordination.

Contrary to popular belief, index of refraction and density need not vary sympathetically, particularly in minerals that contain different principal anions. Minerals of the same or of very similar density may show no correlation with index of refraction (Table 11.4 and Fig. 11.3). When n and ρ are compared for minerals containing one specific anion or for polymorphs or members of isostructural solid solution series, the variation is quite sympathetic. Compare variations of n and ρ for more ionic compounds of F and O with those of the more covalent S, Se, and Te (see Fig. 11.3).

Ionic fluoride minerals have low indices of refraction and relatively high

Figure 11.3. Plot of density (ρ) versus mean index of refraction, $n - 1$, for fluorides, oxides, selenides, sulfides, and carbides. Radians are specific refractivity slopes (k) and show that k increases with covalency of anionic component.

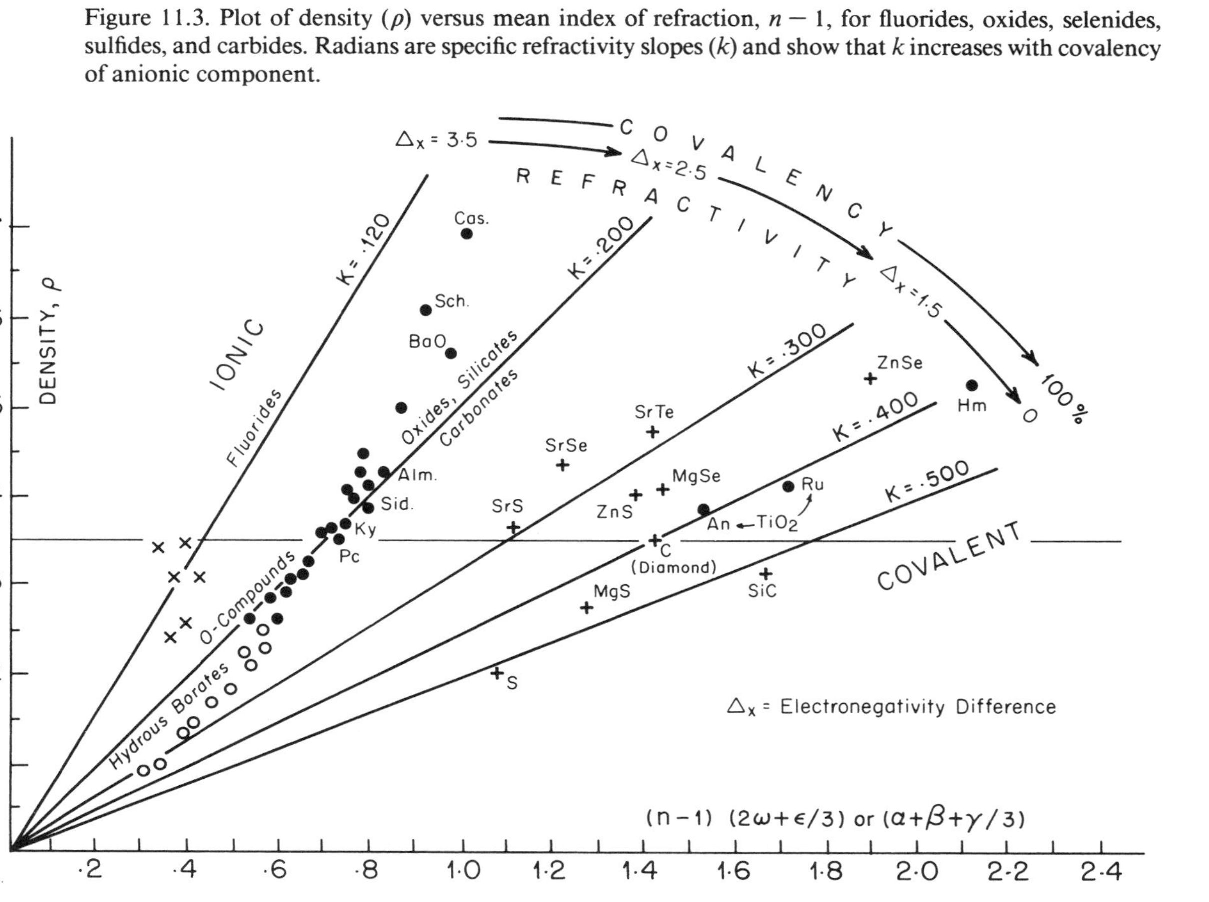

densities, resulting in low specific refractivity values, compared with covalent sulfide and selenide phases that have high indices of refraction and relatively low densities, resulting in high specific refractivity values (Fig. 11.3). Indices of refraction of any species are markedly reduced by the substitution of F for O and/or the substitution of H_2O or (OH) for O; but the low specific refractivity of F ($k = 0.043$) contrasted with the high value for H_2O ($k = 0.340$), results from density contrasts. Note that *K*-line slopes for solid solution series (Fig. 11.4) increase more rapidly for series containing Al^{3+}—Fe^{3+} than for series containing Mg^{2+}—Fe^{2+}. The addition of Fe^{3+} and/or Ti^{4+} to any such mineral will result in a marked increase in the indices of refraction; this expresses the increase in both the number of valency electrons and the covalency.

The practical applications of the law of Gladstone and Dale are many. If n and ρ are measured, it may be used to obtain information on chemical composition. Often the mineralogist has chemical and optical data but cannot measure density, and the simple relation $(n - 1)/K = \rho$ solves the problem in minutes, provided a good table of k values is available. Two such examples cited from Jaffe (1956) and Jaffe et al. (1953) follow.

A spectrographic analysis followed by a preliminary chemical analysis was made of the new mineral sahamalite, $(Mg,Fe)(Ce,La,Nd)_2(CO_3)_4$, named by Jaffe et al. (1953) for the Finnish geochemist Th.G. Sahama. Because only 200 mg of the pure mineral were available, a preliminary chemical analysis was made of a very small quantity, with the following results: $R_2O_3 = 59\%$, $MgO = 6\%$, $FeO = 2\%$, and $CO_2 = 21\%$, for a total of 88 wt.%. After qualitative tests for fluorine and water proved negative, the spectrogram was reexamined. No additional metallic elements could be found. A second spectrogram was made with the same result. By using the measured indices of refraction, the

Table 11.4. *Data showing that index of refraction and density need not vary sympathetically for minerals containing different anions*

Mineral	Formula	Mean refractive index (n)	Density (ρ)
Synthetic	RbBr	1.553	3.41
Synthetic	RbI	1.647	3.48
Synthetic	LiBr	1.784	3.48
Diamond	C	2.412	3.51
Yttrofluorite	(Ca, Y, □)F_2	1.459	3.55
Pyrope	$Mg_3Al_2[SiO_4]_3$	1.714	3.58
Kyanite	$Al_2O[SiO_4]$	1.720	3.60

measured pycnometer density, and the preliminary analytical data, we calculated from $(n = 1)/\rho = K$ that the 12% missing from the analytical summation was a constituent with a specific refractivity k near 0.20. When metallic constituents previously found to be absent by spectrographic analysis were eliminated, this left only CO_2 ($k = 0.217$), Br ($k = 0.214$), I ($k = 0.226$), and O ($k = 0.203$) to account for the missing 12%. The CO_2 analysis seemed to be most suspect and was repeated by two different methods. The correct amount proved to be 31.7% rather than 21%, bringing the analytical summation to 99.+%. The original CO_2 determination was low because the macroabsorption train had recently been set up and was not in equilibrium. Before this was corrected, chemists had argued vehemently that their original CO_2 value could not have been in error!

A second excellent example was provided by a study of the chemically very complex uranium mineral schroeckingerite, described by Jaffe, Sherwood, and Peterson (1948) and Jaffe (1956). Originally, the presence of 2.15% fluorine had gone chemically undetected, leading Larsen and Gonyer (1937) to describe this mineral as a new species they named dakeite. This name was discredited and deleted as a new species when Jaffe et al. (1948) identified the fluorine by spectroscopic analysis and made the necessary chemical analysis. Later, Jaffe (1956) applied the Gladstone and Dale relation to this complex mineral with the excellent result shown in Table 11.5.

Earlier, it was noted that different coordination numbers for a cation would result in different refractivity values. Differences in coordination number, and,

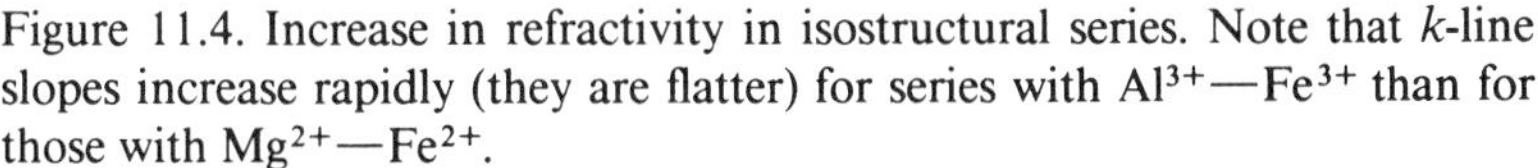

Figure 11.4. Increase in refractivity in isostructural series. Note that k-line slopes increase rapidly (they are flatter) for series with Al^{3+}—Fe^{3+} than for those with Mg^{2+}—Fe^{2+}.

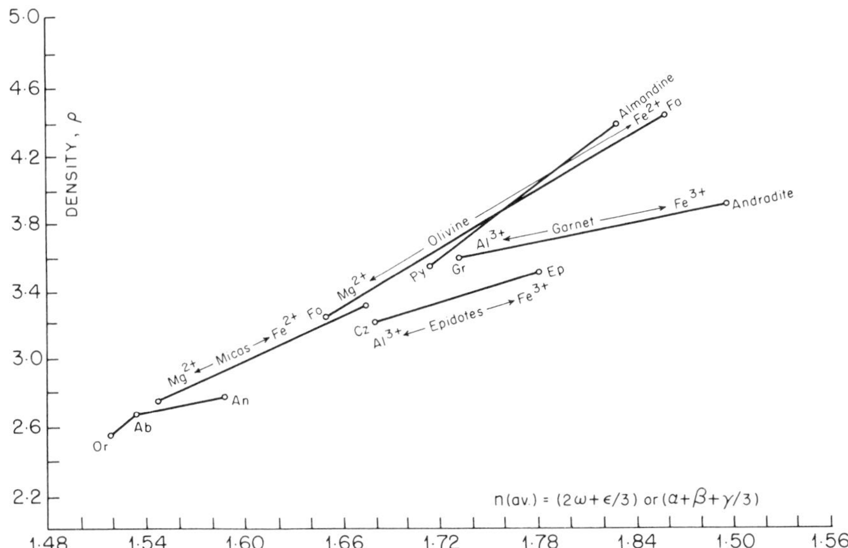

indeed, differences in distortion of a site with the same formal coordination number may be expected to influence a given refractivity value.

Polarizability

Closely related to refractivity is the phenomenon of electronic polarizability, which refers to the temporary displacement of valence electrons in an atom or ion induced by the electric vector of electromagnetic radiation operative at optical frequencies ($\nu_{Na} = 5.09 \times 10^{14}$ cps). After this type of displacement, centers of gravity of the atomic nucleus and the electric charge no longer coincide, and the atom acquires an induced dipole moment μ. The magnitude of such dipole moments summed over all the atoms in a mineral or molecule can be evaluated either by (1) calculation of the molecular electronic polarizability α (Å^3), from measurements of indices of refraction, density, and molar volume, or from (2) measurement of a dielectric constant ϵ. At optical frequency the mean index of refraction n and dielectric constant ϵ are related by $\epsilon = n^2$ or $\sqrt{\epsilon} = n$, as discovered independently by Lorentz and Lorenz in 1880.

Here, an analogy with a marathon runner is appropriate. The runner, moving across a meadow with a speed of 9 mph, is slowed to 6 mph on entering and passing through a spruce forest (Fig. 11.5). Branches of the spruce trees are momentarily elastically displaced by the runner in a manner analogous to the

Table 11.5. *Comparison of measured and calculated mean indices of refraction from $(\alpha + \beta + \gamma)/3$ for schroeckingerite, a chemically very complex uranium mineral*

Oxide	$p/100$		k		$(k)(p/100)$	
CaO	18.93	×	.225	=	.04259	
Na_2O	3.49	×	.181	=	.00632	
UO_3	32.19	×	.134	=	.04313	
CO_2	14.86	×	.217	=	.03225	
SO_3	9.01	×	.177	=	.01595	
H_2O	20.28	×	.340	=	.06895	
F	2.14	×	.043	=	.00092	
	100.90				.21011	
—O=F	.90	×	.203	=	.00183	
	100.00		K	=	.20828	$\alpha = 1.489$
			d	=	2.51	$\beta = 1.542$
			dK	=	.52278	$\gamma = 1.542$

Note: Mean index, $dk + 1 = 1.523$ (n calculated). Mean index, $(\alpha + \beta + \gamma)/3 = 1.524$ (n experimental).

Source: Reprinted with permission from H. W. Jaffe (1956), Application of the rule of Gladstone and Dale to minerals, *Am. Mineral.*, 41: 759.

momentary elastic displacement of electrons in an atom by the electric vector of electromagnetic radiation, or visible light. The thicker the forest, the greater the reduction in the runner's velocity, hence, the greater his or her retardation. Similarly, the higher the electron density and covalency in a mineral, the greater will be the electronic polarizability of its atoms (Fig. 11.5).

Polarizability should increase with numbers of valence electrons affected and by the length of orbitals of atoms of different radii. Within a vertical group of the periodic table, polarizability should increase downward together with both volume and mass in an opposite sense to electronegativity and ionization potential values.

Equations relating specific refractivity *K*, molecular electronic polarizability

Figure 11.5. Relation between polarizability, retardation, and chemical bonding, analogous to a marathon runner entering a spruce forest. Refractivity, *K*, polarizability, $\alpha_{G(Å^3)}/V_{m(Å^3)}$, and index of refraction, *n*, increase with covalent overlap in the sequence: MgF–MgO–MgS.

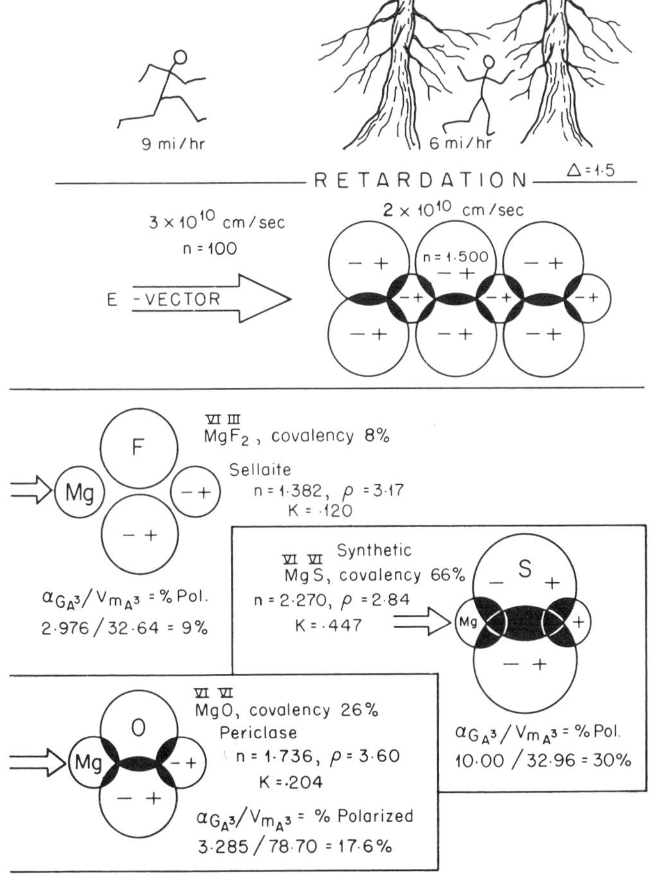

α, and molecular refractivity R to the time-honored formulae of Lorentz–Lorenz and Gladstone and Dale are as follows:

$$\text{Lorentz–Lorenz} \quad \frac{\epsilon - 1}{\epsilon + 2} = \frac{n^2 - 1}{n^2 + 2} \cdot \frac{1}{\rho} = K_L \tag{1}$$

$$\text{Gladstone–Dale} \quad \frac{n^2 - 1}{n + 1} \cdot \frac{1}{\rho} = \frac{n - 1}{\rho} = K_G \tag{2}$$

$$\frac{K \cdot M}{4\pi/3N} = \frac{R\,(\text{cm}^3)}{2.523} = \alpha_L\,(\text{Å}^3) \quad \text{or} \quad \alpha_G\,(\text{Å}^3) \tag{3}$$

depending on whether K_L or K_G is used,

$$\frac{n - 1 \cdot V_m\,(\text{Å}^3)}{4\pi/3} = \frac{R\,(\text{Å}^3)}{4.189} = \alpha_G\,(\text{Å}^3) \tag{4}$$

where K_L = specific refractivity of Lorentz and Lorenz, K_G = specific refractivity of Gladstone and Dale, ϵ = the dielectric constant measured at $\nu_{Na} = 5.09 \times 10^{14}$ cps, n = the mean index of refraction measured for Na light, $\lambda = 5893$ Å, M = mass, V_m = molar volume (in cm^3 or Å^3), R = molar or molecular refractivity (in cm^3 or Å^3), N = Avogadro's number = 6.0228×10^{23} g/mole, $4\pi/3$ is a Lorentz function or factor describing the mutual interaction of atomic dipoles, ρ = density in g/cm^3, α_L and α_G (Å^3) = molecular electronic polarizability at optical frequency. Note that $4\pi/3 = 4.189$ and $4\pi/3N = 2.523$.

Now K_L, K_G, α_L, α_G, R_L, and R_G may be calculated from equations (1), (2), (3), and (4) from the following data for α-quartz, SiO_2:

$$n_\omega = 1.544, \qquad n_\epsilon = 1.553, \qquad n = 1.547$$

$$M = 60.09/\rho = 2.655 = V_m = 22.632 \text{ cm}^3 \text{ or } 37.575 \text{ Å}^3$$

These data give for α-quartz:

$$K_L = 0.1194 \qquad \alpha_L = 2.843 \text{ Å}^3 \qquad R_L = 7.174 \text{ cm}^3\ (11.91 \text{ Å}^3)$$

$$K_G = 0.206 \qquad \alpha_G = 4.907 \text{ Å}^3 \qquad R_G = 12.378 \text{ cm}^3\ (20.551 \text{ Å}^3)$$

In this book we use specific refractivity values K and molecular electronic polarizability values α_G (Å^3) derived from the Gladstone–Dale equation. If the Lorentz–Lorenz equation is used, the calculated polarizability values α_L (Å^3) will yield values comparable to those reported by Lasaga and Cygan as optical polarizabilities.

A list of some of the molecular electronic polarizabilities α_G (Å^3), derived from selected minerals in this study, is presented in Table 11.6. Just as specific refractivity K does not measure index of refraction, neither does the absolute value of polarizability α_G measure index of refraction. What matters is the *polarizability in a given molar volume.*

Table 11.6. *Polarizability of minerals: % mol polarized and relation of the % molar volume polarized to the mean index of refraction*

Mineral	α_G (Å^3) /	V_m (Å^3) =	%	· $4\pi/3$ =	$(n-1)$ /	ρ	= K
Villiaumite	1.956	24.98	.0783	4.189	.328	2.79	.1175
H_2O	2.378	29.91	.0795		.333	1.00	.340
Sellaite	2.976	32.64	.0912		.382	3.17	.1205
Fluorite	4.213	40.76	.1034		.433	3.18	.136
Borax	25.615	231.76	.1105		.463	1.70	.272
Albite	21.05	166.12	.1267		.531	2.62	.2027
Halite	5.83	44.926	.1298		.544	2.16	.252
Quartz	4.907	37.58	.1306		.547	2.655	.206
Anorthite	23.33	167.36	.1394		.584	2.76	.2116
Coesite	4.88	34.17	.1428		.598	2.92	.205
Calcite	8.81	61.32	.1437		.601	2.71	.222
Aragonite	8.49	56.33	.1507		.632	2.95	.214
Andalusite	12.93	85.55	.1511		.633	3.145	.201
Magnesite	7.13	46.98	.1518		.636	2.98	.2134
$MgCl_2$	10.53	68.15	.1545		.647	2.32	.279
Forsterite	11.27	72.56	.1553		.652	3.22	.202
Sillimanite	13.10	82.26	.1581		.662	3.247	.2042
Pyrope	31.84	186.88	.1704		.714	3.582	.1993
Kyanite	12.62	73.41	.1719		.720	3.665	.1965
Grossular	36.42	208.91	.1743		.734	3.594	.2042
Periclase	3.285	18.70	.1757		.736	3.58	.2056
Rhodochrosite	9.15	51.58	.1774		.743	3.70	.2008
Corundum	7.74	42.32	.1829		.766	4.00	.1915
Siderite	9.22	48.58	.1898		.795	3.96	.2008
Spessartine	37.46	196.15	.1910		.800	4.19	.1909
Stishovite	4.46	22.93	.1945		.815	4.35	.1874
Almandine	37.92	191.40	.1981		.830	4.318	.1922
Lime	5.65	27.63	.2045		.838	3.37	.254
SrO	7.11	34.14	.2083		.870	5.04	.173
Andradite	46.29	218.65	.2117		.887	3.859	.2298
BaO	10.41	44.51	.2339		.980	5.72	.1713
Cassiterite	7.86	31.99	.2457		1.029	6.99	.1472
Sulfur	52.67	204.27	.2578		1.080	2.085	.518
MgS	10.00	32.96	.3033		1.271	2.84	.4475
Sphalerite	12.90	39.46	.3269		1.370	4.1	.334
Diamond	1.913	5.676	.3370		1.412	3.51	.4023
Anatase	12.48	34.01	.3670		1.537	3.90	.394
Brookite	12.41	32.04	.3873		1.622	4.14	.392
Rutile	12.60	30.99	.4066		1.708	4.29	.398

The significant relation is

$$\alpha_G / V_m\,(\text{Å}^3) \times 4\pi/3 = n - 1 \qquad (5)$$

For example, borax, $Na_2B_4O_5(OH)_4 \cdot 8H_2O$, has the high Gladstone–Dale polarizability 25.615 Å^3 but the large molar volume 231.76 Å^3; equation (5) yields $n = 1.463$, a low index of refraction. Diamond has a low α_G of 1.913 Å^3, but its very small molar volume, 5.67 Å^3, yields $n = 2.412$ from equation (5) (Table 11.6).

Table 11.7 lists molecular polarizabilities α_G derived for oxides that form their own minerals and that are components of more complex silicate minerals. Table 11.8 illustrates the additivity of several of these oxide polarizabilities to obtain those of more complex silicate minerals, such as kyanite, forsterite,

Table 11.7. *Polarizability of some oxides in minerals*

Oxide & CN	α_G	Mineral example	Oxide & CN	α_G	Mineral example
$Si^{IV}O_2$	4.907	Quartz	$K_2^{IX}O$	7.206	Sanidine
$Al_2^{VI}O_3$	7.76	Corundum	$Fe^{VI}O$	5.346	Siderite
$Al_2^{IV}O_3$	8.26	Sillimanite	$Mn^{VI}O$	4.926	Rhodochrosite
$Mg^{VI}O$	3.28	Periclase	$Fe_2^{VI}O_3$	16.82	Andradite
$Ca^{VI}O$	4.925	Calcite	$Ti^{VI}O_2$	12.615	Rutile
$Ca^{IX}O$	4.655	Aragonite	$Sr^{VI}O$	7.092	SrO
CO_2	3.872	Magnesite	$Ba^{VI}O$	9.89	BaO
$Na_2^{VIII}O$	4.566	Jadeite	$Sn^{VI}O_2$	8.80	Cassiterite
H_2O	2.375	H_2O	(SO_3)	6.11	Anhydrite

Table 11.8. *Some examples of the additivity of polarizabilities of oxides to yield values for silicates*

	α_G		α_G
Corundum	7.76	3 periclase	9.846
+	+	+	+
Quartz	4.907	1 corundum	7.76
"kyanite"	12.667	+	+
Kyanite	12.62	3 quartz	14.721
		"pyrope"	32.327
		Pyrope	32.06
Quartz	4.907		
+	+		
2 periclase	6.56		
"forsterite"	11.467		
Forsterite	11.292		

pyrope, and anorthite. The agreement between the polarizabilities α_G of these minerals and the sums of their constituent oxides is indeed good. The reader will profit from designing some additional examples from the data of Tables 11.7 and 11.8

Electronic polarizabilities have been calculated for several ions by Fajans and Joos (1924), Born and Heisenberg (1924), Pauling (1927), Tessman, Kahn, and Shockley (1953), and Lasaga and Cygan (1982) (Table 11.9). These polarizabilities were determined from the Lorentz–Lorenz equation, and although the numbers obtained will differ from these presented here, the results obtained from each formula are internally consistent.

Optical geochemistry

Perhaps the ultimate application of data on refractivity and polarizability lies in the correlation of the optical properties of minerals with their

Table 11.9. *Electronic polarizabilities of ions in Å^3*

Ion	Tessman, Kahn, & Shockley (1953)	Pauling (1927)	Born & Heisenberg (1924)	Fajans & Joos (1924)	Lasaga & Cygan (1982)
Li^+	.03	.029	.075	.08	.03
Na^+	.41	.179	.21	.196	1.14
K^+	1.33	.83	.87	.88	1.98
Be^{2+}	—	.008	—	.04	.05
Mg^{2+}	—	.094	.012	.12	.48
Ca^{2+}	1.1	.47	—	.51	1.66
Sr^{2+}	1.6	.86	1.42	.86	—
Ba^{2+}	2.5	1.55	—	1.68	—
B^{3+}	—	.003	—	.02	—
Al^{3+}	—	.052	.065	.067	.13
La^{3+}	—	1.04	—	1.3	—
Si^{4+}	—	.016	.043	.04	.08
Ti^{4+}	—	.185	—	.236	—
O^{2-}	.5–3.2	3.88	—	2.75	1.31
S^{2-}	4.8–5.9	10.2	—	8.6	—
Se^{2-}	6.0–7.5	10.5	—	11.2	—
Te^{2-}	8.3–10.2	14.0	—	15.7	—
F^-	.64	1.04	.99	.98	—
Cl^-	2.96	3.66	3.05	3.53	—
Br^-	4.16	4.77	4.17	4.97	—
I^-	6.43	7.10	6.28	7.55	—
$[CO_3]^{2-}$	2.7–5.2	—	—	—	—
$[SO_4]^{2-}$	3.0–5.3	—	—	—	—

crystal chemistry. This new field has been named *optical geochemistry* by this author, who teaches the discipline as a graduate course at the University of Massachusetts.

Bragg (1933) showed that optical character or class and birefringence, *B* (the difference between the maximum and minimum indices of refraction for constant thickness), are related to crystal structure. According to Bragg, planar structures such as calcite (see Fig. 19.6) or sassolite (see Fig. 17.2), with all planar (CO_3) or (BO_3) groups lying *perpendicular* to the *c* crystal axis, are optically negative with high birefringence (Fig. 11.6). Bastnaesite (see Fig. 19.8), (Ce,La,Nd)(CO_3)F, is optically positive because the planar (CO_3) groups are oriented *parallel* to the *c* axis. Bragg also noted that atoms lying in rows staggered parallel to the *c* crystal axis will have optically positive character and high birefringence, for example, rutile (Fig. 11.7; also see Fig. 18.9).

These relations have been studied extensively by this author, who has his graduate students build packing models of minerals from the atomic coordinate crystal structure maps known as unit cell projections. The projections and the models can be used to decipher and relate the travel paths of different velocities of light and the magnitude of their differences in many minerals. A summary of these relations follows.

Figure 11.6. Minerals built of layers perpendicular to the *c* axis are optically negative with high birefringence. Polarizability is high parallel to the layers, low perpendicular to the layers in minerals with $[CO_3]^{2-}$, $[NO_3]^{-}$, and $[BO_3]^{3-}$ groups oriented perpendicular to the *c* axis.

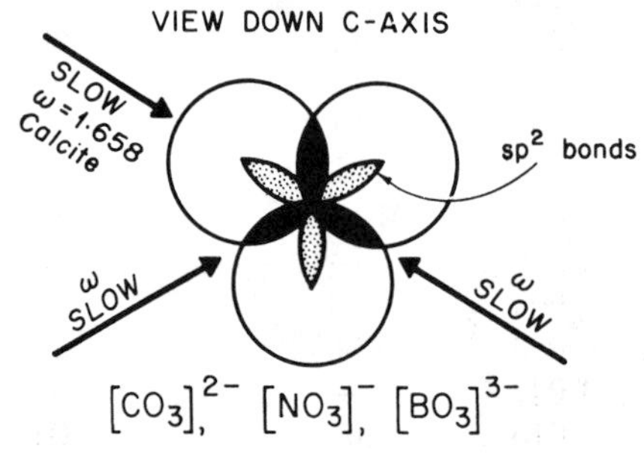

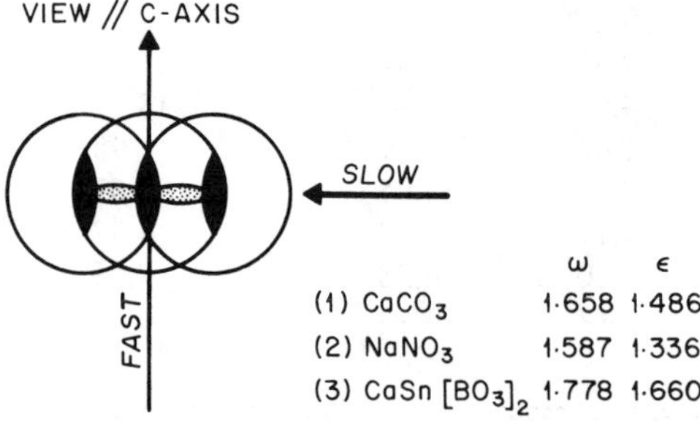

It will be apparent that structural geologic elements (foliation and lineation of strata) may be used as analogues of crystal chemical elements (planar atom and linear atom arrays). Minerals, like rocks, may be regarded as foliated, lineated, or massive, with respect to the degree of anisotropy of electron density conferred by chemical bonding and crystal structure.

Minerals with maximum electron density paths oriented in layers perpendicular to the *c* axis (phyllosilicates, carbonates, nitrates) will be optically negative with a high birefringence *B*. In calcite and soda-niter, covalent hybrid bond sp^2 orbitals impose short O—O distances of 2.22 Å and 2.15 Å in planar (CO_3) and (NO_3) groups, conferring high negative *B* of 0.172 and 0.251, respectively.

Similarly, talc, muscovite, and pyrophyllite are optically negative with high *B*. Chlorite, however, though equally well layered perpendicular to the *c* axis, may be positive or negative around the optic angle (Fig. 11.1), $2V = \pm 0°$ because of the H^+ ion or proton. Brucite and gibbsite are positive with moderately high *B* because the H^+ bonds extensively polarize the O ions parallel to the *c* axis. When the brucite-like layer is combined with an aluminous talc-like layer in chlorite, interlayer tetrahedral–octahedral O—H—O hydrogen bonds parallel to the *c* axis (optically positive) neutralize the (Fe, Mg, Al)—O Si—O layers (optically negative) to yield low *B* and 2V near 0°. With Fe predominant, the octahedral "talc" layer has electron density slightly greater than that of the

Figure 11.7. Minerals with shared-edge-shortened polyhedra lying in staggered rows parallel to the *c* axis will be optically positive with high birefringence, as in rutile, TiO_2. High electron density is linear parallel to *c*, and not in layers. Zircon and anhydrite are similar.

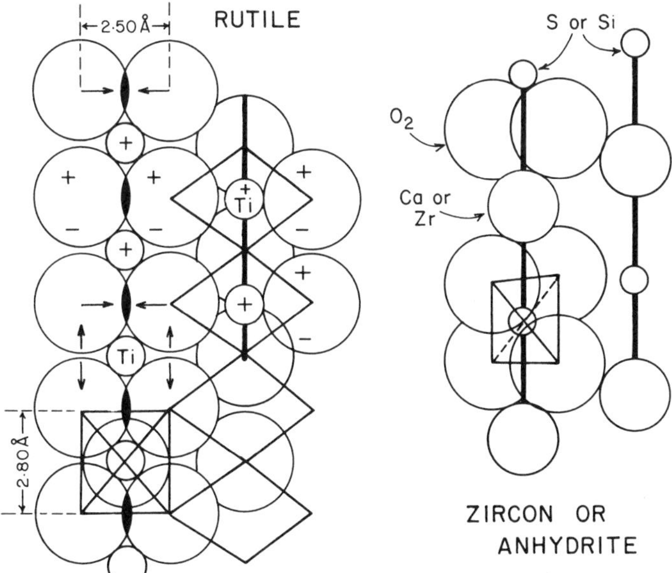

"brucite" layer, and chlorite is optically negative and length slow; with Mg dominant, chlorite is optically positive and length fast (Fig. 11.8).

If atoms and electron density maxima are oriented in staggered rows or spindles parallel to the *c* axis (rutile, anhydrite, zircon), high optically positive *B* will result. In rutile, opposed shared-edge shortening of octahedral edges forms ladders parallel to the *c* axis (lineation) to yield a very high ϵ of 2.901, whereas light vibrating parallel to the *a* axes traverses discontinuous octahedral sites interrupted by voids, with a much lower electron density of $\omega = 2.605$ and $B = 0.296$. In anatase, each octahedron shares four edges in two pairs oriented at right angles to one another, both lying in planes parallel to the *a* axes. In the *c* axis direction, every other octahedral site is empty. Along *a*, the pairs of shared-edge-shortened O—O octahedral edges zigzag or undulate to build up continuous high electron density paths, contrasting with void-interrupted low electron density paths parallel to *c*. Anatase is optically negative with $B = 0.073$ (see Figs. 18.11, 18.12).

Anhydrite is strongly layered with O, S, and Ca atoms lying parallel to the optic plane (100) and with all (SO_4) tetrahedra lying within intersecting (011) layers. Yet anhydrite is optically positive because opposed keels of (SO_4) tetra-

Figure 11.8. In sheet silicates, O^{2-} in tetrahedral and octahedral layers perpendicular to *c* is strongly polarized perpendicular to *c*, but interlayer H^+ strongly polarizes O parallel to *c*. Thus H-bonds parallel to *c* neutralize or counter high polarizability of O^{2-} layers perpendicular to *c*. Layer lattices are normally optically negative, but H bonds can reverse optic sign to positive. In upper series of drawings, octahedral layers (in rectangles) lie between tetrahedral Si_4O_{10} layers. Lower series of drawings shows decreasing ellipticity of optical indicatrix planes and wave velocity contrasts expressed as index of refraction contrasts.

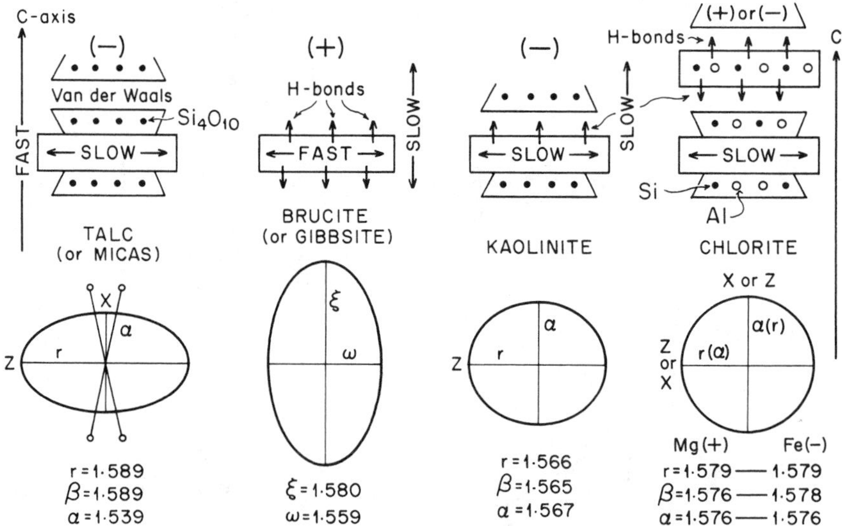

hedra alternate with Ca ions in staggered rows or spindles of high electron density parallel to the *c* axis. These keels form infinite chains of Ca—O_2—S—O_2. In the (100) optic plane, $Z = c$, $X = b$, and $r - \alpha = 1.609 - 1.569 = B = 0.040$. The uniaxial positive zircon structure is similar, with opposed (SiO_4) keels forming, with Zr, infinite spindles of high electron density of Zr—O_2—Si—O_2 parallel to *c* (Figs. 11.7; also see Figs. 12.5, 19.1).

Minerals with equidimensional polyhedral groups, such as tetrahedra, sharing corners, as in tektosilicates (feldspars, feldspathoids, zeolites) are predominantly ionic and do not build up paths of high electron density. These minerals are either isotropic or have low birefringence. In an electric field, bonded electrons cause F, O, and S ions and atoms to behave as a billiard ball, a tennis ball, and a balloon, respectively, resulting in corresponding increases in polarizability, refractivity, and mean index of refraction. The sequence F, O, S is one of decreasing electronegativity and increasing covalency, electron density, and refractivity (Fig. 11.3).

In addition to the effects of orientation of shared-edge-shortened packages, planes, and rows, the effect of the orientation of periodic bond chains (Hartman and Perdok 1955) or unbroken links and chains may be a principal cause of high polarizability in some minerals. For example, in sphene (see Fig. 12.3), the kinked bond string of O—Ti—O—Ti--- correlates with the orientation of the slow vibration direction and maximum index of refraction.

In the descriptive section that follows (Chapters 12–20), crystal chemical analyses of selected minerals are made, which illustrate many of the principles put forth in this chapter and earlier in this book.

Summary

Optical mineralogy is the best single method for identifying transparent minerals. It involves

1. The measurement under the microscope of one, two, or three principal indices of refraction by matching these with calibrated reference liquids
2. A determination of the optic sign and optical class of a given mineral.

There are five optical classes to which all minerals may be assigned, depending on the number of circular sections (CS) or equal velocity planes they contain, and their orientation inside the crystal.

An optic axis (OA) emerges perpendicularly from each circular section. The five optical classes are

1. Omniaxial (CS = ∞; hence OA = ∞)
2. Uniaxial positive and
3. Uniaxial negative (each with one CS and one OA)
4. Biaxial positive and
5. Biaxial negative (each with two CS and two OA).

These classes may be modeled as ellipsoids of revolution from radii equal to indices of refraction, as a

1. Spheroid
2. Prolate spheroid of revolution
3. Oblate spheroid of revolution
4. Triaxial ellipsoid (prolate)
5. Triaxial ellipsoid (oblate)

All isometric minerals and amorphous compounds crystallize in Class 1, tetragonal, hexagonal, and trigonal minerals in Classes 2 or 3, and all orthorhombic, monoclinic, and triclinic minerals in Classes 4 or 5.

The electric vector of electromagnetic radiation of Na or visible light ($\mathbf{E}_{Na}$ or $\mathbf{E}_{vis}$) is retarded and, in most cases, refracted on entering a crystal plate or grain from the air; thus, the velocity c is reduced, the wavelength λ is also reduced, but the frequency ν remains unchanged.

Refractivity and polarizability are phenomena that express the retardation of the **E** vector of light by a mineral of given composition, mass, molar volume, and density.

Specific refractivity $K = (n - 1)/\rho$ measures the rate of retardation of $\mathbf{E}_{Na}$ with density.

Molecular polarizability α_G measures the total of electric dipole moments operative in a crystal, and α_G/V_m (Å^3) $\times 4\pi/3 = n - 1$. Thus, the polarizability in a given molar volume of mutually interacting dipoles is the index of refraction, just as is the product of the specific refractivity and density + unity. One may be derived from the other.

Both α_G and K are reasonably additive when oxide values are combined to sum to complex silicate values if reasonable attention is paid to choosing the correct value for a given CN of a component.

New and revised lists of K and α_G constants or values for use in mineralogical calculations are presented.

The relationships between the vibration directions or vibration velocity vectors (**Z** = slow, **Y** = intermediate, **X** = fast) and their orientation in crystals depends essentially upon the arrays of atoms or ions. Thus, refractivity and crystal chemistry (optical geochemistry) form a rewarding, if somewhat neglected, subject for added study.

Bibliography

Ahrens, L. H. (1959). Variation of refractive index with ionization potential in some isostructural crystals. *Min. Mag.* 31: 929.

Allen, R. D. (1956). A new equation relating index of refraction and specific gravity. *Am. Mineral.,* 41: 245–57.

Batsanov, S. S. (1961). *Refractometry and chemical structure.* Consultants Bureau, New York.

Bloss, F. D. (1961). *An introduction to the methods of optical crystallography.* Holt, Rinehart and Winston, New York.

Born, M., and Heisenberg, K. (1924). Uber den Einfluss der Deformierbarkeit der Ionen auf optische und chemische Konstanten. I. *Z. Phys.*, 23: 388–410.

Bragg, W. L. (1933). *The crystalline state.* Vol. I. *A general survey.* G. Bell, London.

Cygan, R. T., and Lasaga, A. C. (1986). Dielectric and polarization behavior of forsterite at elevated temperatures. *Am. Mineral.*, 71: 758–66.

Dana, E. S., and Ford, W. E. (1932). *A textbook of mineralogy,* 4th. ed. Wiley, New York.

Deer, W. A., Howie, R. A., Zussman, J. (1962–1963). *Rock-forming minerals,* Vols. 1–5. Longman, London.

(1966). *Introduction to the rock-forming minerals.* Longman, London.

Fajans, K., and Joos, G. (1924). Molrefraktion von Ionen und Molekülen im Lichte der Atomstruktur. *Z. Phys.*, 23: 1–46.

Gladstone, J. H., and Dale, T. P. (1864). Researches on the refraction, dispersion, and sensitiveness of liquids. *Roy. Soc. London Philos. Trans.*, 153: 337.

Hartman, P., and Perdok, W. G. (1955). On the relations between structure and morphology of crystals – I. *Acta Crystallogr. Pt. 1* 8: 49–52.

Jaffe, H. W. (1956). Application of the rule of Gladstone and Dale to minerals. *Am. Mineral.*, 41: 757–77.

Jaffe, H. W., Meyrowitz, R., and Evans, H. T. (1953). Sahamalite, a new rare earth carbonate mineral. *Am. Mineral.*, 38: 749.

Jaffe, H. W. and Molinski, V. J. (1962). Spencite, the yttrium analogue of tritomite from Sussex County, New Jersey. *Am. Mineral.* 47: 9–25.

Jaffe, H. W., Sherwood, A. M., and Peterson, M. J. (1948). New data on schroeckingerite. *Am. Mineral.*, 33: 152–7.

Larsen, E. S., Jr., and Berman, H. (1934). The microscopic determination of the nonopaque minerals. U.S. Dept. of Interior, *Geol. Surv. Bull.* 848, 2d ed. (1921, 1st ed.).

Larsen, E. S., Jr., and Gonyer, F. A. (1937). Dakeite, a new uranium mineral from Wyoming. *Am. Mineral.*, 22: 561–3.

Lasaga, A. C., and Cygan, R. T. (1982). Electronic and ionic polarizabilities of silicate minerals. *Am. Mineral.*, 67: 328–34.

Lorentz, H. A. (1880). Ueber die Beziehung zwischen der Fortpflanzungsgeschwindigkeit des Lichtes und der Körperdichte. (On the relations among velocity of propagation of light and the density of solids). *Widem. Ann. Phys.*, IX: 641–65.

Lorenz, L. (1880). Ueber die Refractionsconstante. (On refraction constants). *Widem. Ann. Phys.*, XI: 70–103.

Mandarino, J. A. (1976). The Gladstone–Dale relationship. Part I. Derivation of new constants. *Can. Min.*, 14: 498–502.

(1978). The Gladstone–Dale relationship. Part II. Trends among constants. *Can. Min.*, 16: 169–74.

(1979). The Gladstone–Dale relationship. Part III. Some general applications. *Can. Min.*, 17: 71–76.

(1981). The Gladstone–Dale relationship. Part IV. The compatibility concept and its application. *Can. Min.*, 19: 441–50.

Morse, S. A. (1968). Revised dispersion method for plagioclase. *Am. Mineral.*, 53: 105–16.

Nassau, K. (1983). *The physics and chemistry of color.* Wiley, New York.

Pauling, L. (1927). The theoretical prediction of the physical properties of many-elec-

tron atoms and ions. Mole refraction, diamagnetic susceptibility, and extension in space. *Proc. Roy. Soc. (London) A,* 114: 181–211.
(1960). *The nature of the chemical bond,* 3d ed. Cornell Univ. Press, Ithaca, N.Y., pp. 605–10.

Roberts, S. (1949). Dielectric constants and polarizabilities of ions in simple crystals and barium titanate. *Phys. Rev.,* 76: 1215.

Smith, J. W. (1955). *Electron dipole moments.* Butterworth, London.

Smyth, C. P. (1955). *Dielectric constant and molecular structure.* McGraw-Hill, New York.

Tessman, J. R., Kahn, A. H., and Shockley, W. (1953). Electronic polarizabilities of ions in crystals. *Phys. Rev.,* 92: 890–5.

Tilley, C. E. (1922). Density, refractivity and composition relations of some natural glasses. *Min. Mag.,* 19: 275–94.

Tröger, W. E. (1982). *Optische Bestimmung der gesteinbildenen Minerale, 5. Auflage. Teil 1.* Bestimmungstabellen von Profs. Bambauer, H. U., Taborszky, F., und Trochim, H. D. E. Schweizerbart'sche Verlagsbuchhandlung, Stuttgart.

Van Vleck, J. H. (1932). *The theory of electric and magnetic susceptibilities.* Oxford Univ. Press, London.

Winchell, A. N., and Winchell, H. (1961). *Elements of optical mineralogy,* 4th ed. Wiley, New York; Chapman and Hall, London.

PART II
Descriptive crystal chemistry

12 Silicates: classification and formulation

Because the crust of the earth, and, indeed, much of the mantle, is made up predominantly of oxygen and silicon, the most abundant and diverse mineral species are silicates. Feldspars alone make up roughly 60% of the earth's crust, followed by about 12% each of quartz and pyroxenes, with the remaining 16% represented by thousands of accredited mineral species. Perhaps as many as 50% to 75% of all known mineral species have been found only once or have been found at only one locality.

Silicate minerals are classified on the basis of the degree of polymerization of their $[SiO_4]$ tetrahedra into the seven categories illustrated in Figure 12.1. Note that each solid line boxes in the correct Si—O content of the tetrahedral polymerization unit that is the basic building block of each silicate class. Each class of silicate minerals represents a stacking or packing of these basic units, like the bricks in a chimney. The important aspects of each category and a mineral example of each are summarized in Table 12.1.

The negative electric charge on each $[Si_xO_y]$ unit indicates and restricts the composition of minerals that might be expected to occur. For example, $[Si_2O_7]^{6-}$ represents a close stacking of tetrahedral dumbbells or bow ties linked together by cations having a total positive charge of +6. The simplest way to balance the charge is with two 3+ cations, and the rare mineral thortveitite, $Sc_2[Si_2O_7]$, provides us with the best example. From Figure 5.2B, we note that the bridging oxygen that connects the two tetrahedra of the $[Si_2O_7]$ dumbbell is in contact with two Si^{4+} atoms and has its electrostatic bond strength (EBS) exactly satisfied. All of the remaining oxygens, however, are in contact with one Si^{4+} and two Sc^{3+} to balance their negative charge exactly (Fig. 12.2). The negative charge on the Si—O tetrahedral unit suggests a possible formula. Pauling's first and second rules (see Chapters 4 and 5) allow us to predict coordination numbers and EBS distribution and to limit the number of possible crystal structures that will satisfy these conditions.

In the double chain or band silicate unit, $[Si_4O_{11}]^{6-}$ (Fig. 12.1), each tetrahedron shares alternately two and three of its oxygens along the band axis, equivalent to the *c* axis. This formula is doubled to $[Si_4O_{11}]_2$, or, as most commonly used, $[Si_8O_{22}]^{12-}$, in order to allow the use of integral numbers of charge-balancing cations $[Ca_2Mg_5]^{14+}$ as in the amphibole tremolite, where the excess positive charge of +2 requires the addition of $(OH)_2^{2-}$ to effect charge balance.

All amphiboles and sheet silicates such as micas and talc require the addition of $(OH)^-$ ions to effect charge balance. Both will have twenty-four oxygens in a formula unit: amphiboles $[Si_8O_{22}](OH)_2^{14-}$ and talc $[Si_8O_{20}](OH)_4^{12-}$. Addition of $\square_2Mg_7$ to $Si_8O_{22}(OH)_2$ yields the anthophyllite formula, and addition of $\square_2Mg_6$ to $Si_8O_{20}(OH)_4$ yields the talc formula. The empty box or square preceding each formula denotes one or two empty or vacant lattice sites per doubled formula. Partial or complete occupancy of such a site results in the formation of a different mineral. For example, talc becomes the mica phlogopite by use of the coupled substitution (see Chapter 9) $K_2Al_2 \leftrightarrows \square_2Si_2$ to change talc, $\square_2Mg_6Si_8O_{20}(OH)_4$, to phlogopite, $K_2Mg_6Si_6Al_2O_{20}(OH)_4$.

The location of the $(OH)^-$ ions and the K^+ ions in the void sites, and the determination of the coordination number of K^+ as XII, can best be understood from a close examination of Figures 12.1, 15.17, and 15.18, which show large holes enclosed in the $[Si_4O_{10}]$ sheet silicate unit illustrated. Each such void is bounded in hexagonal outline above and below by six oxygens, and one $(OH)^-$ ion is placed coplanar with each apical tetrahedral oxygen, whereas superposition of two tetrahedral sheets places six oxygens over six others to produce the

Figure 12.1. Ion packing drawings delineating the seven principal silicate polymerization classes. Lines drawn from the center of each large sphere (=O^{2-} anion) will outline $[SiO_4]$ tetrahedra. See Table 12.1 for details.

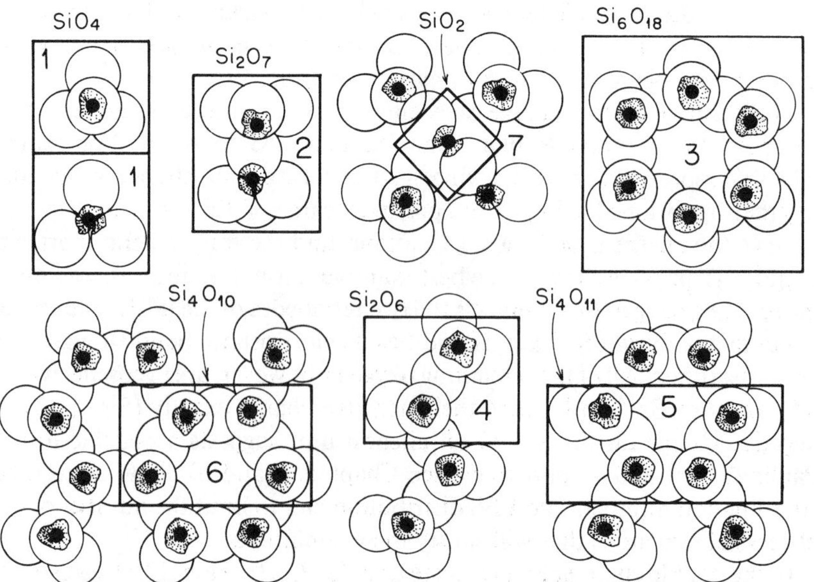

Table 12.1. *Data for tetrahedral polymerization units [Si,Al:O] of Figure 12.1*

Name	[Si,Al:O]	Tetrahedral unit charge	Mineral example	Formula	Shared corners/ tetrahedron
1. Nesosilicate	1:4	4^-	Almandine	$Fe_3Al_2[SiO_4]_3$	0
			Zircon	$Zr[SiO_4]$	0
2. Sorosilicate	2:7	6^-	Thortveitite	$Sc_2[Si_2O_7]$	1
			Åkermanite	$Ca_2Mg[Si_2O_7]$	1
3. Cyclosilicate	6:18	12^-	Beryl	$Al_2Be_3[Si_6O_{18}]$	2
	3:9	6^-	Benitoite	$BaTi[Si_3O_9]$	2
4. Inosilicate	1:3	2^-	Wollastonite	$Ca[SiO_3]$	2
(single chain)	2:6	4^-	Jadeite	$NaAl[Si_2O_6]$	2
5. Inosilicate (double chain = band)	4:11 (8:22)	$6^-(12^-)$	Tremolite	$Ca_2Mg_5[Si_4O_{11}]_2(OH)_2$	2, 3
6. Phyllosilicate	4:10 (8:20)	$4^-(8^-)$	Talc	$Mg_3[Si_4O_{10}](OH)_2$	3
			Phlogopite	$KMg_3[Si_3AlO_{10}](OH)_2$	3
7. Tektosilicate	1:2	0	Quartz	SiO_2	4
	4:8	2^-	Anorthite	$Ca[Al_2Si_2O_8]$	4

large XII-coordinated hole or site that is occupied by K^+. Such large sites can also accommodate the rarer alkali ions in micas (e.g., Rb^+) and can also act as cages enclosing the gas ^{40}Ar derived from the radioactive decay of the ^{40}K isotope.

There are two problems associated with this silicate classification system: first, the presence of appreciable amounts of Al^{3+} proxying for Si^{4+} in tetrahedra, and second, the presence of O^{2-}, $(OH)^-$, or F^- anions in excess of those needed to build tetrahedral groups. For example, sphene (titanite) has the formula $Ca(Ti,Al)[SiO_4](O,F)$: it is not a simple island or nesosilicate. The fifth O^{2-} anion (O-3 of Fig. 12.3) does not contact an Si^{4+} ion but is used to complete octahedral (VI) coordination sites around Ti^{4+}. Introduction of F^- into this site, replacing the O^{2-}, will require a compensating cation substitution, such as $Al^{3+} \leftrightarrows Ti^{4+}$, as well as a coupled substitution $Al^{3+}F^- \leftrightarrows Ti^{4+}O^{2-}$.

Figure 12.2. Packing model (above) and polyhedral analogue (below) of thortveitite, $Sc_2[Si_2O_7]$.

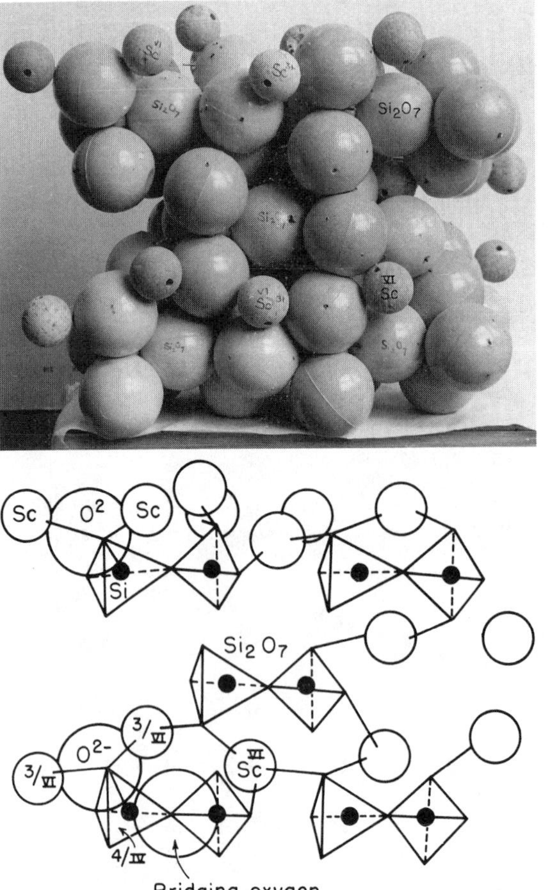

Other minerals classed as silicates contain far more Al^{3+} than Si^{4+} and are best classed as silico*aluminates* rather than as alumino*silicates,* with respect to the principal polymerization of their polyhedral elements.

Chains or bands of octahedral composition, Al[O,OH]$_4$, are present in minerals such as sillimanite, kyanite, andalusite, epidote, and lawsonite, and are illustrated in Figure 12.4. In sillimanite, chains of Al—O octahedra polymerize along the *c* axis with composition [AlO_4] and provide more information on the structure of the mineral than do the isolated [SiO_4] tetrahedra. Thus, for sillimanite, the choice of formulae may be

$$Al_2O_3 \cdot SiO_2 \quad (1)$$
$$Al_2SiO_5 \quad (2)$$
$$Al_2O[SiO_4] \quad (3)$$
$$Al^{IV}O^{II}Si^{IV}[Al^{VI}O_4] \quad (4)$$
$$Al^{VI}[Al^{IV}SI^{IV}O_5] \quad (5)$$

Formulae (1) and (2) ignore all aspects of the crystal structure; formula (3) lumps the two different coordination sites of Al and provides no information on polymerization. Formula (4) indicates that octahedral Al—O chains of composition [AlO_4] are propagated in the *c* axis direction, and these chains are in turn cross-linked to tetrahedral Al^{IV} and Si^{IV} oxygens via a fifth oxygen, O^{II}. Formula (5) elegantly represents the mineral as perfectly ordered tetrahedral bow ties $AlSiO_5$ (Fig. 12.4; also see Fig. 12.10) propagated along the *c* axis and connected solely by Al^{VI} ions. It thus ignores the octahedral chain polymerization of formula (4).

Figure 12.3. Packing model (left) and polyhedral analogue (right) of part of the structure of monoclinic sphene (=titanite), CaTi(③,F)[SiO_4]. View is approximately along *c* sin *β*, (001) projection.

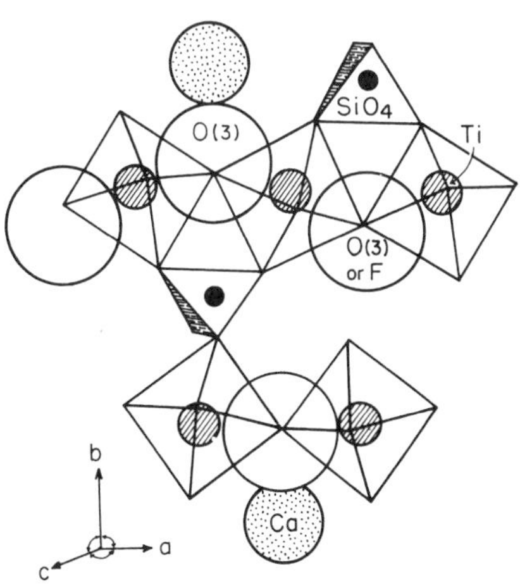

Formula (4) is probably the best choice, because it can more readily be related to formulae of andalusite and kyanite, polymorphous with sillimanite. Crystal chemists should always choose the formula that gives the maximum possible information on polyhedral polymerization, and not restrict themselves to a less informative formula based on an imperfect classification scheme.

A descriptive crystal chemical analysis of one or more minerals of each silicate class follows. No attempt is made to present structural data for a large number of species [for which see the volumes of Wyckoff (1963–8)].

Nesosilicates = island silicates = orthosilicates

Nesosilicates (Greek νεσος = island) are the most difficult of all silicate groups to visualize from either packing, polyhedral, or ball-and-stick models, because of the absence of $[SiO_4]$ polymerization and the resulting closeness of packing of large ions such as O^{2-}. Accordingly, use of cation–anion polymerization of ions other than Si with O (e.g., Al—O) will greatly facilitate visual and crystal chemical comprehension of these minerals.

Nesosilicates may be divided into two categories: simple *(tetraoxy type)* and complex *(peroxy type)* as follows:

Simple tetraoxy silicates

1. Isolated $[SiO_4]_n$ groups account for all the anions in the structure.
2. Each O^{-2} anion has an identical cation array and EBS distribution (Pauling Rule 2A; see Chapter 5).

Figure 12.4. Bow tie chains of Si and Al tetrahedra of composition $[Si,AlO_5]$ as in sillimanite (A), $[AlO_4]$ octahedral chains as in sillimanite, andalusite, kyanite, and epidote (B), and $[AlO_3(OH)]$ octahedral chains as in lawsonite and epidote (C). Chains are infinitely polymerized parallel to *c* in sillimanite, andalusite, kyanite, and parallel to *b* in lawsonite and epidote.

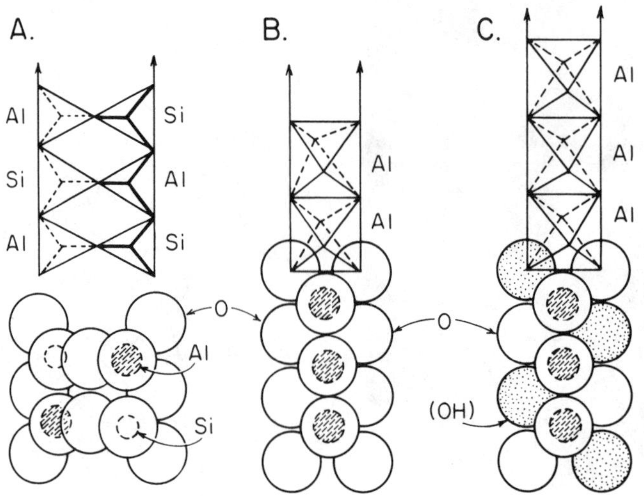

Examples: olivine, garnet, zircon.

Complex peroxy silicates

1. Isolated $[SiO_4]_n$ groups do not account for all anions in the structure; there is an excess of O^{2-}, $(OH)^-$, or F^- anions.
2. Each anion does not have an identical cation array, and more complex EBS solutions apply (Pauling Rule 2C; see Chapter 5).

Examples: sphene, sillimanite.

Simple (tetraoxy) nesosilicates

Although Si^{4+} is the principal tetrahedral cation, it is understood that Al^{3+}, Fe^{3+}, P^{5+}, and B^{3+} may proxy for Si. In the case of a complete proxy for Si, not usually met with in nature, the new mineral may be other than a silicate. The groups of minerals listed in Table 12.2 are all isostructural. Note that they may show complete solid-solution relationships or none whatsoever.

Zircon

$Zr^{VIII}[Si^{IV}O_4^{III}]$ or $4/Zr^{VIII}O_2^{III}Si^{IV}O_2^{III}$

Tetragonal $I_{4_1}\underset{(001)}{a}\ \underset{(100)}{m}\ \underset{(110)}{d}$

$a = 6.63$ Å, $c = 6.02$ Å, $V_{uc} = 264.62$ Å^3

Table 12.2. *Isostructural mineral groups that show complete to no solid solution.*

VI IV IV	Mineral	Comment
Mg_2 $[Si\ O_4]$	Forsterite	No tetrahedral substitution;
Fe_2 $[Si\ O_4]$	Fayalite	Mg—Fe solid solution complete
MnLi $[P\ O_4]$	Lithiophilite	Tetrahedral proxy complete but no
MgAl $[B\ O_4]$	Sinhalite	solid solution
Al_2 $[Be\ O_4]$	Chrysoberyl	
VIII VI IV IV		
Mn_3 Al_2 $[Si\ O_4]_3$	Spessartine	Complete substitution of tetrahe-
Y_3 Al_2 $[Al\ O_4]_3$	Yttrogarnet (YAG)	dral Si by Al and/or Fe^{3+} in syn-
Y_3 Fe_2 $[Fe\ O_4]_3$	Yttrium–iron garnet (YIG)	thetic systems. Limited substitu-
		tion in nature.
VIII IV III		
Zr $[Si\ O_4]$	Zircon	Extensive solid solution between
Th $[Si\ O_4]$	Thorite	zircon and thorite, moderate sub-
Y $[P\ O_4]$	Xenotime	stitution of Si by P in natural min-
U $[Si\ O_4]$	Coffinite	erals

$M/\rho = V_m = 183.31/4.60 = 39.85\ cm^3$
Electronegativity: $\Delta\chi_{Zr-O} = 2.1$ (67% ionic bond)
$+\Delta\chi_{Si-O} = 1.7$ (51% ionic bond)
Integrated $\Delta\chi$ for O—Zr—Zr—Si = 1.97 (62% ionic bond)
Hardness: 7.5 Luster: adamantine
Color: pale yellow, pink, purple, brown-black
CN: Zr^{VIII} – triangular dodecahedron
Si^{IV} – tetragonally distorted tetrahedron
O^{III} – triangle
A—X distances: Zr—O = 2.15 Å(4), 2.29 Å(4); Si—O = 1.61 Å
O—O (shared edge) = 2.43 Å, O—O (unshared edge) = 2.73 Å
Specific refractivity: $K = 0.204$
Molecular polarizability: $\alpha_G = 14.82$ Å^3
Optical properties: $\epsilon = 1.968$, $\omega = 1.923$, $\epsilon - \omega = 0.045$
uniaxial (+); colorless in transmitted light, and nonpleochroic

Discussion. Crystals are well-formed four-sided prisms with a square cross section; they show high relief and are often concentrically zoned. The optical indicatrix model is a prolate spheroid of revolution.

Figure 12.5. Packing and polyhedral representation of the eightfold (triangular dodecahedral) coordination of Zr^{4+} by oxygen anions, O^{2-}, in zircon. Each Zr cation links six SiO_4 tetrahedra: T-1 and T-2, edge sharing; T-3, T-4, T-5, and T-6, corner sharing.

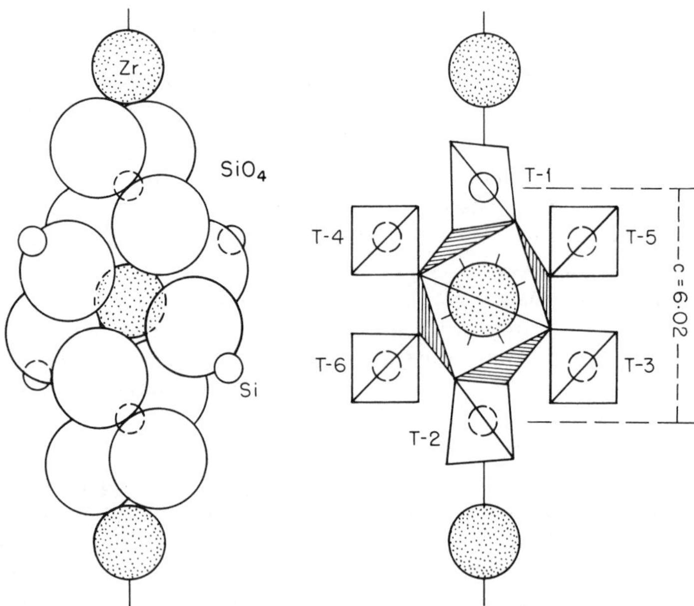

Zircon consists of isolated [SiO_4] tetrahedra linked by Zr^{4+} ions. Each Zr^{4+} ion bonds to four O^{2-} ions that form opposed keels of adjoining [SiO_4] tetrahedra, and to four additional O^{2-} ions that form corners of four other adjoining [SiO_4] tetrahedra (T-3, T-4, T-5, T-6, Fig. 12.5). The eight oxygens that surround and coordinate each Zr^{4+} ion lie at the corners of a very closely packed and deformed triangular dodecahedron (Fig. 12.5). Dodecahedra and tetrahedra polymerize into infinite twisted link chains or spindles running parallel to the *c* crystal axis. Zr^{4+} and Si^{4+} ions straddle pairs of O^{2-} ions that form opposed keels of [SiO_4] tetrahedra, giving each twisted link chain the composition ZrO_2SiO_2. Each chain is staggered *c*/4 from and rotated from adjoining chains by the 4_1 and 4_3 screw axes (Figs. 12.5, 12.6). Mirror planes (100) bisect

Figure 12.6. (A) The structure of zircon viewed along an *a* axis, showing the edge and corner linkages of ZrO_8 and SiO_4 polyhedra, and their translation along an *a* axis by axial glide planes *a* (001). (B) Projection of the SiO_4 tetrahedra of zircon along the *c* axis, showing the relation of all such tetrahedra to mirror planes (*m*), diamond glides (*d*), and 4_1 and 4_3 screw axes. These combine to yield space group *I*4_1/*amd*.

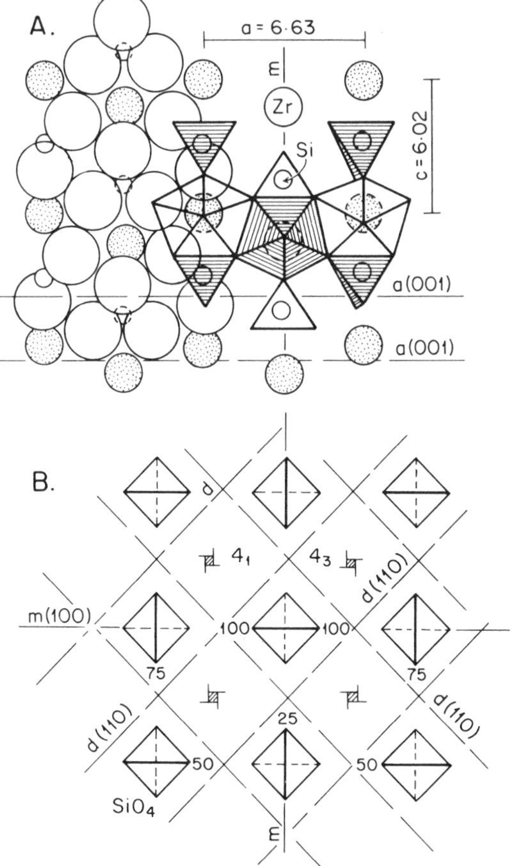

the unit cell, and axial glides parallel to (001) translate atoms $a/2$ (Fig. 12.6). Diamond glides (d) are oriented in planes parallel to (110) and translate atoms $a/4 + c/4$ (110) as in Figure 12.6. The staggering of ZrO_2SiO_2 chains parallel to c, the shortened-shared-edge overlap of O—O = 2.43 Å (T-1 and T-2; see Fig. 12.5), the very tight packing, and the large valence of both Zr and Si combine to produce a high electron density parallel to c and an optically positive crystal (see Chapter 11).

Virtually all zircons are radioactive and contain from about 0.5 to 1000 ppm of radiogenic lead produced from the decay of uranium and thorium substituted as U^{4+} and Th^{4+} in Zr^{4+} sites in the crystal lattice. Zonation of U and Th during crystal growth results in alternating thin zones of different Pb content. When Pb content exceeds 125 ppm, zircon becomes gradually metamict because of α-particle bombardment and atomic recoil over time. The indices of refraction and the birefringence become reduced, more so in U- and Th-rich zones, less so in U- and Th-poor zones, giving crystals a concentrically zoned appearance. Continuous α-particle bombardment over time also deepens color from colorless to hyacinth purple and ultimately to dark brown or even black as the structure undergoes extensive damage. Zircon occurs in all granitic igneous rocks, and its U, Th, and Pb content permit its use as a geochronometer. ^{238}U decays to $^{206}Pb + 8 \times {}^{4}He$, ^{235}U to $^{207}Pb + 7 \times {}^{4}He$, and ^{232}Th to $^{208}Pb + 6 \times {}^{4}He$. Because these isotopes decay at different rates through time, each is a separate geochronometer. When all three decay schemes yield the same age ($\pm$ 10%) or a concordance of values, the age is considered reliable.

Thorite, $ThSiO_4$, and xenotime, YPO_4, are isostructural with zircon. Extensive solid solution occurs among these minerals, as well as with coffinite, $USiO_4$, which is reported by some investigators to contain OH (Stieff, Stern, and Sherwood 1956; see also Table 12.2).

Olivine

The structure of olivine is adequately covered in Chapter 8 and need not be repeated here.

Garnet

The garnets represent simple (*tetraoxy-type*) nesosilicates and cover a wide range of chemical composition. All garnets are isometric,

$$\text{Space group: } \underset{(100)}{I\,a} \quad \underset{(111)}{3} \quad \underset{(110)}{d}$$

Garnets have the structural formula

$$8/A_3^{VIII}B_2^{VI}[T^{IV}O_4^{IV}]_3$$

where

A^{VIII} = Mg, Fe^{2+}, Mn^{2+}, Ca^{2+} in the more common silicate varieties, as well as Na^+, Y^{3+}, Nd^{3+}, Sm^{3+}, Dy^{3+}, and other rare earth elements.

B^{VI} = Al, Fe^{3+}, Cr^{3+} in the more common silicate garnets, as well as V^{3+}, Zr^{4+}, and Ti^{4+}.

T^{IV} = Si^{4+} in the more common silicate garnets, and Al, Fe^{3+}, Li^+, As^{5+}, H_4, and Ge^{4+}.

O^{IV} is principally O^{2-}, but may be proxied by $(OH)^-$ and F^-.

Discussion. All garnets have but one type of oxygen in which EBS is satisfied by two A^{VIII} cations, one B^{VI} cation, and one T^{IV} cation around each oxygen.

The CN VIII site of garnet is very much like that described for zircon – a triangular dodecahedron. The CN VI site is a distorted octahedron, and the CN IV site is a tetragonally distorted tetrahedron (Fig. 12.7).

Figure 12.7. Polyhedral drawing of part of the garnet structure, showing distorted triangular dodecahedral (CN VIII) polyhedra around the large cation, distorted octahedra (CN VI) enclosing Al, and Si in tetrahedra (CN IV). Each O^{2-} anion is a corner common to two dodecahedra, one octahedron, and one tetrahedron. Key relates locations of Figures 12.7 and 12.8.

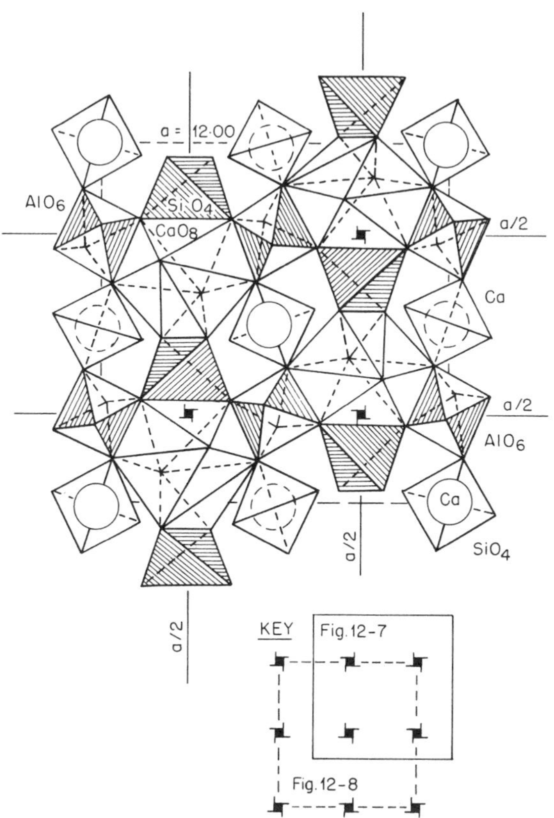

Because garnets have two CaO_8 triangular dodecahedra, one AlO_6 octahedron, and one SiO_4 tetrahedron all joined at corners common to every oxygen, the structure is extremely difficult to visualize. In Figures 12.7 and 12.8, those parts of the structure of garnet are shown that best illustrate some of its complex crystal chemistry. Note that staggered chains or spindles Ca—O_2—Si—O_2—Ca—O_2—Si—O_2 (Fig. 12.7) are similar to (and twice as long per unit cell edge) those of ZrO_2—Si—O_2 of zircon (see Fig. 12.5).

In garnet these chains are cross-linked in parallel by chains of Ca—O—Al—Si—O—Al—O (Fig. 12.7). The electron density and index of

Table 12.3. *Varieties of garnet and synthetic analogues*

	A_3^{VIII}	B_2^{VI}	$[T^{IV}$	$X_4^{IV}]_3$	a (Å)	ρ	n
Natural							
Pyrope	Mg	Al	Si	O	11.46	3.58	1.714
Almandine	Fe	Al	Si	O	11.53	4.32	1.830
Spessartine	Mn	Al	Si	O	11.62	4.19	1.800
Grossular	Ca	Al	Si	O	11.85	3.59	1.734
Andradite	Ca	Fe^{3+}	Si	O	12.05	3.86	1.887
Uvarovite	Ca	Cr^{3+}	Si	O	12.00	3.90	1.860
Schorlomite	Ca	Ti	(Fe,Al,Si)	O		3.85	1.98
Hydrogrossular	Ca	Al	H_4	O	12.16	3.13	1.675
Goldmanite	Ca	V^{3+}	Si	O	12.07	3.76	1.834
Kimseyite	Ca	(Zr,Ti)	(Al,Fe,Si)	O	12.56	3.94	1.94
Berzeliite	NaCa	Mg	As	O	12.37	4.08	1.707
Cryolithionite	Na	Al	Li	F	12.16	2.77	1.339
Yttrian spessartine where x = 0.3	$(Mn_{3-x}Y_x)$	Al	$(Si_{3-x}Al_x)$	O	11.88	4.22	1.820
Synthetic							
YAG	Y	Al	Al	O	12.01	4.55	1.823
YIG	Y	Fe^{3+}	Fe^{3+}	O	12.38		
	Tb	Al	Al	O	12.07	6.06	1.872
	Dy	Al	Al	O	12.04	6.19	1.867
	Ho	Al	Al	O	12.01	6.30	1.858
	Er	Al	Al	O	11.98	6.40	1.857
	Tm	Al	Al	O	11.96	6.48	1.855
	Ca	(Zr,Co)	Ge	O	12.54		
	Y	Fe^{2+}	Al	O	12.16		
	Ca	(Ti,Mg)	Ge	O	12.35		
	Y	Ga	Ga	O	12.28		
	Nd	Fe	Fe	O	12.60		
	Sm	Fe	Fe	O	12.53		
	Dy	Fe	Fe	O	12.41		
	Ca	Zr	Fe	O	12.62		

refraction of garnets will be relatively high (Table 12.3), but the three-dimensional polymerization of the cross-linked chains results in an isometric mineral represented by a spherical indicatrix of $r = n$ (see Fig. 11.1).

Following the substitution scheme suggested by Jaffe (1951), Yoder and Keith (1951) synthesized the first pure yttrium aluminum end member garnet (YAG) $Y_3Al_2[AlO_4]_3$. Soon after, numerous analogues of garnet were synthesized (Table 12.3), which have had many commercial applications (see Chapter 9).

All garnets crystallize in space group

$$\underset{(100)}{I\ a}\ \underset{(111)}{3}\ \underset{(110)}{d}$$

and, in addition to the *a* axial glides and diamond glides *d*, they also have 4_1 and 4_3 screw axes alternating through cube faces of the unit cell. Figure 12.8 illustrates many of these symmetry elements and shows particularly well the counterclockwise or sinistral rotation and translation of all polyhedra, $a/4$, around

Figure 12.8. A part of the garnet structure viewed along an a_3 axis. Unit cell is shown centered on a 4_3 left-handed screw axis. Many polyhedra are omitted for clarity. Elevations show that tetrahedra, octahedra and triangular dodecahedra rotate counterclockwise around the central 4_3 screw axis of the drawing, and translate polyhedra one-fourth the length of c with each 90° rotation.

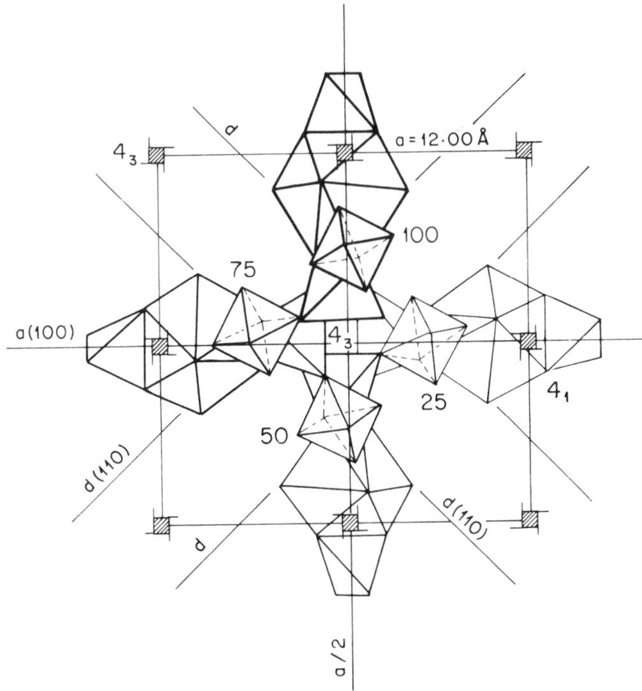

the 4_3 screw axis at the center of the figure. Additional CaO_8 dodecahedra above and below the SiO_4 tetrahedra at the center of the figure are not shown because they would severely clutter up the relations. The a axial glides $a/2$ (100) and less obvious diamond glides $a/4$ and $c/4$ (110) are also illustrated. The inclined cant of the dodecahedral and octahedral axes increases the difficulty of visualizing this elegant but complex structure.

Complex peroxysilicates

Datolite

$4/CaB(OH)[SiO_4] = 4/Ca^{VIII}B^{IV}(OH)^{III}[Si^{IV}①_2^{III}②^{IV}④^{III}]$
Monoclinic (pseudoorthorhombic) $P2_1/a$
$a = 9.62$ Å $b = 7.60$ Å $c = 4.84$ Å $\wedge\beta = 90°09'$
$V_{uc} = 353.79$ Å^3
$M/\rho = V_m = 160.03/3.003 = 53.27$ cm^3
Electronegativity: $\Delta\chi_{Ca-O} = 2.5$ (79% ionic bond)
$\Delta\chi_{B-O} = 1.5$ (43% ionic bond)
$\Delta\chi_{Si-O} = 1.7$ (51% ionic bond)
$\Delta\chi_{H-O} = 1.3$ (34% ionic bond)
Integrated $\Delta\chi$ for datolite $= 2.01$ (63% ionic bond)
Hardness: 5.5 Luster: vitreous
Color: white–pale yellow
CN: Si, B^{IV} – tetrahedra
Ca^{VIII} – square antiprism
O^2 – ①,③,④ – triangular
② – tetrahedral
(note ③ = (OH))
A—X distances: Si—O = 1.52–1.69 Å, B—O = 1.44–1.47 Å,
B—(OH) = 1.56 Å
Specific refractivity: $K = 0.2152$
Molecular polarizability: $\alpha_G = 13.64$ Å^3
Optical properties: $\gamma = 1.666$, $\beta = 1.649$, $\alpha = 1.622$
$\gamma - \alpha = 0.044$, biaxial (−), $2V = 75°$,
$\mathbf{Z} \wedge a = 2°$, $\mathbf{Y} = b$, $\mathbf{X} \wedge c = 2°$
Colorless and nonpleochroic (transmitted light)
Optical indicatrix model: oblate triaxial ellipsoid

Discussion. Datolite is a complex peroxy-type nesosilicate and borosilicate at the same time. $[SiO_4]$ and $[BO_3OH]$ tetrahedra are present in equal proportions. Although it is classed as a nesosilicate, this mineral displays aspects of both inosilicate and phyllosilicate polymerization. Study of Figure 12.9 shows tetrahedral chains of alternating $[SiO_4] - [BO_3(OH)]$ undulating parallel to the a axis, but these are cross-linked almost perfectly into $(BSiO_4OH)$ sheets parallel to (001). These sheets are held together by Ca^{2+} ions that occupy VIII-fold square antiprismatic voids in the sheets. Octahedral CN VI voids are located on corners and in the center of the primitive unit cell. The single monoclinic symmetry axis is a 2_1 screw axis parallel to the *b* crystal axis, and this is perpendicular to an axial glide *a*/2 parallel to the (010) plane (Fig. 12.9).

Figure 12.9. Polyhedral and packing models of datolite, CaB(OH)[SiO_4], in (001) projection. Note the edge and corner polymerization of polyhedra CaO_8, SiO_4, and BO_4 into sheets parallel to (001).

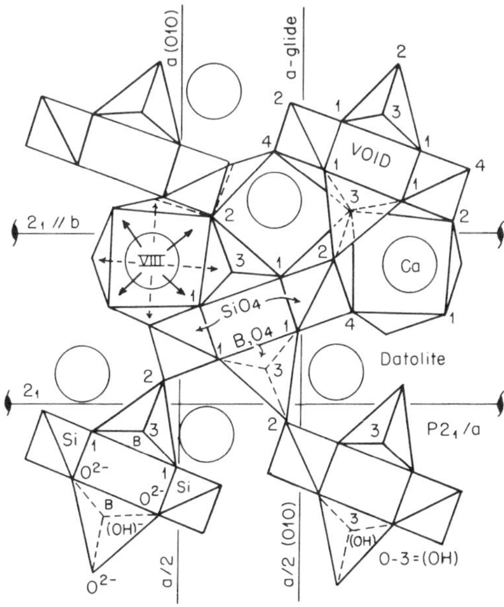

The fairly high birefringence, $\gamma - \alpha = 0.044$, and the optically negative sign conform to the tenet that minerals with their atoms in planar sheets perpendicular to the c axis will have higher electron density parallel to the sheets than perpendicular to them, and thus will be optically negative with high birefringence (large double refraction).

Isostructural with datolite are herderite, $CaBe(PO_4)F$, and possibly gadolinite, $FeY_2Be_2O_2Si_2O_8$. The Fe^{2+} ions in gadolinite occupy the octahedral voids described for datolite, which are located at the center and corners of the unit cell (Fig. 12.9).

Sillimanite

Orthorhombic *Pbnm*
$4/Al^{IV}O^{II}Si^{IV}[Al^{VI}O_4^{III}]$ or $Al^{VI}[Al^{IV}Si^{IV}O_5^{(II,III)}]$
$a = 7.49$ Å $b = 7.67$ Å $c = 5.77$ Å $V_{uc} = 331.47$ Å^3
$M/\rho = V_m = 162.05/3.246 = 49.92$ cm^3
Electronegativity difference: sillimanite, integrated = 1.9 (59% ionic)
Hardness: 7 Luster: vitreous
Color: white, in needles, radial clusters of needles, and fibers elongated with c
CN: Al^{VI} octahedral and Al^{IV} tetrahedral
Si^{IV} tetrahedral
O^{III} triangular in AlO_4 chains parallel to c
O^{II} in bow ties that connect AlO_4 chains
A—X distances: Al—O (octahedral) = 1.91 Å
Al—O (tetrahedral) = 1.78 Å
Si—O = 1.62 Å
Specific refractivity: $K = 0.205$
Molecular polarizability: $\alpha_G = 13.17$ Å^3
Optical properties: $\gamma = 1.680$ $\beta = 1.660$ $\alpha = 1.659$ $\mathbf{Z} = c$
$\mathbf{Y} = b$ $\mathbf{X} = a$ $\gamma - \alpha = 0.021$ biaxial (+)
$2V = 20°$
Color (transmitted light): colorless, nonpleochroic
Optical indicatrix model: prolate triaxial ellipsoid

Discussion. The maximum polarizability is along c, parallel to and controlled by the direction of the staggering of $[AlO_4]$ chains and Al—O—Si bow ties (Fig. 12.10).

Andalusite

Orthorhombic *Pnnm*
$4/Al^{V}O^{III}Si^{IV}[Al^{VI}O_4^{III}]$
$a = 7.79$ Å $b = 7.90$ Å $c = 5.56$ Å $V_{uc} = 342.16$ Å^3

$M/\rho = V_m = 162.05/3.145 = 51.53\ cm^3$
Electronegativity difference: andalusite,
integrated = 1.9 (59% ionic)
Hardness: 7 Luster: vitreous
Color: white, in prisms of square cross section often enclosing a black cross formed by inclusions in cavities (var. chiastolite)
CN: Al^{V} – triangular dipyramid
Al^{VI} – octahedron
Si^{IV} – tetrahedron
O^{III} – triangular, in chains and shared edges of paired dipyramids that connect $[AlO_4]$ chains

Figure 12.10. Packing model (upper) and polyhedral model (lower) of sillimanite, $Al^{IV}—O—Si[Al^{VI}O_4]$, projected along *c*, (001) projection. $Al^{IV}—O^{II}—Si^{IV}$ bow ties connect with their corners to $[Al^{VI}O_4]$ chains running parallel to *c*, up toward viewer.

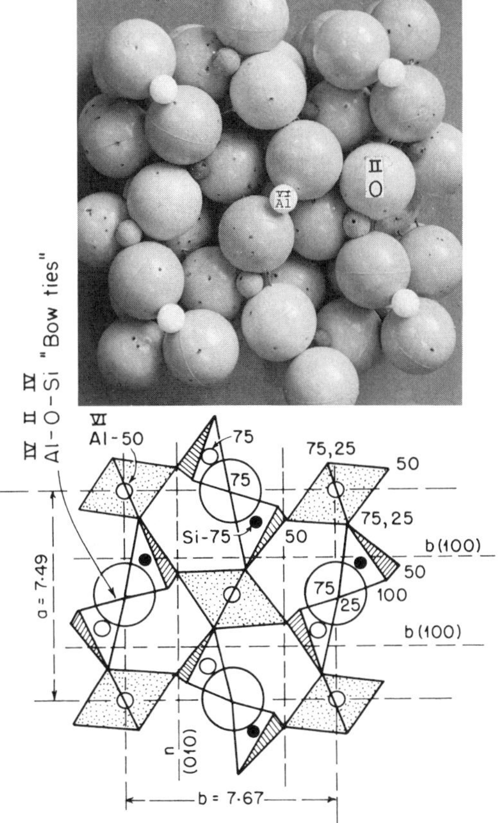

A—X distances: Al—O (dipyramid) = 1.84 Å
Al—O (octahedral chains) = 1.90 (1.83–2.09) Å
Si—O = 1.63 Å
Specific refractivity: $K = 0.2014$
Molecular polarizability: $\alpha_G = 12.94$ Å³
Optical properties: $\gamma = 1.639$ $\beta = 1.635$ $\alpha = 1.627$ $\mathbf{Z} = a$
$\mathbf{Y} = b$ $\mathbf{X} = c$ $\gamma - \alpha = 0.012$ biaxial (−)
2V = 80°
Color (transmitted light): colorless, weak pleochroism
X colorless, **Y** rose, **Z** pale greenish yellow
Optical indicatrix model: oblate triaxial ellipsoid

Discussion. In andalusite, the *minimum* rather than the *maximum* polarizability direction lies along *c*, the chain-axis direction. Evidently the Al triangular dipyramid pairs, each containing three oxygens, which are all coplanar at 0- and 50-levels, around one Al^{V} (Fig. 12.11), mimic the planar $[CO_3]$ groups of common carbonates in which these triangles are oriented in planes perpendicular to *c*. Thus andalusite has the γ and β indices of refraction (**Z** and **Y** vibration directions) in the *a-b* (001) plane, and α (**X** vibration direction) parallel to *c*. Because γ and β are closer together than β and α, the indicatrix model describes an optically negative mineral. It is interesting that appreciable substitution of Fe^{3+} and/or Mn^{3+} ions for Al reverses these optical relations, so this impure andalusite becomes optically positive with orientation as in sillimanite. Evidently Fe^{3+} and Mn^{3+} build covalent orbitals with O^{2-}, along *c*, that are sufficient to reverse the vibration velocities. Because the contrasts of vibration velocity along **Z**, **Y**, and **X** are small, it does not take much to change their relative orientations, hence the optic sign.

Isostructural with

andalusite $Al^{V}O^{III}Si^{IV}[Al^{VI}O_4^{III}]$

are

libethenite $Cu^{V}OP[Cu^{VI}O_3(OH)]$
adamite $Zn^{V}OAs[Zn^{VI}O_3(OH)]$
olivenite $Cu^{V}OAs[Cu^{VI}O_3(OH)]$

Thus all of these minerals retain in their structures the unusual five-coordination polyhedron, a triangular dipyramid, of oxygen atoms around one of their two atoms of Al, Cu, or Zn. The other Cu atom in libethenite occupies the octahedral chain site analogous to that of Al in andalusite and shows an unusual distortion. The octahedron around this Cu atom is markedly prolate. The four equatorial oxygens of the octahedron are drawn in close to the central Cu atom, and the two apical oxygens have moved to larger distances. This distortion suggests that the Cu atom uses a dsp^2 square planar hybrid orbital in bonding to the four oxygens. This crystal chemical feature has been observed in many minerals containing Cu^{2+} coordinated by oxygen atoms (see Chapter 3).

Kyanite

Triclinic $P\bar{1}$
$4/Al^{VI}O^{III}Si^{IV}[Al^{VI}O_3^{III}O^{IV}]$
$a = 7.104$ Å $b = 7.74$ Å $c = 5.57$ Å $V_{uc} = 298.95$ Å^3
$\wedge\alpha = 89.9\text{–}90.5°$ $\wedge\beta = 101.2°$ $\wedge\gamma = 105.4\text{–}106.0°$
$M/\rho = V_m = 162.05/3.60 = 45.01$ cm^3
Electronegativity difference: kyanite,
integrated = 1.9 (59% ionic)
Hardness: Unusual, in that it differs on crystallographic planes; parallel to c on (100), H = 4–5, whereas on (010), H = 7
Luster: vitreous

Figure 12.11. (A) Polyhedral projection on (001) of andalusite, Al^VO^{III}-$SI^{IV}[Al^{VI}O_4]$. Bow ties of sillimanite structure are replaced by paired, irregular, five-coordinated triangular dipyramids containing only Al. These corners connect to $[Al^{VI}O_4]$ chains along c. (B) Same projection as in (A) (001), showing the locations of large voids in the oxygen-packing assemblage. These result from the open packing of the paired CN V polyhedra, and their filling by amorphous to poorly crystalline matter gives rise to the chiastolite crosses that characterize andalusite in many occurrences.

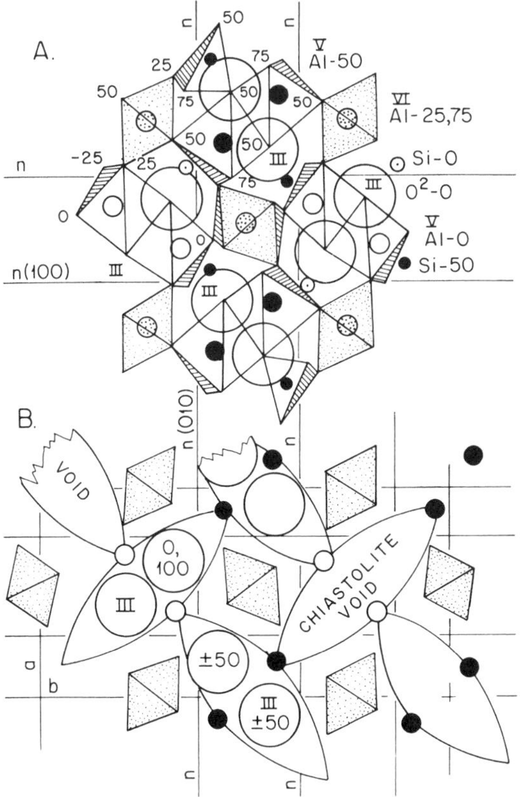

Color: pale to deep blue, less commonly pale green or colorless; in flat elongate bladed crystals showing two or three cleavages
A—X distances: Al—O (all octahedral) = 1.92 Å (1.81–2.00 Å)
Si—O = 1.63 Å
Specific refractivity: $K = 0.200$
Molecular polarizability: $\alpha_G = 12.86$ Å^3
Optical properties: $\gamma = 1.728$ $\beta = 1.721$ $\alpha = 1.713$ $\mathbf{Z} \wedge c = -13°$ on (100) $\mathbf{Y} \wedge b = 30°$ $\mathbf{X}' \wedge a = 1°$ (001)
$\gamma - \alpha = 0.015$ biaxial (−) 2V = 84°
Color (transmitted light): colorless to pale blue, pleochroism nil to weak
Optical indicatrix model: oblate triaxial ellipsoid

Discussion. In kyanite the **Z** (slow) vibration direction is predicted to be parallel to *c*, the direction of staggered [$Al^{VI}O_3^{III}O^{IV}$] chains (Fig. 12.12), but the reason for the angular orientation of −13° (**Z** ∧ *c*) is not apparent. Optical properties of triclinic minerals always present problems.

Comparison of the aluminosilicate polymorphs

The three principal aluminum silicate polymorphs consist chemically of 1 mole each of $Al_2O_3 + SiO_2$, with mass 162.05. Sillimanite is the high-

Figure 12.12. Polyhedral drawing of kyanite viewed along the *b* axis, showing staggered [$Al^{VI}O_3O$] chains parallel to the *c* axis. Note that these chains contain oxygen, O^{IV}, which is in contact with four Al atoms and does not touch Si.

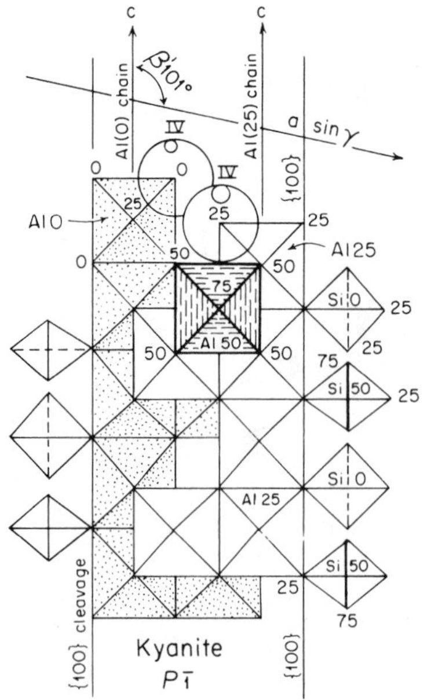

temperature form; kyanite, the high pressure form; and andalusite, the low pressure form (Fig. 12.13). All three polymorphs grow by propagation of $[Al^{VI}O_4]$ octahedral chains that share O—O octahedral edges or bridges along the *c* axis.

In sillimanite these octahedral chains are joined by bow ties of composition Al^{IV}—O^{II}—Si^{IV} that translate along *b* as axial glide *b*/2 (100), and also as diagonal or *n* glides along *a*, as $a/2 + c/2$ (see Fig. 12.10). The *n* glides and Figure 12.4A give the further information that Al^{IV} and Si^{IV} alternate in a perfectly ordered sequence along *c*.

In andalusite the $[AlO_4]$ chains are cross-linked by aluminum in pairs of unusual triangular oxygen dipyramids (see Fig. 12.11) and by silicon in the customary tetrahedron. The linkage is now described as Al^{V}—O^{III}—Si^{IV}. Pairs of these polyhedra at 0- and 50-levels translate as diagonal *n* glides along (100) and (010).

In kyanite, the dense high pressure form, the oxygens are in cubic close packing, so all coordination sites are octahedra and tetrahedra (see Fig. 4.3). Now the octahedral chains parallel to *c* have the composition $[AlO_3^{III}O^{IV}]$ and are cross-linked by a nonchain aluminum octahedron and silicon tetrahedron of composition $Al^{VI}O^{III}Si^{IV}$ (see Fig. 12.12). Although the oxygens are in a cubic close-packed array, the site occupancies and their repeat distances impose a triclinic symmetry on kyanite (Fig. 12.14).

Summing up these differences in structural formulae leads to

Sillimanite	$Al^{IV}O^{II}Si^{IV}[Al^{VI}O_4^{III}]$
Andalusite	$Al^{V}O^{III}Si^{IV}[Al^{VI}O_4^{III}]$
Kyanite	$Al^{VI}O^{III}Si^{IV}[Al^{VI}O_3^{III}O^{IV}]$

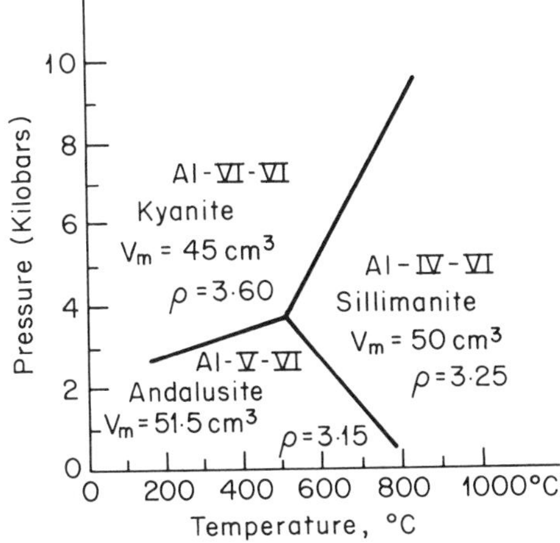

Figure 12.13. *P-T* diagram for stability of Al-silicate polymorphs, after Holdaway (1971).

It is important to underscore the presence, in each polymorph, of *one unique oxygen* as follows:

Sillimanite: O^{II} is nonchain and forms the bow tie knots of Al^{IV}—Si^{IV} pairs (see Fig. 12.10).
Andalusite: O^{III} is also nonchain and joins pairs of Al^{V} to Si^{IV} polyhedra (see Fig. 12.11).
Kyanite: O^{IV} is part of an $[AlO_3O_4^{IV}]$ chain and is unique in that it joins four Al—O octahedra and does not contact Si at all (see Fig. 12.12).

The packing models (see Figs. 12.14, 12.10) show that the oxygens in kyanite are extremely close packed, whereas those in sillimanite are somewhat less so, which might be predicted from their respective Al^{VI}-versus-Al^{IV} coordination sites. In andalusite (photo not shown), the placement of the paired Al^{V}—O polyhedra leads to the presence of large, gaping holes or voids in the structure, and this accounts for the anomaly of higher CN (Al^{V}) and lower density for andalusite, compared with lower CN (Al^{IV}) and higher density for sillimanite. Therefore the slope bounding the stability fields of andalusite and sillimanite on the *P-T* diagram is negative, whereas the boundary curves for kyanite–andalusite and kyanite–sillimanite are positive. The higher temperature forms have the larger volumes and the lower densities (see Fig. 12.13). It is the filling of those gaping voids in andalusite with inclusions of amorphous carbon, poorly crystallized clays, and so on, that leads to the black crosses of the chiastolite variety of andalusite (see Fig. 12.11B).

Electrostatic bond strength (EBS) is satisfied in a different way in each of these polymorphs, which is obvious from the differing coordinations of Al.

In kyanite, EBS increments are all either +1 (Si^{4+}/IV) or 0.5 (Al^{3+}/VI), and

Figure 12.14. Packing model of kyanite showing cubic close-packed oxygens and outline of triclinic unit cell imposed by atom periodicity.

four of the five oxygens of the formula unit contact 1 Si + 2 Al = +2.00. The fifth (unique) oxygen of the chain contacts 4 Al + 0 Si = +2.00.

In sillimanite, EBS relations are more complex. There are increments of +1 (Si), +0.5 (Al^{VI}), and +0.75 (Al^{3+}/IV).

In andalusite, EBS relations become even more complex. There are increments of +1 (Si), +0.5 (Al^{VI}), and +0.6 (Al^{3+}/V).

The disposition of the EBS increments among all five oxygens of each polymorph is as follows:

	Sillimanite	Andalusite	Kyanite
O(1)	+1.75	+1.60	+2.00
O(2)	+2.00	+2.00	+2.00
O(3)	+2.25	+2.10	+2.00
O(4)	+2.25	+2.10	+2.00
O(5)	+1.75	+2.20	+2.00
	+10.00	+10.00	+10.00

Staurolite

Monoclinic, pseudo-orthorhombic $C2/m$
$2/[Fe_2Al_{8.67}O_6(OH)_2][SiO_4]_4$ or $2/[Fe_2Al_9O_7(OH)][SiO_4]_4$
$a = 7.9$ Å $b = 16.7$ Å $c = 5.6$ Å $\beta =$ about 90°
$V_{uc} = 738.18$ Å^3
$M/\rho = V_m = 843.9/3.79 = 222.61$ cm^3
Electronegativity difference: staurolite,
integrated = 1.8 (55% ionic bonding)
Hardness: 7 Luster: vitreous
Color: brown, in prismatic crystals that often show characteristic "sawhorse" penetration twins (or cruciform twins)
CN Fe^{IV} – tetrahedral, a very unusual crystal chemical feature
Al^{VI} – octahedral
Si^{IV} – tetrahedral
O^{IV} – tetrahedral
H – uncertain.
A—X distances: Fe—O = 2.020 Å Al—O = 1.907–1.992 Å
Si—O = 1.641 Å O—O = 2.86–3.25 Å
Specific refractivity: $K = 0.1989$
Molecular polarizability: $\alpha_G = 66.52$ Å^3
Optical properties: $\gamma = 1.761$ $\beta = 1.753$ $\alpha = 1.747$ $\mathbf{Z} = c$
$\mathbf{Y} = a$ $\mathbf{X} = b$ $\gamma - \alpha = 0.014$ biaxial (+)
2V= 80–90°
Color (transmitted light): characteristically golden yellow

Pleochroism: weak to moderate in shades of yellow with normal absorption scheme **Z** > **Y** > **X**; typically **Z** golden yellow, **Y** pale yellow, **X** colorless
Optical indicatrix model: slightly prolate triaxial ellipsoid

Discussion. Staurolite is a common mineral that frequently occurs with kyanite in Al-rich parent rocks that have been metamorphosed and folded under conditions of considerable depth in the earth with moderately high temperature, on the order of 15–30 km and 400–600°C.

The structure of this interesting silicate is shown in Figure 12.15 in (001) projection. Staurolite contains kyanite-like regions or zones that alternate with wüstite- and diaspore-like zones. In Figure 12.15, AlO_6 octahedral groups attached to one another and to SiO_4 tetrahedra represent the kyanite-like zones. The tetrahedral Fe associated with O in staurolite, however, does not mimic the octahedral Fe found in wüstite and most other silicates. The formula of staurolite may be represented chemically as 2 wüstite + 1 diaspore + 4 kyanite, as follows:

$$(FeO)_2AlO(OH)[Al_2O(SiO_4)]_4$$

The location of Fe in a tetrahedral site next to an empty octahedral site results from a better bond strength distribution rather than from packing considerations. Because CN is expected to increase with pressure or depth in the earth, under conditions where staurolite grows, the uncommonly low CN IV for Fe^{2+} is indeed an enigma. Zn atoms commonly replace Fe in tetrahedra of staurolite, which perhaps indicates that sp^3 hybrid bond orbitals favored by Zn—O play an important role in bonding Fe and Zn to four O atoms. Optically, staurolite has the same low birefringence as kyanite, perhaps confirming a structural relationship. With an addition of Fe, the indices of refraction of kyanite would increase to values very close to those of staurolite.

The chemistry and structure of staurolite have yet to be definitively solved. Although the unit cell is known to contain forty-eight oxygen atoms, the H content varies and appears to be an important crystal chemical component that stabilizes growth of this mineral.

Chloritoid

Monoclinic $C2/c$

$$4/[Fe_2Al(OH)_4O_2]Si_2[\underset{(1)}{O_6}\underset{(2)}{Al_3}\underset{(3)}{O_2}\square]$$

$a = 9.52$ Å $b = 5.47$ Å $c = 18.19$ Å $\beta = 101.6°$
$V_{uc} = 927.90$ Å^3
$M/\rho = V_m = 503.88/3.606 = 139.73$ cm^3
Electronegativity difference: chloritoid, integrated = 1.95 (61% ionic bonding)
Hardness: 6.5 Luster: dull or vitreous

Figure 12.15. Polyhedral (001) projection of staurolite, emphasizing isolated SiO_4 tetrahedra, joined to AlO_6 in octahedral chains simulating kyanite-like parts of structure. Noteworthy is the occurrence of Fe^{2+} (black spheres) and Zn^{2+} (spotted spheres) in expanded tetrahedra.

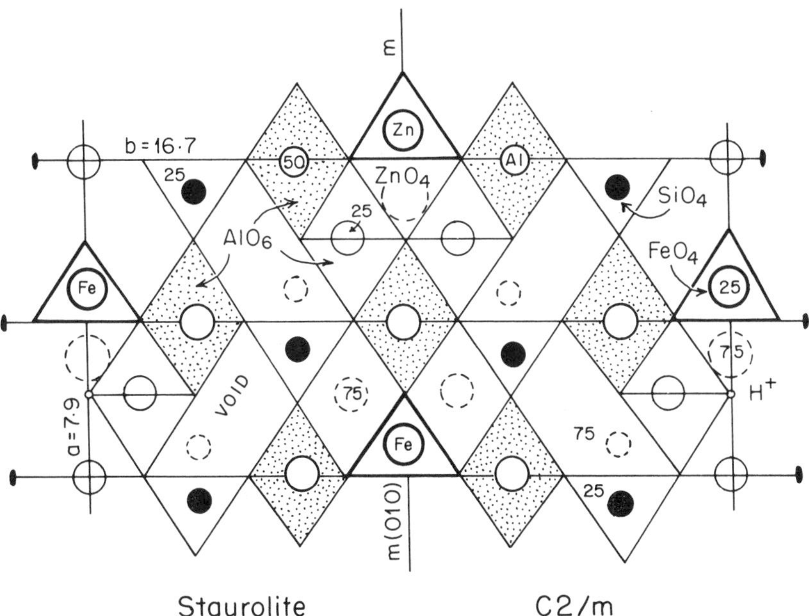

Color: light green, brown, black
CN: Fe^{VI} – distorted octahedron
Al^{VI} – distorted octahedron
Si^{IV} – tetrahedron
A—X distances: Fe—O = 2.22 Å Al—O = 1.91–1.96 Å
Si—O = 1.630 Å O—O = 2.59–3.23 Å with a mean of 2.74 Å
Specific refractivity: $K = 0.2017$
Molecular polarizability: $\alpha_G = 40.30$ Å^3
Optical properties: $\gamma = 1.732$ $\beta = 1.726$ $\alpha = 1.725$ $\mathbf{Z} \wedge c = 15°$
$\mathbf{Y} \wedge a = 15°$ $\mathbf{X} = b$ $\gamma - \alpha = 0.007$ biaxial (+) 2V = 45°
Optical color (transmitted light): pale bluish green

Figure 12.16. Polyhedral (left) and packing model (right) of chloritoid in (100) projection illustrates c axial glide $c/2$ (010) and layered nature of structure. Arrows locate H^+ bonds.

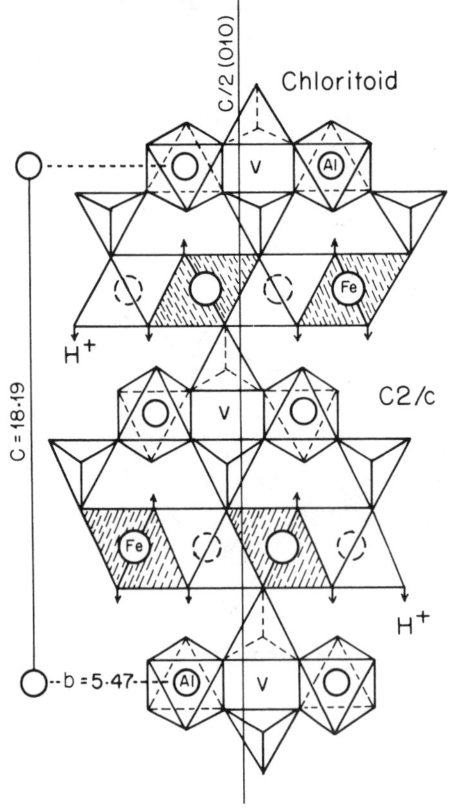

Pleochroism: marked with **Y** = blue, **X** = light green, **Z** = pale yellow, with absorption scheme **Y** > **X** > **Z**, an unusual pattern; lamellar twinning common

Optical indicatrix model: slightly prolate triaxial ellipsoid

Discussion. Chloritoid is an interesting mineral that occurs as slender hexagonal prisms in argillaceous parent rocks, such as shale, that have been thermally

Figure 12.17. Packing model (upper) and polyhedral drawing (lower) of chloritoid in (001) projection emphasizing brucite-like layers of $Fe_2AlO_2(OH)_4$. Compare with Figure 12.16.

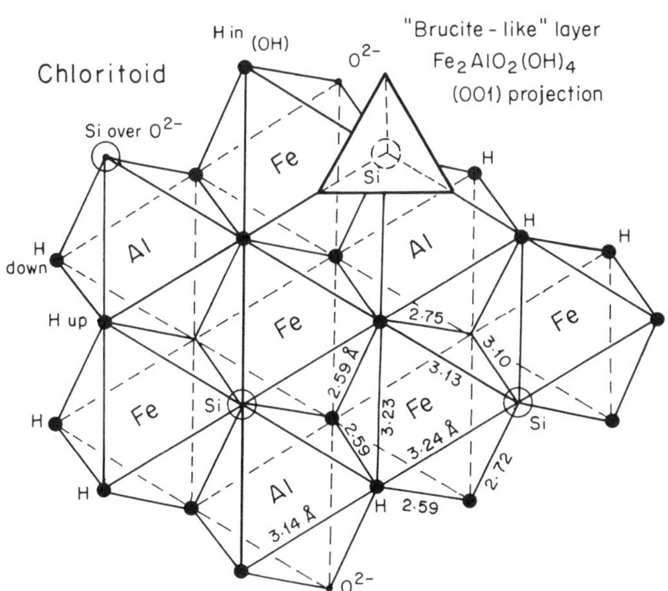

metamorphosed by proximity to a magma emplaced at a shallow level in the earth's crust. Because of its pronounced basal (001) cleavage, it has often been classed with micas or phyllosilicates. Although it possesses an octahedral layer similar to that of the micas, there is no tetrahedral layer. It is replaced by a corundum-like or spinel-like layer. From the formula given earlier and from Figures 12.16 and 12.17, it is seen that chloritoid may best be visualized as consisting of a brucite-like layer, $Fe_2Al(OH)_4O_2$, linked by tetrahedral Si_2 to a spinel- or corundum-like layer, $O_6Al_3O_2\square$. Pairs of upward- and downward-pointing SiO_4 tetrahedra along (010) are translated $c/2$ by an axial glide plane (Fig. 12.16). An (001) projection of the octahedral or brucite-like layer (Fig. 12.17) shows the edge sharing sheet of distorted Fe and Al octahedra, with the black dots representing the H^+ ions or protons that point or orient in up-and-down fashion along *c*. See also the arrows illustrating this feature in Figure 12.16.

Because spinel is isotropic and corundum nearly so, it is the brucite-like layer that controls the optical properties, and more specifically the preferred orientations of the H^+ ions or protons along *c*. Brucite itself, $Mg(OH)_2$, is optically positive, as are some chlorites, as a function of the marked orientation of the H^+ ions perpendicular to (0001) or (001) layers (see Fig. 11.8). As with the optically positive chlorites, the polarizability induced by H—O—H--- along *c* exceeds that along the layer, and chloritoid is optically positive.

Bibliography

Bragg, L., Claringbull, G. F., and Taylor, W. H. (1965). *Crystal structures of minerals. The crystalline state,* Vol. IV. Cornell Univ. Press, Ithaca, N.Y.

Belov, N. V., Belova, E. N., Andrianova, N. N., and Smirova, R. F. (1951). Parameters of olivine. *Dokl. Akad. Nauk SSSR,* 81: 399.

Bertaut, E. F., and Forrat, F. (1965). Structure des ferrites ferromagnétiques des terres rares, *C.R. Acad. Sci. Paris,* 242: 382–4.

Bertaut, E. F., Forrat, F., Herpin, A., and Mériel, P. (1956). Etude par diffraction de neutrons du grenat ferromagnétique, *C.R. Acad. Sci. Paris,* 243: 898–901.

Brown, G. E., Jr. (1982). Olivines and silicate spinels. *Mineral. Soc. Am. Rev. Mineral.,* 5: 275–381.

Burnham, C. W. (1963a). Refinement of the structure of sillimanite. *Z. Krist.,* 118: 127–148.

(1963b). Refinement of the crystal structure of kyanite. *Z. Krist.,* 118: 337–60.

Burnham, C. W., and Buerger, M. J. (1961). Refinement of the crystal structure of andalusite. *Z. Krist.,* 115: 269–90.

Cruikshank, D. W. J., Lynton, H., and Barclay, G. A. (1962). A reinvestigation of the crystal structure of thortveitite, $Sc_2Si_2O_7$. *Acta Crystallogr.,* 15: *Pt. 5,* 549.

Deganello, S. (1978). The crystal structure of triphylite after oxidation at 670°C. *Neues Jahrb. Mineral.,* 128–34.

Fang, J. H., and Newnham, R. E. (1965). The crystal structure of sinhalite. *Min. Mag.,* 35: 196–9.

Farrell, E. F., Fang, J. H., and Newnham, R. E. (1963). Refinement of the chrysoberyl structure. *Am. Mineral.,* 48: 804.

Finger, L. W. (1974). Refinement of the crystal structure of zircon. Carnegie Inst. Yearbook 73, Geophys. Lab., Washington, D.C., pp. 544–7.

Finger, L. W., and Rapp, G. R., Jr. (1970). Refinement of the crystal structure of triphylite. Carnegie Inst. Yearbook 68, Geophys. Lab., Washington, D.C., pp. 290–5.

Fleischer, M., Wilcox, R. E., and Matzko, J. J. (1984). Microscopic determination of the nonopaque minerals. U.S. Geol. Survey Bull. 1627 [revision of U.S.G.S. Bull. 848 of E. S. Larsen, Jr. and H. Berman (1934)], U.S. Department of the Interior, Geol. Survey, Washington, D.C.

Fuchs, L. H., and Hoekstra, H. R. (1959). The preparation and properties of uranium (IV) silicate. *Am. Mineral.,* 44: 1057–63.

Geller, S. (1967). Crystal chemistry of the garnets. *Z. Krist.,* 125: 1–47.

Geller, S., and Durand, J. L. (1960). Refinement of the structure of $LiMnPO_4$. *Acta Crystallogr.,* 13: *Pt. 4,* 325–31.

Halferdahl, L. B. (1957). Chloritoid. Carnegie Inst. Yearbook 56, Geophys. Lab., Washington, D.C., pp. 225–8.

(1961). Chloritoid: its composition, X-ray, and optical properties, stability, and occurrence. *J. Petrol.,* 2: 49–135.

Hanscom, R. H. (1975). Refinement of the crystal structure of monoclinic chloritoid. *Acta Crystallogr., Sect. B,* 31: 780–4.

Harrison, G. W., and Brindley, G. W. (1957). The crystal structure of chloritoid. *Acta Crystallogr.,* 10: *Pt. 1,* 77–82.

Hey, J. S., and Taylor, W. H. (1931). The coordination number of aluminum in the alumino-silicates. *Z. Krist.,* 80: 428.

Holdaway, M. J. (1971). Stability of andalusite and the aluminum silicate phase diagram. *Am. J. Sci.,* 271: 97–131.

Holdaway, M. J., Dutrow, B. L., Burthwick, J., Shore, P., and Harmon, R. S. (1986). H-content of staurolite as determined by H-extraction line and ion microprobe. *Am. Mineral.,* 71: 1135–41.

Holdaway, M. J., Dutrow, B. L., and Shore, P. (1986). A model for the crystal chemistry of staurolite. *Am. Mineral.,* 71: 1142–59.

Ito, T., and Mori, H. (1953). The crystal structure of datolite. *Acta Crystallogr.,* 6: 24.

Jaffe, H. W. (1947). Reexamination of sphene (titanite). *Am. Mineral.,* 32: 637–42.

(1951). The role of yttrium and other minor elements in the garnet group. *Am. Mineral.,* 36: 133–55.

Kristanovic′, I. R. (1958). Redetermination of the oxygen parameter in zircon. *Acta Crystallogr.,* 11: *Pt. 12,* 896–7.

Liebau, F. (1985). *Structural chemistry of silicates.* Springer-Verlag, New York.

Meagher, E. P. (1982). Silicate garnets. *Mineral. Soc. Am. Rev. Mineral.,* 5: 25–66.

Menzer, G. (1926). Die Kristallstruktur von Granat. *Z. Krist.,* 63: 157–8.

Náray-Szabó, St. v., and Sasvari, K. (1958). On the structure of staurolite, $HFe_2Al_9Si_{14}O_{24}$. *Acta Crystallogr.,* 11: *Pt. 12,* 862–5.

Novak, G., and Gibbs, G. V. (1971). The crystal chemistry of the silicate garnets. *Am. Mineral.,* 56: 791–825.

Ribbe, P. H. (1982a). Kyanite, andalusite and other aluminum silicates. *Mineral. Soc. Am. Rev. Mineral.,* 5: 189–214.

(1982b). Staurolite. *Mineral. Soc. Am. Rev. Mineral.,* 5: 171–88.

Ed. (1982c). *Mineral. Soc. Am. Rev. Mineral.,* 5: 1–450.

Robinson, K., Gibbs, G. V., and Ribbe, P. H. (1971). The structure of zircon: a comparison with garnet. *Am. Mineral.,* 56: 782–90.

Sahama, Th. G. (1946). On the chemistry of the mineral, titanite. *Bull. Comm. Gêol. Finlande,* 24: 88–118.

Smith, J. V. (1968). The crystal structure of staurolite. *Am. Mineral.,* 53: 1139–55.

Speer, J. A. (1982). Zircon. *Mineral. Soc. Am. Rev. Mineral.,* 5: 67–135.

Speer, J. A., and Gibbs, G. V. (1976). The crystal structure of synthetic titanite, $CaTiSiO_4$, and the domain textures of natural titanites. *Am. Mineral.,* 61: 238–47.

Stieff, L. R., Stern, T. W., and Sherwood, A. M. (1956). Coffinite, a uranous silicate with hydroxyl substitution: a new mineral. *Am. Mineral.,* 41: 675–88.

Taylor, W. H., and Jackson, W. W. (1928). The structure of cyanite, Al_2SiO_5. *Proc. Roy. Soc. London A,* 119: 132.

Winchell, A. N., and Winchell, H. W. (1964). *The microscopical characters of artificial inorganic solid substances: optical properties of artificial minerals.* Academic Press, New York.

Wyckoff, R. W. G. (1963–8). *Crystal structures,* Vols. 1–4, 2d ed. Wiley, New York.

Wyckoff, R. W. G., and Hendricks, S. B. (1927). The crystal structure of zircon and criteria for the special positions in tetragonal space groups. *Z. Krist.,* 66: 73.

Yoder, H. S., Jr., and Keith, M. L. (1951). Complete substitution of aluminum for silicon: the system $3\,MnO \cdot 3SiO_2$—$3Y_2O_3 \cdot 5Al_2O_3$. *Am. Mineral.,* 36: 519–33.

Zachariasen, W. H. (1930). The crystal structure of thortveitite $Sc_2Si_2O_7$. *Z. Krist.,* 73: 1.

13 Sorosilicates – dumbbell silicates – pyrosilicates

Sorosilicates (Greek *σορος* = coupled) generally consist of a close packing of $[Si_2O_7]$ dumbbells. In different sorosilicates, the SiO_4 halves of each dumbbell may be oriented in parallel or in opposition, or be canted at different angles: for example, åkermanite, $Ca_2Mg[Si_2O_7]$, parallel (see Fig. 13.3); thortveitite, $Sc_2[Si_2O_7]$, opposed; and lawsonite, $CaAl_2(OH)_2[Si_2O_7]H_2O$, angled (Fig. 13.1).

All sorosilicates have one bridging oxygen, which rarely is in contact with more than the two Si^{4+} ions that exactly satisfy its bond strength. In lawsonite, contact with an additional Ca^{2+} ion overcharges the bridging oxygen, ① (see the following section). The close packing of dumbbells is favored in high pressure environments: β-$Mg_2SiO_4 \times 2 = Mg_4[Si_2O_7]O$ is favored at 400 km in the low velocity zone of the earth's mantle, and lawsonite is favored in subduction zone environments with high tectonic pressure. Contradictory data, however, are supplied by the relatively high T, low P occurrences of melilite in alkalic magmatic rocks.

Lawsonite

$4/Ca^{V}Al_2^{VI}(OH)_2^{II}[Si_2^{IV}O_7^{III}]H_2O$

Orthorhombic: $\underset{}{C}\ \underset{(100)}{c}\ \underset{(010)}{m}\ \underset{(001)}{m}$

$a = 8.78$ Å $b = 5.85$ Å $c = 13.13$ Å $V_{uc} = 674.40$ Å^2
$M/\rho = V_m = 314.24/3.094 = 101.56\ cm^3$
Electronegativity: $\Delta\chi_{Ca-O} = 2.5$, $\Delta\chi_{Al-O} = 2.0$, $\Delta\chi_{Si-O} = 1.7$
Integrated $\Delta\chi$ for lawsonite = 1.9 (59% ionic bonding)
Hardness: 6 Luster: vitreous
Color: white
CN Ca^V – tetragonal pyramid ($=\frac{1}{2}$ octahedron)
Al^{VI} – octahedron
Si^{IV} – tetrahedron

A—X distances: Ca—O = 2.36–2.53 Å, Al—O = 1.92–1.95 Å
Si—O = 1.65–1.69 Å
Specific refractivity: $K = 0.218$
Polarizability: $\alpha_G = 27.16$ Å^3
Optical properties: $\gamma = 1.685$ $\beta = 1.674$ $\alpha = 1.665$
$\gamma - \alpha = 0.20$ optically positive 2V = 85°
X = b **Y** = a **Z** = c
Optical indicatrix model: prolate triaxial ellipsoid

Discussion. Assignment of crystal axes as *a*, *b*, and *c* is arbitrary in orthorhombic minerals, and the *b* and *c* axes of lawsonite are interchanged by different investigators. Here we choose the long *c* axis in place of a long *b* axis. This leads to infinite [$AlO_3(OH)$] octahedral chains (Fig. 13.1; also see Fig.

Figure 13.1. Packing model (upper) and polyhedral drawing (lower) of lawsonite, 4/$CaAl_2(OH)_2[Si_2O_7]\cdot H_2O$ with [Si_2O_7] tetrahedral groups rotated on *c* axis (turned toward viewer) for clarity. Note arrows locating orientation of H^+ ions. Plane of projection is (010).

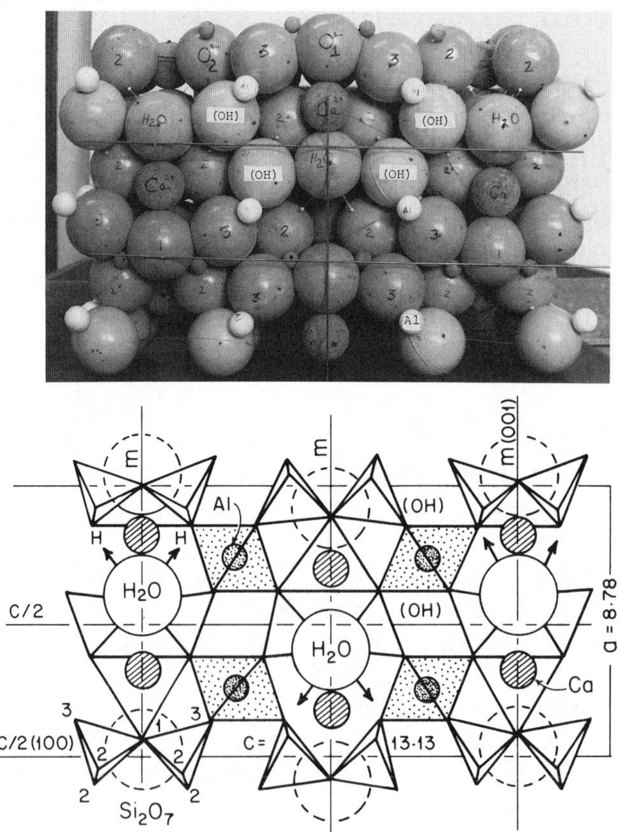

12.4C) parallel to the *b* axis, as in epidote, and space group *Ccmm*. These chains of $[AlO_3(OH)]$ tetrahedra are cross-linked to $[Si_2O_7]$ groups, resulting in the rare CN V for Ca^{2+}. If the Ca is considered to be in octahedral coordination using one apical oxygen ① and one H_2O along with 4 × ② oxygens (Fig. 13.1), the total bond strength to 7 O and 2 (OH) is 15.665 or 0.335 underbonded. If the H_2O is considered to be extraneous, the CN V polyhedron is a tetragonal pyramid made up of one apical ① oxygen centered over a four-oxygen basal square. The bond strength of Ca then increases from $2^+/VI = 0.333$ to $2^+/V = 0.400$, and the total bond strength to 7 O + 2 (OH) = 16.00. From Figure 13.1, it can be seen, with some study, that

①	touches	Ca^{2+}/V, Si^{4+}/IV, $Si^{4+}/IV = 2.40 \times 1$
②	touches	Ca^{2+}/V, Si^{4+}/IV, $Al^{3+}/VI = 1.90 \times 4$
③	touches	Al^{3+}/VI, Al^{3+}/VI, $Si^{4+}/IV = 2.00 \times 2$
(OH)	touches	Al^{3+}/VI, $Al^{3+}/VI = 1.00 \times 2$

Now noting that in $[Si_2O_7](OH)_2$ there are one ①, four ②, two ③, and two (OH), multiplying out bond strength to each anion in proportion to their stoichiometry by 1, 4, 2, and 2 yields exactly 16.00+ for 16.00−. The H_2O and (OH) do, however, contribute to the bonding. It is apparent in Figure 13.1 that each H_2O orients its two H^+ protons toward the underbonded ② anion, and each (OH) also orients its single H^+ proton toward underbonded ② oxygens. Perhaps more striking is the way in which H_2 proton pairs on each H_2O preserve the *c*/2 axial glide planes parallel to (100) by orienting in opposite directions across the *c* glide planes. In the photograph and in the polyhedral drawing, the reader will better be able to visualize the structure by understanding that each ${}^{H}\backslash_{H_2O}/{}^{H}$ strip parallel to the *c* axis lies directly over ${}^{3}\backslash_{1}/{}^{3}$ of an $[Si_2O_7]$ group, and that each (OH) anion forming a corner of an $[AlO_3(OH)]$ octahedron directly overlies oxygen ③ of an $[Si_2O_7]$ group. A cardinal rule of crystal chemistry is obeyed here, in that (OH) anions are never in contact with or coordinate Si^{4+} ions, because a large electrostatic overcharge would result.

That the γ index of refraction and the **Z** vibration direction (slow velocity) run parallel to *c* confirms that electron density is highest along the string of $[Si_2O_7]$ groups that form a lineation parallel to this crystal axis (Fig. 13.2). This increase in electron density would presumably be enhanced by $d\pi$ covalent Si—O bonds, in addition to the sp^3 hybrid component of $Si—O_4$ bonds. The intermediate velocity **Y** and the β index of refraction parallel to *a* are associated with chains of $O_3—Ca—H_2O$ oriented in staggered rows parallel to the mirror planes (001). The association of the fast vibration direction **X** and its low index of refraction α parallel to *b* and the octahedral chains $AlO(OH)_3$ is contrary to its effect in sillimanite, where **Z** is parallel to *c* and to octahedral chains of $[AlO_6]$. In lawsonite, the orientation of the H^+ protons into a plane near (010) enhances the indices of refraction in this plane that contains β and α values and overcomes the effects of shared-edge shortening and high index of refraction parallel to the *b* chain axis.

Lawsonite occurs in low temperature regional metamorphic environments often associated with high tectonic pressures resulting from subduction of oceanic crustal plates. It is here a common component of blueschists that contain low T, high P minerals, such as glaucophane (a blue Al^{VI}-rich amphibole; see Chapter 15) and aragonite, the high pressure polymorph of $CaCO_3$ (see Chapter 19).

Åkermanite

$Ca_2MgSi_2O_7$
Tetragonal $P\bar{4}2_1m$
$a = 7.82$ Å $c = 5.02$ Å $V_{uc} = 306.98$ Å^3
$M/\rho = V_m = 272.66/2.95 = 92.43$ cm^3
Electronegativity: $\Delta\chi_{Ca-O} = 2.5$ $\Delta\chi_{Mg-O} = 2.0$ $\Delta\chi_{Si-O} = 1.7$
Integrated $\Delta\chi$ for melilite = 2.5 (79% ionic bond)
Hardness: 5 Luster: vitreous
Color: pale yellow–brown
CN polyhedra: Ca^{VIII} – very distorted square antiprism
Mg^{IV} – fairly regular tetrahedron
Si^{IV} – tetrahedron
A—X distances: Ca—O = 2.4(4), 2.7(4) Å Mg—O = 1.87 Å
Si—O = 1.63 Å

Figure 13.2. Oxygen anion packing drawing, (010) projection, of lawsonite emphasizing the linear orientation of $[Si_2O_7]$ groups and $d\pi$ bonds stacked parallel to c, the direction of the slow velocity vibration vector **Z**, along which is measured the maximum index of refraction γ.

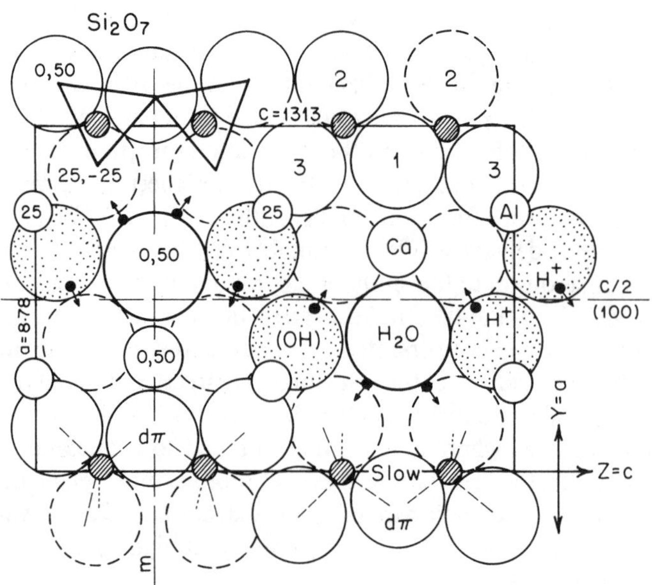

Specific refractivity: $K = 0.2145$
Molecular polarizability: $\alpha_G = 23.18$ Å^3
Optical properties: $\omega = 1.630$ $\epsilon = 1.638$ $\epsilon - \omega = 0.008$
uniaxial (+), colorless and nonpleochroic in transmitted light
Indicatrix: prolate spheroid of revolution, approaching a sphere

Discussion. Åkermanite is a member of the melilite mineral group that consists of $Ca_2Mg[Si_2O_7]$, åkermanite, and $Ca_2Al[AlSiO_7]$, gehlenite. They form a solid solution series and crystallize in magmatic rocks at high temperatures. Noteworthy is the occurrence of the Mg^{2+} ion in tetrahedral coordination (CN IV), indicative of a high temperature occurrence, in contrast with its dodecahedral (CN VIII) occurrence in pyrope garnet, a high pressure mineral. In most silicate and oxide minerals, Mg^{2+} is in octahedral (CN VI) coordination.

The space group $P\bar{4}2_1m$ is illustrated in Figure 13.3, where the c inversion axis is shown by the opposed orientation of alternating pairs of $[Si_2O_7]$ groups. Each pair of $[Si_2O_7]$ groups is bisected by a mirror plane $m(110)$, and 2_1 screw axes pass through the tetrahedral $[Si_2O_7]$ group corners. The presence of a glide plane (110)/2 is unusual. It reflects $[Si_2O_7]$ groups $\frac{1}{2}$ (110) and translates them (110)/2. Note that mirror planes, and screw axes as well, need not pass through the center of the crystal. Here they do not (see also the olivine structure, Figure 6.7, where they also do not).

The EBS solution conforms with Pauling Rule 2C (see Chapter 5), and is interesting because it illustrates the averaging of unequal bond strength of three different oxygens over the entire $[Si_2O_7]$ group. Because each $[Si_2O_7]$ group is made up of two oxygens O-1, four oxygens O-2, and one oxygen O-3, the bond strength distribution is as follows (Fig. 13.3):

$$\begin{aligned} \text{O-1} = \text{EBS} = 1.75 \times 2 &= \quad 3.50 \\ \text{O-2} = \text{EBS} = 2.00 \times 4 &= \quad 8.00 \\ \text{O-3} = \text{EBS} = 2.50 \times 1 &= \quad \underline{2.50} \\ &\quad +14.00 \\ O^{2-} \times 7 &= -14.00 \end{aligned}$$

Thus, the unequal bond strength must be averaged over the formula ratio or stoichiometry of the three different oxygen ions.

Although the structure has been described as a sheet-like structure held together by Ca^{2+} ions, this does not agree with refractivity data. Structures with sheets perpendicular to the c axis should be optically negative with high birefringence ($\omega \gg \epsilon$).

Most minerals of the melilite family have low birefringence, and some are nearly isotropic. This agrees better with the idea that the structure closely approximates a tetrahedral framework. The $[MgO_4]$ tetrahedron of åkermanite (center of Fig. 13.3) shares all of its four corners with $[SiO_4]$ tetrahedra of $[Si_2O_7]$ groups. Three of every four corners of $[Si_2O_7]$ groups are shared with other tetrahedra. Only the O-1 or apical oxygens are not shared with other tetrahedra. Important to the comprehension of the structure is the stacking of the $[Si_2O_7]$ groups along the *c* axis (vertical in Fig. 13.3). All $[Si_2O_7]$ groups in a single column parallel to *c* point in a common direction. All such groups are oriented with their apical O-1 oxygens up *c* in one column, and alternate with those that have apical O-1 oxygens pointing down *c*. $[Si_2O_7]$ groups turned at 90° to one another are schematically oriented as in Figure 13.3.

Figure 13.3. Packing model (upper) and polyhedral drawing (lower) of the structure of åkermanite, $Ca_2Mg[Si_2O_7]$, and EBS distribution around oxygens O-1, O-2, and O-3. Unusual feature is MgO_4 tetrahedron formed by linkage of corners of four $[Si_2O_7]$ groups.

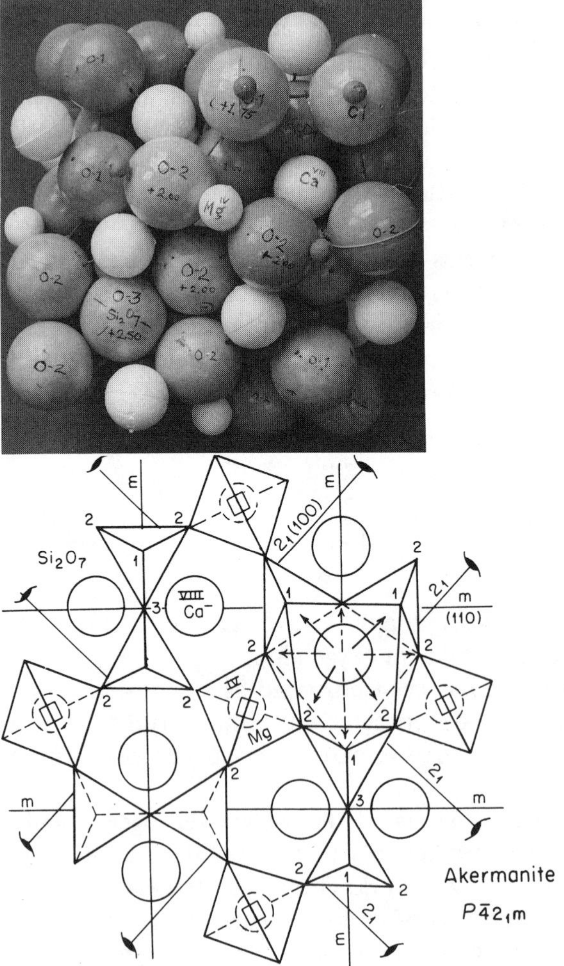

Minerals isostructural with

åkermanite, $Ca_2Mg[Si_2O_7]$

include

gehlenite, $Ca_2Al[Si,AlO_7]$
hardystonite, $Ca_2Zn[Si_2O_7]$
meliphanite, $(Ca,Na)_2Be[(Si,Al)_2(F,O)_7]$

Synthetic analogues include

$Ca_2Be[Si_2O_7]$
$Ca_2Co[Si_2O_7]$
$Sr_2Mg[Si_2O_7]$
$Sr_2Fe[Si_2O_7]$
$Ba_2Mn[Si_2O_7]$

Epidote

Monoclinic $P2_1/m$
$2/Ca_2^{VIII}Fe^{3+VI}O^{III}Al_2^{VI}(OH)^{III}[Si^{IV}O_4^{III}][Si_2^{IV}O_7^{IV,II}]$
$a = 8.96$ Å $b = 5.63$ Å $c = 10.30$ Å $\beta = 115.4°$
$V_{uc} = 469.35$ Å^3
$M/\rho = V_m = 483.26/3.42 = 141.30$ cm^3
Electronegativity differences: Ca—0 = 2.5 Fe—O = 1.7
Al—O = 2.0 Si—O = 1.7
H—O = 1.4
Epidote integrated = 1.8 (55% ionic bonding)
Hardness: 7 Luster: vitreous
Color: pistachio green
CN: Ca^{VII} or VIII – very irregular polyhedron
Al^{VI} – octahedron
Fe^{VI} – octahedron
Si^{IV} – tetrahedron
A—X distances: Ca—O = 2.20–3.00 Å Fe—O = 1.92–2.22 Å
Al—O = 1.85–1.95 Å Si—O = 1.51–1.73 Å
O—O = 2.54–3.28 Å
Specific refractivity: $K = 0.224$
Molecular polarizability: $\alpha_G = 42.84$ Å
Optical properties: $\gamma = 1.787$ $\beta = 1.768$ $\alpha = 1.740$
$\mathbf{X} \wedge c = -10°$ $\mathbf{Y} = b$ $\mathbf{Z} \wedge a = 30°$
$\gamma - \alpha = 0.047$ 2V = 80 – 90° (–)
Color (transmitted light): pale yellow-green, weakly pleochroic
Optical indicatrix model: oblate triaxial ellipsoid

Discussion. Epidote is neither a nesosilicate nor a sorosilicate, but has characteristics of both. Silicon is located in $[SiO_4]$ tetrahedra, and in $[Si_2O_7]$ dumb-

Figure 13.4. Polyhedral drawing of epidote in (100) projection: small drawing (upper) and (010) projection; large drawing (lower). Note the presence of both $[Si_2O_7]$ and isolated $[SiO_4]$ groups that join $[Al^{VI}O_3OH]$ chains running parallel to *b*.

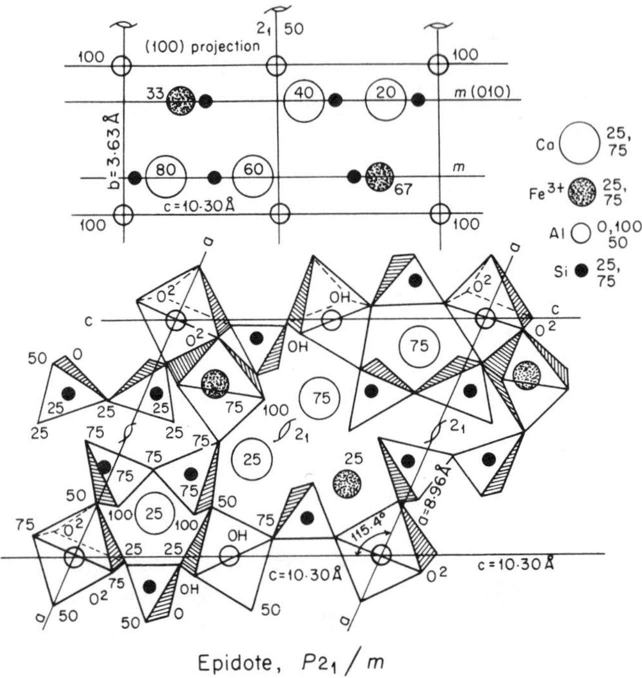

Figure 13.5. Packing model of epidote in the same (010) projection as Figure 13.4. Vertical rod transecting the monoclinic cell outline represents the slow (Z) vibration velocity vector associated with the densest alignment of atoms.

bells each one is isolated from another. Edge sharing chains of octahedra of composition [AlO_4] that run parallel to the *b* axis are linked along *c* by [SiO_4] and [Si_2O_7] groups via corner sharing (Fig. 13.4). These linked [AlO_4], [SiO_4], and [Si_2O_7] groups are, in turn, joined along *a* by FeO_6 octahedra that share edges with [AlO_4] octahedra of chains, and corners with [SiO_4] and [Si_2O_7]. Calcium ions at levels 25 and 75 occupy large holes or voids in this structure.

The lattice is primitive, and 2_1 screw axes parallel to *b* emerge from corners, in the center, and midway along edges of the (010) projection of the unit cell (Fig. 13.4). Mirror planes (010) perpendicular to 2_1 axes, located at levels 25 and 75 are difficult to see in the (010) projection, but are readily apparent in the (100) upper projection of Figure 13.4. The slow-velocity vibration direction **Z** lies at about 30° to the *a* axis and is represented by the rod that emerges at the upper end of Figure 13.5. This direction runs along a line of densely packed atoms that includes Fe^{3+}—O—H—O—Si, and polarizability along this line exceeds that developed along the shared O_2—Al—O_2—Al - - - edges parallel to *b*. Fe^{3+} and H^+ atoms aligned with O^{2-} atoms always induce and propagate large dipole moments and, hence, large polarizability.

Bibliography

Baur, W. H. (1978). Crystal structure refinement of lawsonite. *Am. Mineral.*, 63: 311–5.

Belov, N. V., and Rumanova, I. M. (1953). The crystalline structure of epidote. *Dokl. Akad. Nauk SSSR*, 89: 853.

Horiuchi, H., and Sawamoto, H. (1981). β-Mg_2SiO_4: Single-crystal X-ray diffraction study. *Am. Mineral.*, 66: 568–75.

Ito, T., Morimoto, N., and Sadanaga, N. (1954). The structure of epidote. *Acta Crystallogr.*, 7 *Pt. 1*: 53.

Moore, P. B., and Smith, J. V. (1970). Crystal structure of β-Mg_2SiO_4: Crystal chemical and geophysical implications. *Phys. of the Earth and Planetary Interiors*, 3: 166–77.

Rumanova, I. M., and Skipetrova, T. I. (1959). Crystal structure of lawsonite. *Dokl. Akad. Nauk. SSSR*, 124: 324–7.

Smith, J. V. (1953). Re-examination of the crystal structure of melilite. *Am. Mineral.*, 38 643.

Smyth, J. R. (1987). β-Mg_2SiO_4: A potential host for water in the mantle? *Am. Mineral.* 72: 1051–55.

Warren, B. E. (1930). The structure of melilite $(Ca,Na)_2(Mg,Al)(Si,Al)_2O_7$. *Z. Krist.*, 14: 131.

Wyckoff, R. G. W. (1963–8). *Crystal structures IV*, 2d ed. Wiley, New York, pp. 232–3.

14 Cyclosilicates

Ring structures or cyclosilicates (Greek κυκλος = cycle) are built of SiO_4 tetrahedra joined in the ratio Si_nO_{3n}. They form units or building blocks such as $[Si_3O_9]^{6-}$, $[Si_6O_{18}]^{12-}$, $[Si_{12}O_{30}]^{12-}$, and may be associated with additional rings of $[BO_3]_3^{9-}$. These Si_nO_{3n} groups are not directly joined to one another but are packed together as distinct structural units joined together by other cations. For example, in beryl, hexagonal $[Si_6O_{18}]^{12-}$ rings are joined by Al and Be cations. In benitoite, $[Si_3O_9]^{6-}$ rings, triangular in outline, are linked by Ba and Ti. When the ratio of Si : O is 1 : 3 in a mineral, each SiO_4 tetrahedron will always share two of its corners with adjoining tetrahedra.

This ratio Si_nO_{3n} is not in itself diagnostic of the silicate structural type. The pyroxenes have the same Si_nO_{3n} ratio, but the two shared corners of each tetrahedron are polymerized into infinite chains of composition $[Si_2O_6]^{4-}$ parallel to the *c* crystal axis. Wadeite, $K_2Zr[Si_3O_9]$, for example, is a cyclosilicate built of rings of $[Si_3O_9]^{6-}$, whereas wollastonite, $Ca_3[Si_3O_9]$, is an inosilicate (single chain silicate) built of chains of $[Si_3O_9]^{6-}$ infinitely polymerized parallel to the *b* crystal axis, along which SiO_4 tetrahedra repeat in patterns of three tetrahedra (see Fig. 15.2C).

In some cyclosilicates, rings of three, four, or six tetrahedra form mirror planes (Fig. 14.1), whereas in others, three oxygens of each tetrahedron in a ring form a base capped by one apical oxygen per tetrahedron. All such oxygens point in the same direction, and the structure becomes polar, as in tourmaline (Fig. 14.1). Thus, groups of $[Si_6O_{18}]^{6-}$ in beryl and in tourmaline produce markedly different structures. In beryl, inner rings of Si_6O_6 form mirror planes parallel to the basal plane (0001), whereas in tourmaline rings of Si_6O_{12} are parallel to (0001) and are capped by an O_6 apical ring that induces polarity.

Another feature of the crystal chemistry of cyclosilicates is the contrast between the polyhedral and the packing models. The void space at the ring centers is grossly exaggerated by polyhedral representation. Note, for instance, that the very small void at the center of the Si_3O_9 ring in the packing model is, in the polyhedral analogue, represented by an empty triangle, apparently a large void (Fig. 14.1).

In some ring structures, Si and Al in additional, non-ring tetrahedra cannot be distinguished from those that form the rings, and an added complication in classification arises. In cordierite, the formula $\square Mg_2Al_3[Si_5AlO_{18}]$ is that of a cyclosilicate. If, however, all of the tetrahedral cations are grouped together, $\square Mg_2[Si_5Al_4O_{18}]$, the formula is that of a tektosilicate, in which the cation–anion ratio in tetrahedra is always $(Si,Al)_nO_{2n}$. Accordingly, cordierite is listed as a cyclosilicate by some investigators, and as a tektosilicate by others.

Beryl

$2/\square Al_2Be_3[Si_6①_6②_{12}]$ or $2/Al_2Be_3[Si_6O_{18}]$
Hexagonal: $P6/mcc$
$a = 9.20$ Å $c = 9.15$ Å $V_{uc} = 670.99$ Å^3
$M/\rho = V_m = 537.2/2.66 = 201.95$ cm^3

Figure 14.1. Polyhedral and packing drawings of cyclosilicate rings of composition $[Si_3O_9]^{6-}$, $[Si_6O_{18}]^{12-}$, and $[Si_6O_{18}]^{12-}$ polar, shown perpendicular to and parallel to the c axes.

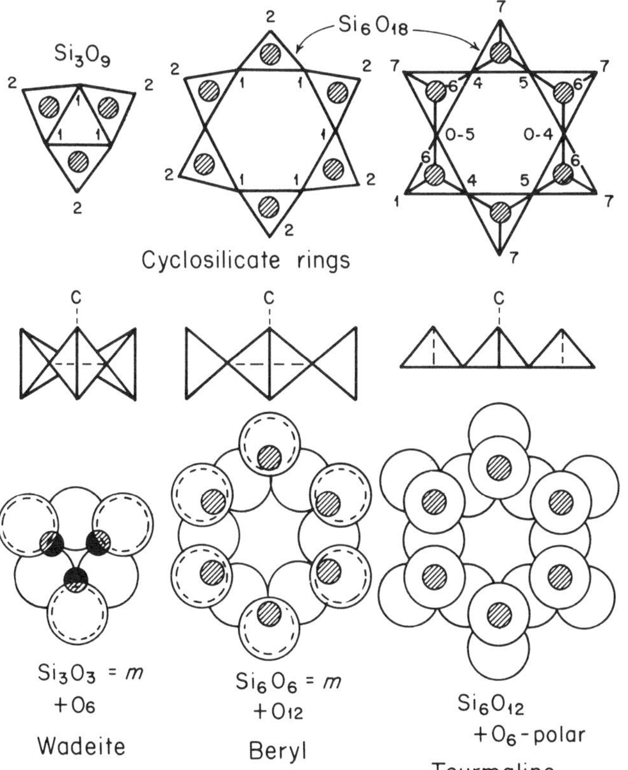

Electronegativity: $\Delta\chi_{Al-O} = 2.0$ $\Delta\chi_{Be-O} = 2.0$ $\Delta\chi_{Si-O} = 1.7$
Integrated $\Delta\chi$ for beryl = 1.8 (55% ionic bonding)
Hardness: 8 Luster: vitreous
Color: blue-green, yellow, white
CN: Al^{VI}-octahedron
Be^{IV}-distorted tetrahedron
Si^{IV}-tetrahedron
Voids □ XII-superposed VI rings rotated 30° from each other form tunnels parallel to the *c* axis
A—X distances (Å): Si—O = 1.594–1.620 Al—O = 1.903
Be—O = 1.660 O—O = 2.36–3.03
Specific refractivity: $K = 0.213$
Polarizability: $\alpha_G = 45.41$ Å
Optical properties: $\omega = 1.568$ $\epsilon = 1.564$ $\omega - \epsilon = 0.004$
uniaxial (−), colorless, nonpleochroic in transmitted light
Optical indicatrix model: oblate spheroid of revolution, almost a sphere

Discussion. Beryl is built of hexagonal rings of composition Si_6O_{18} that are stacked up the *c* axis in hexagonal columns; each ring is rotated 30° from the one below. The rings and the columns they build are linked to one another by Al^{3+} and Be^{2+} ions. Linkage of these columns coordinates each Al into an octahedron and each Be into a tetrahedron (Fig. 14.2). Thus each Al joins six

Figure 14.2. Polyhedral drawing of the structure of beryl, $2/\square Al_2Be_3$-[$Si_6$①$_6$②$_{12}$], (0001) projection.

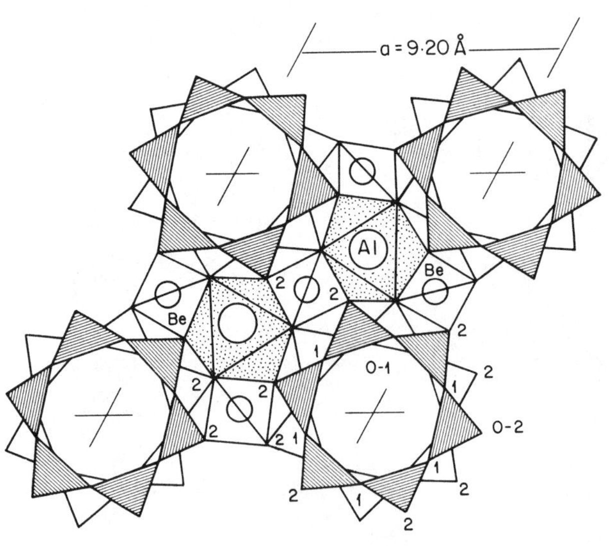

Si_6O_{18} rings, and each Be ion joins four such rings. Each AlO_6 octahedron shares three of its edges with BeO_4 tetrahedra, and each SiO_4 tetrahedron shares two corners with other SiO_4 tetrahedra. Hexagonal Si_6O_{18} rings consist of six inner oxygens, O-1, surrounded by twelve outer oxygens, O-2. Oxygens labeled O-1 contact two Si atoms, whereas oxygens O-2 contact one each of Al, Be, and Si atoms (Figs. 14.1, 14.2; also see Fig. 12.1).

The large tunnels parallel to *c*, although commonly void, may contain large ions of Cs^+ or Rb^+, molecules of H_2O, and smaller ions of Li^+. These alkali ions in the tunnels evidently serve to satisfy a positive electric charge deficiency caused by a vacancy somewhere in the Be—Al—Si—O structure. The presence of nine tetrahedral cations Be_3Si_6 to eighteen oxygens relate beryl to the framework (tekto) silicates that are characterized by low birefringence, such as is found in quartz and feldspars. Structures built mostly of tetrahedra tend to have low birefringence, because their electron density is uniformly distributed rather than being concentrated in planes or rods.

Cordierite

Orthorhombic *Cccm*

$4/(Mg,Fe)_2(Al_2,Si)[Al_2Si_4O_{18}]\cdot nH_2O\cdot nNa$

$a = 17.08$ Å $b = 9.74$ Å $c = 9.34$ Å $V_{uc} = 1539.80$ Å^3

$M/\rho = V_m = 609.78/2.63 = 231.85\ cm^3$

Electronegativity differences: Mg—O = 2.3 Fe—O = 1.7
Al—O = 2.0 Si—O = 1.7

Integrated = 1.9 (59% ionic bonding)

Hardness: 7 Luster: vitreous

Color: deep blue to purplish-blue

CN: $(Mg,Fe)^{VI}$ – distorted octahedra

Al^{IV} – distorted tetrahedra, both within the tetrahedral ring and within the octahedral–tetrahedral ring (Fig. 14.3)

Si^{IV} – fairly regular tetrahedra within the tetrahedral ring, but also as distorted tetrahedra within the tetrahedral–octahedral ring

O^{2-} – There are six different types, labeled on polyhedral corners as 1 through 6 (Fig. 14.3). The EBS summed over these six O^{2-} anions is $+11.749$. Perhaps this slight positive charge deficiency is the reason why small but significant amounts of Na^+ ions occupy some sites in the *c* axis tunnels. It is directly analogous to the occupancy of similarly located Cs^+ ions in the analogous tunnels of beryl (Fig. 14.2). Cordierite (orthorhombic) and beryl (hexagonal) have very similar crystal structures, and high temperature cordierite is hexagonal and isostructural with beryl. Some investigators class cordierite as a tektosilicate. The Al and Si are disordered over the structure and the tetrahedral sites so that adding both together gives $Al_4Si_5O_{18}$, the T–O ratio of a tektosilicate. Either classification is tenable. This author prefers the analogy with beryl because the two structures are strikingly similar. Therefore cordierite, in this book, is restored to its original niche as a cyclosilicate or ring structure.

A—X distances: Mg—O = 2.01–2.12 Å Al—O = 1.748 Å
Si—O = 1.614 Å O—O = 2.50–2.80 Å shortest to longest.

Specific refractivity: $K = 0.209$

Molecular polarizability: $\alpha_G = 50.51$ Å^3

Figure 14.3. Packing model (upper) and polyhedral drawing (lower) of cordierite, $4/(Mg,Fe)_2(Al_2Si)[Al_2Si_4O_{18}]\cdot n Na \cdot n H_2O$, (001) projection. Similarity of structure to that of beryl (Fig. 14.2) is apparent.

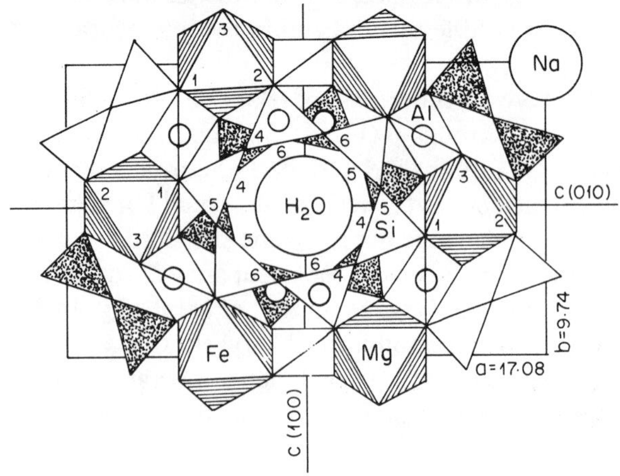

Optical properties: $\gamma = 1.556$ $\beta = 1.550$ $\alpha = 1.544$ $\mathbf{Z} = b$ $\mathbf{Y} = a$ $\mathbf{X} = c$ $\gamma - \alpha = 0.012$, with all properties varying with the Mg—Fe ratio. Cordierite may be either optically positive or negative, because the differences $\gamma - \beta$ and $\beta - \alpha$ are virtually equal. The optical indicatrix model is thus a triaxial ellipsoid not far from being a sphere.

Color (transmitted light): colorless or very weakly pleochroic in blue and purple, increasing with Fe content; minute particles of a radioactive species often present as inclusions are bordered by a characteristic yellow pleochroic halo, seen in grains under the microscope.

Discussion. Cordierite, because of its blue color, vitreous luster, and conchoidal fracture, is sometimes mistaken in rocks for blue quartz. Because of a similarity in both indices of refraction, and low birefringence, it is also commonly mistaken for quartz or feldspar under the microscope. The yellow pleochroic halos already described are, when present, diagnostic. Cordierite is commonly found in argillaceous sedimentary parental rocks that have been metamorphosed by high temperature caused by the proximity of igneous magmas intruded at shallow crustal levels. It also occurs, however, in rocks metamorphosed at considerable depth in the earth and, less often, as spectacular blue crystals in veins with anthophyllite, as well as in some pegmatites.

It was pointed out in Chapter 11 that tektosilicates and ring structures composed largely of equidimensional grouping of tetrahedra and/or octahedra tend to have low birefringence or to be isotropic. Quartz, feldspars, beryl, and cordierite all fall into this category or optical grab bag. All of these minerals lack continuous planes or rows of atoms disposed along particular axial directions; hence electron density is more isotropic than directed in distribution.

Wadeite

$2/K_2^{IX}Zr^{VI}[Si_3^{IV}\textcircled{1}_3^{III}\textcircled{2}_6^{III}]$ or $2/K_2Zr[Si_3O_9]$

Hexagonal $P6_3/m$

$a = 6.89$ Å $c = 10.18$ Å $V_{uc} = 4.18.54$ Å^3

$M/\rho = V_m = 397.69/3.15 = 126.25$ cm^3

Electronegativity: $\Delta\chi_{K-O} = 2.7$ $\Delta\chi_{Zr-O} = 2.1$ $\Delta\chi_{Si-O} = 1.7$

Integrated $\Delta\chi = 2.0$ (63% ionic)

Hardness: 6 Luster: vitreous

Color: colorless

Interatomic distances: Si—O = 1.60 Å Zr—O = 2.07 Å
K—O = 2.82–3.44 Å

CN: K^+ ions lie along the c axis in distorted trigonal antiprisms formed of nine oxygens (Fig. 14.4). Each such antiprism is made up of parts of two Si_3O_9 units.

Zr^{VI} is centered in a regular octahedron that links six Si_3O_9 groups.
Si^{IV} forms tetrahedra.

Specific refractivity: $K = 0.204$

Polarizability: $\alpha_G = 32.22$ Å^3

Optical properties: $\omega = 1.624$ $\epsilon = 1.653$ $\epsilon - \omega = 0.029$
uniaxial (+), colorless and nonpleochroic in transmitted light

Optical indicatrix model: prolate spheroid of revolution

Discussion. The symmetry of the wadeite structure (Fig. 14.4) shows that each Zr atom lies on a 6_3 screw axis, which, when rotated either clockwise or counterclockwise, translates each KO_9 and Si_3O_9 unit half the distance of the *c* axis (*c*/2) with each rotation of 60° (note elevations). Each K atom lies on a $\bar{6}$ or sixfold inversion axis (equivalent to a threefold symmetry axis perpendicular to

Figure 14.4. Packing model (upper) and polyhedral drawing (lower) of wadeite, $2/K_2Zr[Si_3O_9]$, (0001) projection.

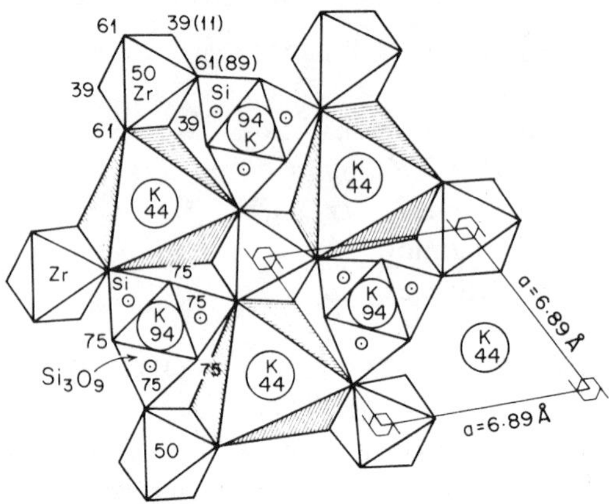

a mirror plane, or $\bar{6} = 3/m$). Twofold, 2_1, screw axes are located on diagonals and edges of the hexagonal unit cell (Fig. 14.4). A mirror plane parallel to (0001), the plane of the illustration, is not apparent. Each such m plane will bisect each Si_3O_9 group parallel to (0001) (Figs. 14.4, 14.5).

Each Si_3O_9 group is made up of three oxygens, O-1, that lie on reflection planes along with the three Si atoms. The six other oxygens, O-2, form the corners of the Si_3O_9 group and are disposed in triads above and below the mirror plane.

The EBS is satisfied as shown in Fig. 14.5:

O-1 is coordinated by 2 K^+/IX and 2 Si^{4+}/IV $= +2.22$
O-2 is coordinated by 2 K^+/IX, 1 Si^{4+}/IV, 1 Zr^{4+}/VI $= +1.88$

O-1	$3 \times 2.22 =$	$+6.66$
O-2	$6 \times 1.88 =$	$+11.33$
	$9\ O^{2-}$	$+18.00$

The staggering of the $[Si_3O_9]$ groups parallel to the c axis, and their corner linkages to Zr^{4+} ions, results in an increased electron density and high index of refraction parallel to this axis. This explains the optically positive sign and the moderate birefringence of wadeite (maximum $n = \epsilon \parallel c$ axis, minimum $n = \omega \perp c$ axis).

A high pressure isostructural analogue of wadeite, $K_2Si^{VI}[Si_3^{IV}O_9]$, in which Si^{VI} proxies for Zr^{VI} of wadeite, has been synthesized by Kinomura, Kume, and Koizumi (1975) and by Swanson and Prewitt (1983). The presence of Si in both CN VI and CN IV in the same compound, synthesized at $P = 24$ kbars (~ 84 km depth) and $T = 900\,^{\circ}$C suggests that this phase may be an important constituent of the mantle of the earth.

Figure 14.5. Part of the structure of wadeite, in approximate (1120) projection, illustrating irregular CN IX coordination of K^+ and electrostatic bonds to O-1 and O-2 of $[Si_3①_3②_6] = [Si_3O_9]$ groups.

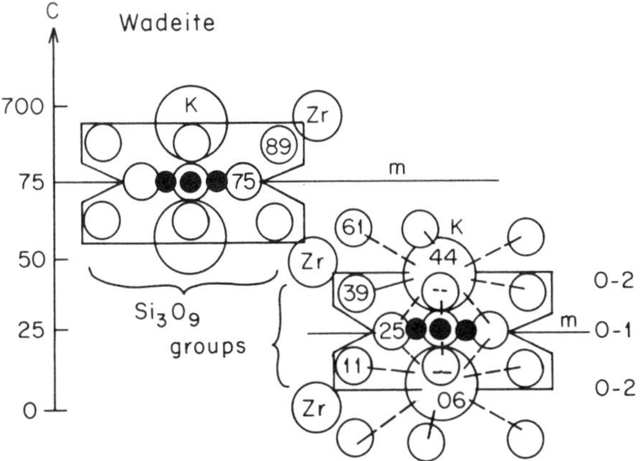

Tourmaline

*R*3*m*

Na

$\square$ $Si_6O_{12} \cdot (Mg,Fe)_3(OH)O_6 \cdot (BO_3)_3(OH)_3$

Al_6

$a_{rh} = 9.54$ Å $\alpha_{rh} = 116°$ $Z = 1$ $V_{uc\,rh} = 532.1$ Å^3
$a_{hex} = 16.00$ Å $c = 7.20$ Å $Z = 3$ $V_{uc\,hex} = 1596.2$ Å^3
$M/\rho = V_m = 1006.16/3.14 = 320.43$ cm^3
Electronegativity: $\Delta\chi_{Na—O} = 2.6$ $\Delta\chi_{Mg—O} = 2.3$ $\Delta\chi_{Al—O} = 2.0$ $\Delta\chi_{B—O} = 1.5$ $\Delta\chi_{Si—O} = 1.7$
Integrated $\Delta\chi = 1.9$ (59% ionic)
Interatomic distances: Si—O = 1.59–1.72 Å Al—O = 1.97–2.08 Å Na—O = 2.64–2.88 Å B—O = 1.37–1.47 Å Mg—O = 1.95–2.04 Å

There are eight different oxygen environments that make up coordination polyhedra (Figs. 14.6, 14.7, Table 14.1).

CN: Na^+ lies in an irregular CN IX cavity above three oxygens (O-2) of $(BO_3)_3$ groups and below six oxygens (3 × O-4, 3 × O-5) of Si_6O_{12} rings (Figs. 14.6, 14.7)

$(Mg,Fe)^{VI}$ lie in fairly regular octahedra of oxygens (2 × O-2, 2 × O-6) and two hydroxyls (O-1 and O-3) (Figs. 14.6, 14.7)

Figure 14.6. "Three-ring circus" of tourmaline in packing models showing rings A, $[Si_6O_{12}]$, overlain by rings B, $Al_6Mg_3(OH)O_6$, in turn overlain by rings C, $Na(BO_3)_3(OH)_3$. Stacking of rings forms columns of sequences ABC, CAB, and BCA parallel to *c*. Each vertical column and each lateral row parallel to a mirror plane give the complex formula of tourmaline. Oxygens are numbered from 1 to 8.

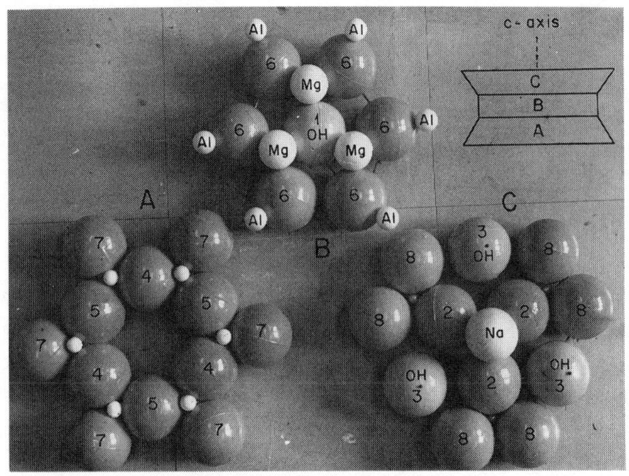

Table 14.1. *Electrostatic bond strength distribution over the eight different anions in tourmaline (thirty-one per formula)*

Anion	Cation neighbors	EBS	No. of atoms/ formula	=	EBS/ formula
O-1(OH)	Mg^2/VI, Mg, Mg	1.000	1		1.000
O-2	Na^1/IX, Mg, Mg, B^3/III	1.778	3		5.334
O-3(OH)	Mg, Al^3/VI, Al	1.333	3		4.000
O-4	Si^4/IV, Si, Na	2.111	3		6.333
O-5	Si, Si, Na	2.111	3		6.333
O-6	Si, Mg, Al	1.833	6		11.000
O-7	Si, Al, Al	2.000	6		12.000
O-8	B^3/III, Al, Al	2.000	6		12.000
			EBS	=	+58.000
			$O_{27}(OH)_4$	=	−58.000

Figure 14.7. Polyhedral drawing, (0001) projection, of tourmaline showing edge sharing of Al and Mg octahedra, corner sharing of (BO_3) and $[Si_6O_{18}]$ groups with octahedra, and unusual polar ninefold coordination of Na^+. Note the large CN IX voids parallel to *c*. Oxygens are numbered as in Figure 14.6.

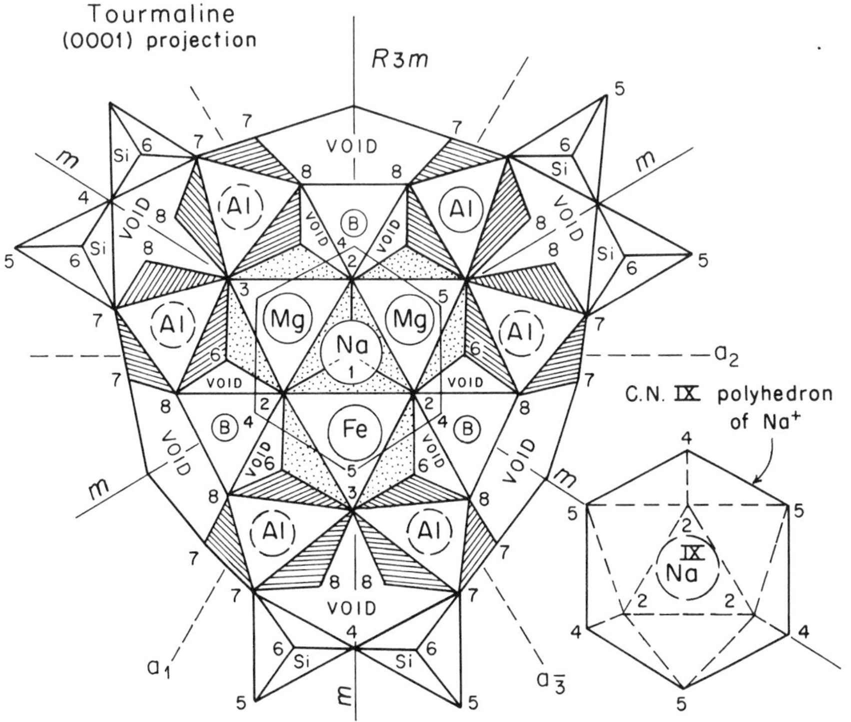

Al^{VI} lies in very irregular octahedra of five oxygens (O-6, 2 × O-7, 2 × O-8) and one hydroxyl (O-3) (Figs. 14.6, 14.7)
Si^{IV} lies in tetrahedra of four oxygens (O-4, O-5, O-6, and O-7) (Figs. 14.6, 14.7)

Specific refractivity: $K = 0.1996$
Polarizability: $\alpha_G = 79.58$ Å^3
Optical properties: $\omega = 1.634$ $\epsilon = 1.612$ (for dravite, Mg-rich variety, increasing rapidly with substitution of Fe) $\omega - \epsilon = 0.022$ uniaxial (−)
Color and pleochroism in transmitted light: colorless to pale yellow, pleochroic yellow to yellow-brown ‖**O** to colorless ‖**E** (dravite); Fe-rich varieties are strongly pleochroic in deep green and brown
Indicatrix model: oblate spheroid of revolution
Substitutions: LiAl ⇌ MgMg in the variety elbaite
FeMn ⇌ MgMg in the variety schorl
Ca□ ⇌ NaNa
F ⇌ (OH)

Discussion. Tourmaline is one of the most complex as well as one of the most elegant of all crystal structures. The complexity of its chemical composition produces an equally complex structure that contains five different cation coordination sites (IX, VI, VI, IV, III), thus violating Pauling's Rule 5, the law of parsimony (see Chapter 4). These sites are enclosed by the eight different anions numbered O-1 to O-8 as in Figure 14.6 and Table 14.2. Tourmaline is a three-ring circus built of successive rings: Si_6O_{12} capped by $Al_6Mg_3(OH)O_6$, which in turn is capped by $(BO_3)_3(OH)_3$ (Fig. 14.6). Na^+ in CN IX links Si_6O_{12} to $(BO_3)_3(OH)$ rings, and Al^{3+} in CN VI links five rings: $2 \times Si_6O_{12}$, $1 \times Mg(OH)O_6$, and $2 \times (BO_3)_3(OH)_3$. These three rings, designated A, B, and C, respectively (Fig. 14.6) are stacked in columns along the *c* axis in the alternat-

Table 14.2. *Makeup of the five different coordination polyhedra and the additional void sites along mirror planes in tourmaline*

Atom	Polyhedron	CN	Nos. of bonding O, (OH) anions
Na	Trigonal antiprism	IX	2,2,2,4,4,4,5,5,5
Mg	Octahedron	VI	1,2,2,3,6,6
Al	Very distorted octahedron	VI	3,6,7,7,8,8
Si	Tetrahedron	IV	4,5,6,7
B	Triangles	III	8,8,2
□	Irregular trigonal polyhedron	IX	3,4,5,6,6,7,7,8,8

ing sequences A, B, C; C, A, B; B, C, A and are cross-linked by Al^{3+} ions into slightly warped sheets that extend infinitely along a_1, a_2, and a_3 axes, forming a plane of high electron density parallel to (0001). These warped sheets are linked by interlayer Na^+ ions, such as are found in the micas of the phyllosilicate family. Thus, although tourmaline is a cyclosilicate, it has much in common structurally with the phyllosilicates. It is, in fact, this layering that causes the marked retardation of the electric vector of light parallel to (0001) and perpendicular to *c*, giving rise to the optically negative character and moderate birefringence of tourmaline. It is fascinating to note that this complex chemistry and symmetry combine to yield one complete formula in each vertical column, A, B, C; C, A, B; or B, C, A (Fig. 14.6) as well as laterally, along C, B, A; B, A, C; A, C, B, along each of the three intersecting mirror planes (Fig. 14.8). The

Figure 14.8. Packing model of tourmaline, showing relation of the true, smaller rhombohedral unit cell to the larger, hexagonal unit cell. Oxygens are numbered as in Figure 14.6.

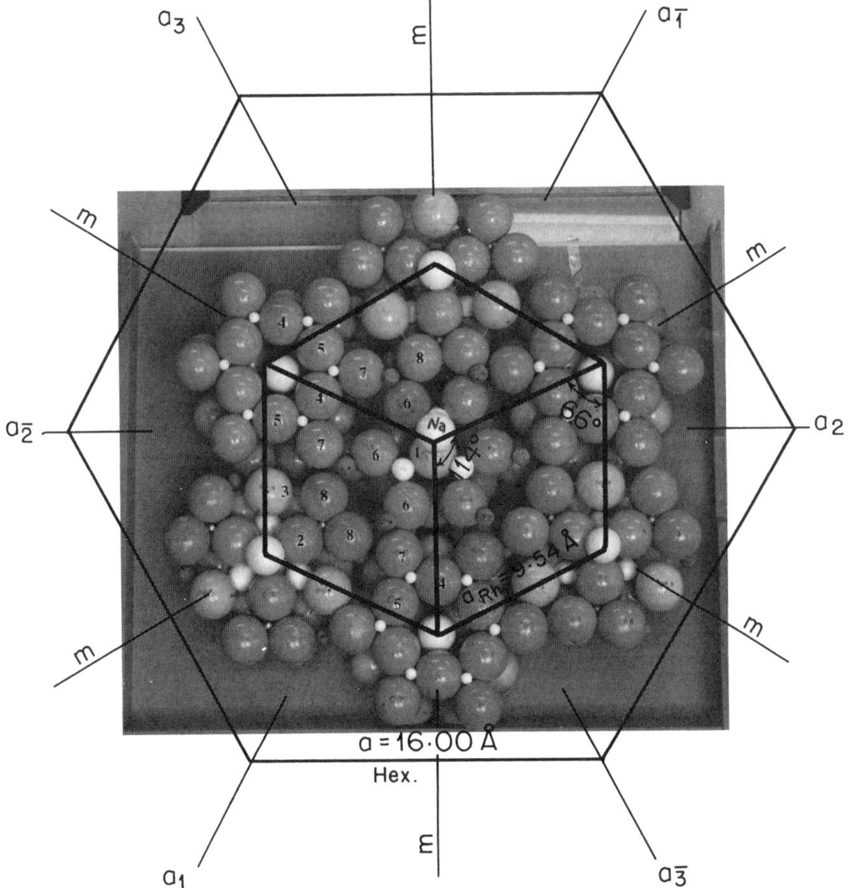

relation of the smaller (truer) unit cell, the rhombohedral cell ($a = 9.54$ Å, $\alpha = 66°$, $Z = 1$) to the larger hexagonal cell ($a = 16.00$ Å, $c = 7.20$ Å, $Z = 3$) is illustrated in Figure 14.8.

The unit rhombohedron, with Na^+ atoms on corners (Fig. 14.8), has a polar or interfacial angle of 66°, which results in a marked flattening of the unit rhombohedron parallel to c (Fig. 14.8). Note that Na^+ and $(OH)^{1-}$ are centered on each vertical column of rings.

One-third of the atoms in the hexagonal unit cell are projected to the sixty-four-anion layer level in Fig. 14.9. The atoms illustrated will repeat at intervals $c/3$ ($Z = 3$ per hexagonal unit cell).

Tourmaline has a primitive lattice and crystallizes in space group $R3m$ with threefold, 3, rotational axes of symmetry oriented parallel to c that lie along the center line of each column of rings and are centered on Na^+ and $(OH)^{1-}$ atoms. Writing the symbol as $3m$ rather than as $3/m$ indicates that the mirror planes are parallel to the threefold axes of symmetry.

The coordination chemistry of tourmaline is actually less complicated than might be expected, but the EBS distribution over the eight different anions, O-1 to O-8, is more so. The makeup of the five different coordination polyhedra is

Figure 14.9. (0001) projection of one-third of the atoms in one hexagonal unit cell of tourmaline to the 64-levels above the base of the unit cell. Can you locate and count up the nineteen cations and thirty-one anions in the unit formula?

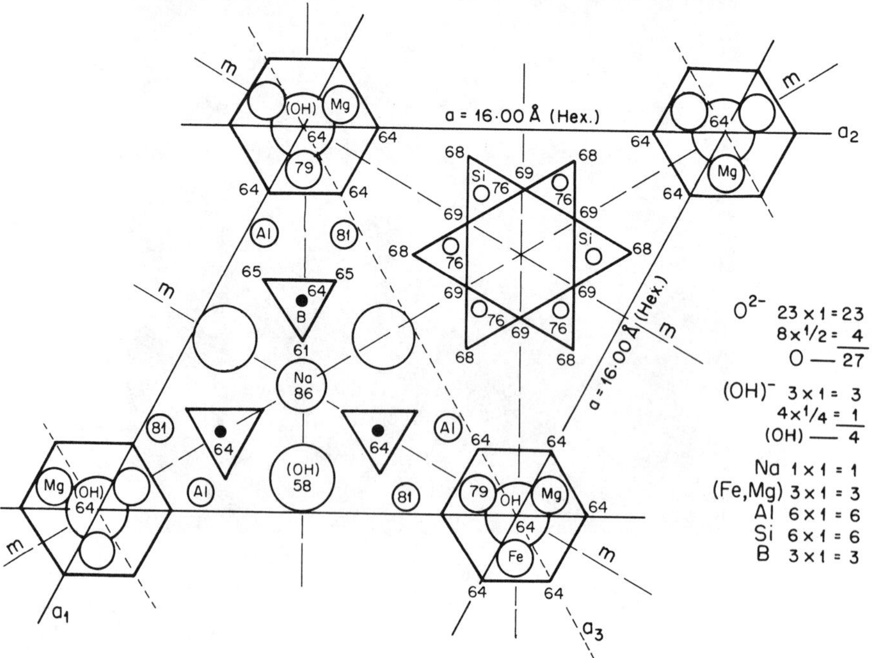

given in Table 14.2 and best illustrated in Figures 14.7, 14.8. Note, in addition, the void sites □ with CN IX located at triple ring junctions along mirror planes (see Fig. 14.7). No studies have yet been made of these void sites to indicate whether any of them may be occupied by cations in some varieties of tourmaline. The sharing of edges of (Mg, Fe)—O and Al—O(OH) octahedra (see Fig. 14.7) forms a microcosm of the infinite octahedral layers in micas. Lateral linkage of these rings, to form warped sheets parallel to *a* axes (thus perpendicular to the *c* axis), endows tourmaline, like micas, with the high electron density and consequent high index of refraction that is found when the electric vector of light vibrates parallel to the sheets. Thus hexagonal (rhombohedral) tourmaline, like pseudohexagonal (monoclinic) micas, is optically negative.

Another important feature of the crystal chemistry of tourmaline is the marked polarity of its structure. All the silica tetrahedra are oriented in the same sense, pointing up the *c* axis, as are all Na^+O_9 polyhedra and Mg and Al—O(OH) octahedra. Hydrogen ions on $(OH)^-$ ions also preserve polarity by pointing up the *c* axis on (OH)-3 ions, although they point down it on (OH)-1 ions. This parallelism of H^+ in (OH)-1 ions will increase the polarizability and electron density parallel to *c*, thus reducing the contrast between ω and ϵ indices of refraction and lowering the birefringence.

Tourmaline provides an excellent example of Pauling's Rule 2C (see Chapter 5), whereby EBS is satisfied only when averaged over all thirty-one anions, twenty-seven O and four (OH), of the unit formula (see Table 14.1). Only O-7, O-8, and O-1(OH) have their bond strength satisfied locally: $\Sigma EBS = +2.000$, $+2.000$, and $+1.000$, respectively. O-2 at the center of borate rings, $\Sigma EBS = +1.778$, and O-6 at the corners of the Mg—O_6 octahedral ring, $\Sigma EBS = +1.833$, are markedly underbonded. The remainder of the anions are overbonded, ΣEBS O-4 and O-5 $= +2.111$, and O-3(OH) $= +1.333$.

In accordance with Zachariasen's elaboration of Pauling's Rule 2 (see Chapter 5), cations will move in closer than normal to underbonded oxygens, and move out further than normal from overbonded oxygens. In tourmaline, underbonded O-2 at the centers of borate rings will pull the rings tightly together, and underbonded O-6 ions will hold the octahedral rings to the tetrahedral rings, because O-6 also forms the apex of each tetrahedron. The underbonded O-6 oxygen, along with the overbonded O-3(OH) hydroxyl, also induce the marked distortion of the Al—O_5(OH) octahedra of tourmaline. Conversely, the hexagonal disposition of underbonded O-6 around Mg ions constrains the Mg—O(OH) octahedra to a regular undistorted symmetry.

The apparent kinship of tourmaline with the phyllosilicates, because of its warped layers of rings, loses its validity when cleavage is considered. Micas have perfect basal cleavage parallel to the layers, because K^+ ions lie parallel to these. In tourmaline, although the stacking of the three-rings induces layering, it also places the Na^+ ions at intervals $c/3$, thus inhibiting the development of a basal cleavage.

Bibliography

Armbruster, T. (1986). Role of Na in the structure of low-cordierite: A single-crystal X-ray study. *Am. Mineral.,* 71: 746–57.

Bragg, W. L., and West, J. (1926). The structure of beryl, $Be_3Al_2Si_6O_{18}$. *Proc. Roy. Soc. London A,* 111: 691.

Buerger, M. J., Burnham, C. W., and Peacor, D. R. (1962). Assessment of the several structures proposed for tourmaline. *Acta Crystallogr.* 15: *Pt. 6* 583.

Bystrom, A. (1942). The crystal structure of cordierite. *Ark. Kemi. Mineral. Geol. B,* 15: 1–5.

Gibbs, G. V. (1966). The polymorphism of cordierite. I. The crystal structure of low cordierite. *Am. Mineral.,* 51: 1068–87.

Henshaw, D. E. (1955). The structure of wadeite. *Min. Mag.,* 30: 585.

Ito, T., and Sadanaga, R. (1951). A Fourier analysis of the structure of tourmaline. *Acta Crystallogr.,* 4: 385.

Kinomura, N., Kume, S., and Koizumi, M. (1975). Synthesis of $K_2SiSi_3O_9$ with silicon in 4^- and 6^- coordination. *Min. Mag.,* 40: 401–4.

Swanson, D. K., and Prewitt, C. T. (1983). The crystal structure of $K_2Si^{VI}Si_3^{IV}O_9$. *Am. Mineral.,* 68: 581–5.

15 Inosilicates and phyllosilicates

Inosilicates (Greek ἰνος = chain) are customarily divided into two groups: single chain silicates, in which tetrahedral polymerization is based upon $[SiO_3]$, as in the pyroxene jadeite, $NaAl[Si_2O_6]$, and double chain silicates with tetrahedral polymerization $[SiO_{2.75}]$, as found in the amphibole glaucophane, $\square Na_2Mg_3Al_2[Si_8O_{22}](OH)_2$. Closely related to these chain silicates are the phyllosilicates (Greek φυλλον = leaf), sheet or layer-lattice silicates with two-dimensional tetrahedral polymerization $[SiO_{2.5}]$, as exemplified by micas and, most simply, in talc, $\square Mg_3[Si_4O_{10}](OH)_2$. Crystal chemical similarities of parts of these structures, and simple arithmetic, show that glaucophane may be represented as a sandwich of one slice of talc between two layers of jadeite, for example,

	Na Al	Si_2O_6	Jadeite
+	$\square$	$Mg_3Si_4O_{10}(OH)_2$	Talc
+	Na Al	Si_2O_6	Jadeite
=	$\square Na_2Mg_3Al_2$	$Si_8O_{22}(OH)_2$	Glaucophane

Similarly, one layer of talc between two of diopside yields tremolite:

	Ca Mg Si_2O_6	Diopside
+ $\square$	$Mg_3Si_4O_{10}(OH)_2$	Talc
+	Ca Mg Si_2O_6	Diopside
=	$\square Ca_2Mg_5Si_8O_{22}(OH)_2$	Tremolite

The similarities in crystal chemical segments of these structures are illustrated in Figure 15.1. Here, unit cells of the pyroxene diopside and the mica phlogopite occupy similar two-dimensional areas on the page, as do the larger unit cells of the amphibole pargasite and the mica muscovite. Diopside and phlogopite each have twenty-four oxygens in one unit cell, and both have similar unit cell dimensions of about $9 \times 9 \times 5$ Å for diopside, and $10 \times 9 \times 5$ Å for phlogopite. Each will therefore occupy a very similar volume in three-dimensional space. Similarly, unit cells of pargasite and muscovite each contain forty-eight oxygens and have unit cell dimensions of about 18 × 10 × 5 Å for pargasite, and 20 × 9 × 5 Å for muscovite, and these too will occupy similar volumes in three-dimensional space. This structural similarity is even more apparent when the reversals of the *a* and *c* crystal axes are disregarded. Thus, double chain minerals, such as the amphiboles, contain mica-like phyllosilicate segments sandwiched between pyroxene-like inosilicate segments. In addition, the large CN XII void □, which houses the K^+ ions that hold mica sheets together, should be and is present in the amphiboles, such as pargasite, where it is occupied by Na^+ (Fig. 15.1).

Because of the importance of these three mineral groups, pyroxenes, amphiboles, and micas, each one will be considered separately.

Single chain silicates

In single chain silicates, $[SiO_3]_n$, SiO_4 tetrahedra, each sharing two oxygens (bridging oxygens), build infinite chains parallel to a crystal axis, most

Figure 15.1. Similarities in the structure of phlogopite (A) and muscovite (B) (projected along *a*) with diopside (C) and pargasite (D) (projected along *c*). Amphiboles contain structural elements of both pyroxenes and micas.

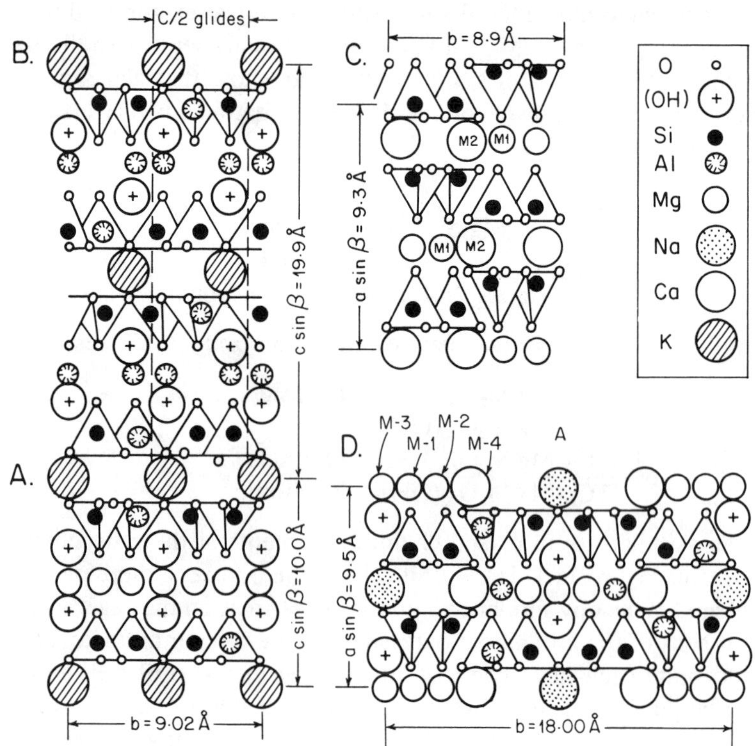

commonly the *c* axis, as in pyroxenes, but often along the *b* axis, as in wollastonite. Chains are stacked in parallel or in bundles that are held together by larger cations which occupy CN VI or VIII sites created by the stacking of chains and the negative charge on each chain. Single chains may repeat their tetrahedral growth or translation pattern in multiples of two (diopside), three (wollastonite), five (rhodonite), or seven (pyroxmangite), as illustrated in Figure 15.2.

The best known and most widely occurring single chain silicates of the pyroxene family are numerous and varied in their complex chemistry, but all are based on a common tetrahedral polymerization scheme of $[Si_2O_6]^{4-}$. Space

Figure 15.2. Single chain inosilicate periodic repeat patterns in multiples of two, three, five, and seven tetrahedra. Reprinted with permission from F. Liebau, *Structural chemistry of silicates,* Springer-Verlag, 1985.

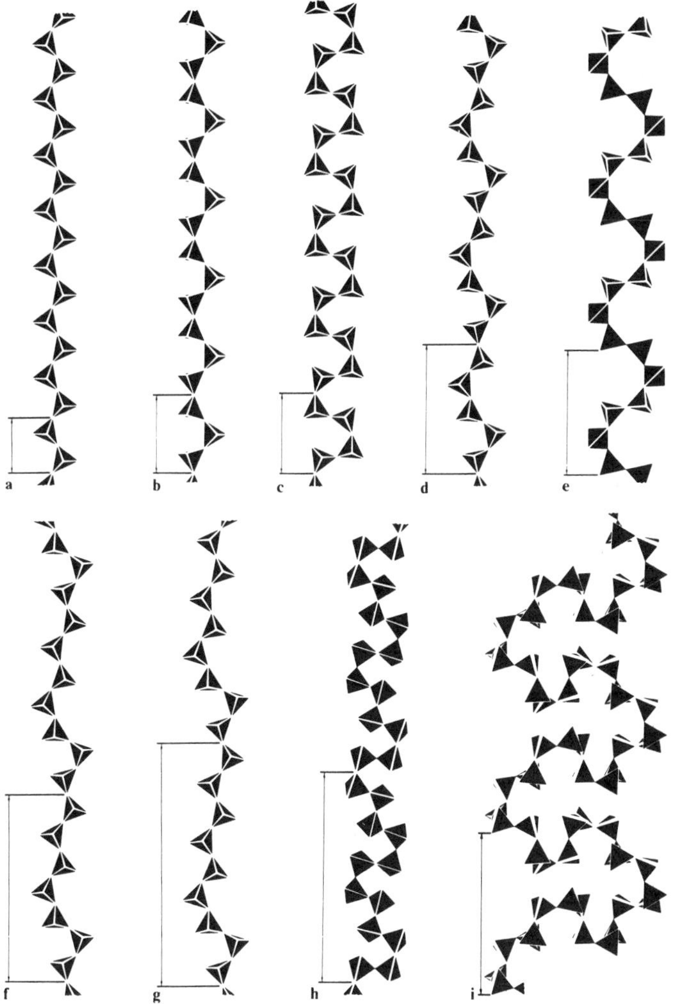

does not permit an examination of all pyroxenes here, but data for a representative group are given in Table 15.1, and crystal chemical details of one of the most common members, diopside, $CaMg[Si_2O_6]$, are presented in Figures 15.3–15.5 and analyzed in the text (also see Table 5.1, Figs. 5.6, 8.1).

Basically, $[Si_2O_6]$ pyroxene chains stack together with tetrahedral chains base to base, alternating with chains stacked apex to apex. This stacking creates cation voids of two sizes: a larger site classed as the M-2 site, and a smaller site classed as the M-1 site. In all pyroxenes, the M-1 site is a distorted octahedron, CN VI, but the larger M-2 site may be a CN VIII distorted square antiprism in some pyroxenes (see Fig. 8.1), or a CN VI distorted octahedron in others. Where both M-1 and M-2 sites are octahedra, M-2 remains the larger. This is analogous to what we have seen in the olivine structure (see Fig. 8.2).

The charge and radius of the larger cation occupying the M-2 site affect the expansion or contraction of the $[Si_2O_6]$ chains and result in the formation of different pyroxene structures. Although they are all based on $[Si_2O_6]$, pyroxenes may crystallize with monoclinic symmetry in space groups $C2/c$, $C2$, $P2$, and $P2_1/c$, as well as with orthorhombic symmetry in space group *Pbca*. Representative examples of each are listed in Table 15.1. For additional details of the crystallography and properties of pyroxenes, see the *Short Course* volume on pyroxenes published by the Mineralogical Society of America (1981).

Table 15.1. *Crystal chemical, unit cell, and optical data for selected pyroxenes*

$C2/c$	$4/\text{M-2}^{\text{VIII}}$	M-1^{VI}	$[T_2^{IV}O_6]$	a (Å)	b (Å)	c (Å)	β	V_{uc} (Å^3)
Diopside	Ca	Mg	Si	9.75	8.90	5.25	105.6°	438.6
Hedenbergite	Ca	Fe^{2+}	Si	9.84	9.02	5.24	104.7°	450.6
Johannsenite	Ca	Mn^{2+}	Si	9.98	9.16	5.29	105.5°	466.0
Augite	$(Na_{0.2}Ca_{0.8})$	(Fe,Mg,Al,Ti)	(Si,Al)	9.71	8.59	5.27	106.5°	434.8
Jadeite	Na	Al	Si	9.42	8.56	5.22	107.6°	401.2
Ureyite	Na	Cr^{3+}	Si	9.55	8.71	5.27	107.4°	418.6
Acmite	Na	Fe^{3+}	Si	9.66	8.79	5.29	107.4°	429.1
$C2$	$4/\text{M-2}^{\text{VI}}$	M-1^{VI}	$[T_2^{IV}O_6]$					
Spodumene	Li	Al	Si	9.45	8.39	5.21	110.1°	388.1
$P2_1/c$	$4/\text{M-2}^{\text{VI}}$	M-1^{VI}	$[T_2^{IV}O_6]$					
Pigeonite	$(Fe_{0.8}Ca_{0.2})$	(Fe,Mg,Ti)	$(Si_{1.9}Al_{0.1})$	9.71	8.96	5.24	108.6°	432.2
$P2$	$4/\text{M-2}^{\text{VIII}}$	M-1^{VI}	$[T_2^{IV}O_6]$					
Omphacite	$(Na_{0.5}Ca_{0.5})$	$(Al_{0.5}Mg_{0.5})$	Si	9.55	8.75	5.25	106.9°	420.2
Pbca	$8/\text{M-2}^{\text{VI}}$	M-1^{VI}	$[T_2^{IV}O_6]$					
Enstatite	Mg	Mg	Si	18.21	8.81	5.18		831.4
Orthoferrosilite	Fe	Fe	Si	18.42	9.08	5.24		875.6
Hypersthene	(Mg,Fe)	(Mg,Fe,Al)	(Si,Al)	18.31	8.93	5.23		854.2

Diopside

$Ca^{VIII}Mg^{VI}[Si_2^{IV}O_6]$: $C2/c$

Perhaps the best-known crystal chemical example of a single chain silicate is the monoclinic pyroxene, diopside. The closely related $C2/c$ pyroxenes jadeite, hedenbergite, augite, and johannsenite (Table 15.1) differ only in minor details as to crystal structure.

In diopside (Fig. 15.3) each tetrahedron has an Si atom enclosed by oxygens labeled O-1, O-2, and pairs of O-3 (the bridging oxygens). When upward- and downward-pointing chains (Fig. 15.3) are stacked together, they enclose M-2, CN VIII, sites, each formed of O-1 (two each), O-2 (two each), and O-3 (four each). An equal number of M-1 octahedral sites are each bounded by O-1 (four each) and O-2 (two each). The *C* face is centered; there is one twofold symmetry axis, the *b* crystal axis, and this is perpendicular to a *c* axis glide plane parallel to (010).

Although diopside may be viewed along different axes in many different crystallographic projections, the most informative is the (100) projection or that projected along $a \sin \beta$ (Fig. 15-3). Here, the *c* axis, the chain axis, runs vertically, and the *b* axis horizontally across the figure. The twofold symmetry axis (symbol ⬮) parallel to the *b* crystal axis may be readily discerned in various parts of the figure. The *c* axial glide that reflects and translates each tetrahedron,

Table 15.1. *(cont.)*

g/cm³	γ	β	α	2V°	sign	$Z \wedge c$	k	α_G (Å³)
3.28	1.694	1.672	1.665	62	+	38°	.206	17.72
3.345	1.771	1.753	1.745	52	+	41°	.226	20.33
3.52	1.729	1.708	1.700	70	+	48°	.202	19.80
3.31	1.730	1.710	1.704	52	+	43°	.216	18.54
3.345	1.652	1.645	1.640	67	+	33°	.193	15.46
3.60	1.762	1.756	1.740	65	−	76°	.209	18.80
3.58	1.830	1.812	1.770	58	−	84°	.225	20.59
3.18	1.677	1.659	1.653	66	+	25°	.208	15.36
3.44	1.721	1.698	1.696	25	+	41°	.205	18.18
3.32	1.718	1.699	1.689	64	+	40°	.211	17.60
3.22	1.658	1.653	1.650	59	+	—	.203	16.22
3.96	1.789	1.780	1.772	86	+	—	.197	20.39
3.60	1.728	1.724	1.714	58	−	—	.200	18.40

each octahedron, and each distorted square antiprism one-half the distance of the *c* axis repeat is well illustrated in Figure 15.3, in which each of the three types of coordination polyhedron may be seen zigzagging left and right up the *c* axis in a $c/2$ repeat pattern. The *c* axis repeat distance, usually given as $[SiO_3]_2$, may be measured along *c* from every second Si but just as well from every second Mg, every second Ca, or from every second like oxygen, designated O-1, O-2, or O-3. All atoms are related to like atoms by the $c/2$ axial glide.

A careful study of Figure 15.3 will show that Si is situated in a somewhat distorted tetrahedron (T-site) and that the $[Si_2O_6]$ chain running along *c* is slightly twisted. Magnesium ions are located in a very distorted octahedral site (M-1), and calcium ions in CN VIII polyhedra (M-2 sites) that represent very distorted square antiprisms (see Fig. 8.7). Note that along chains, SiO_4 tetrahedra share only corners, whereas MgO_6 octahedra share edges, as do CaO_8

Figure 15.3. Polyhedral representation of the structure of diopside projected along $a \sin \beta$ (100) projection. Note that tetrahedral chains Si_2O_6 parallel to *c* are mimicked by Mg octahedral and Ca square antiprismatic chains. For example, the *c* unit cell dimension may be measured between like pairs of Si, Mg, Ca, or O atoms.

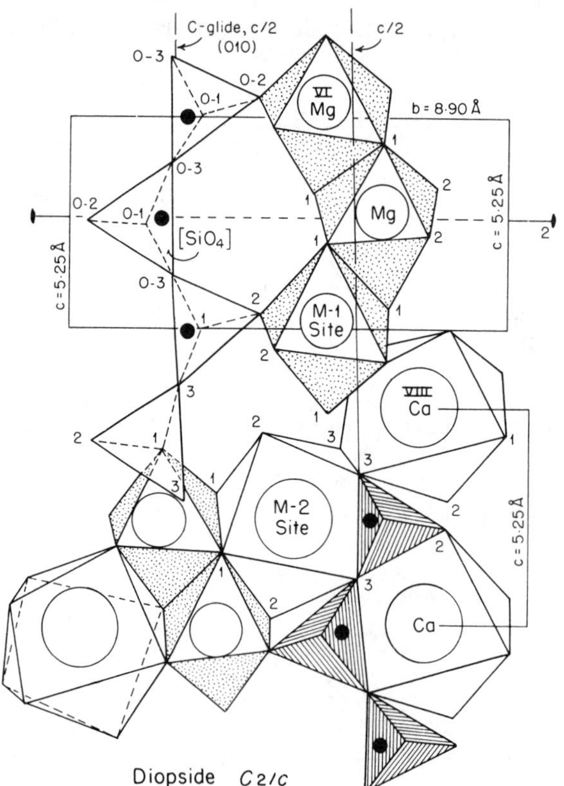

polyhedra. SiO_4 tetrahedra link to corners of MgO_6 octahedra, but also form common edges with CaO_8 polyhedra. MgO_6 and CaO_8 also share edges. Over the entire structure, a large number of all polyhedral edges present are shared.

Note that the three different oxygens, O-1, O-2, and O-3, are overcharged (overbonded) or undercharged (underbonded) as has already been shown (see Fig. 5.6, Table 5.1). Although electrostatic bond strengths (EBS) for O-1 = +1.93, O-2 = +1.58, and O-3 = +2.50 average out to +2.00 per six formula-unit oxygens, it is this overbonding and underbonding that induces the polyhedral distortion observed and described in Chapter 5.

Pyroxenes, and amphiboles as well, are often represented in (010) projection along the *b* crystal axis (Fig. 15.4) to show the linkage of up- and down-facing

Figure 15.4. Packing model (upper) and polyhedral representation (lower) of the diopside structure in (010) projection. Note monoclinic $a \wedge c$ angle, $\beta = 105°$, tetrahedral chains parallel to *c*, and Mg octahedra that alternate with Ca polyhedra.

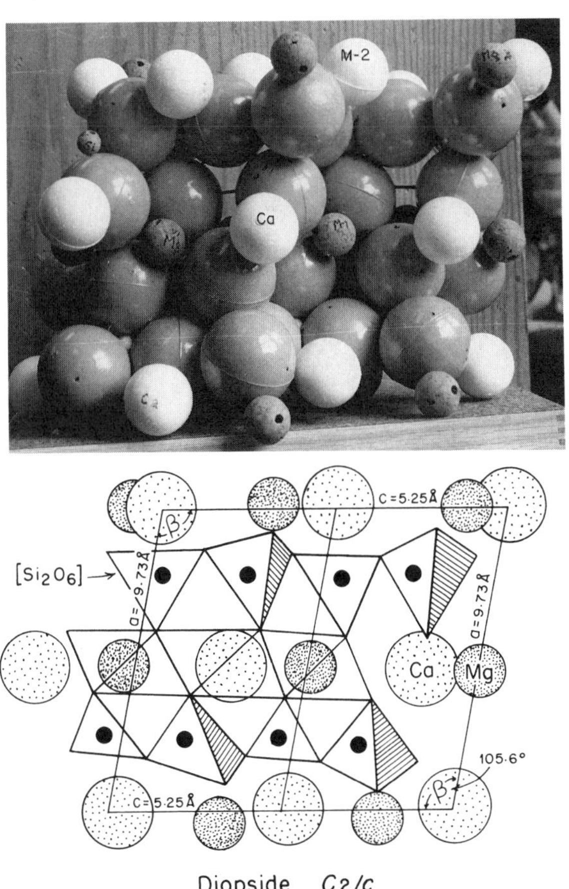

tetrahedral (Si_2O_6) chains by Mg^{VI} and Ca^{VIII} cations. Another widely used representation is the *c* axis projection, viewed along the zigzag chains, in which case the structure is visualized as a packing of units analogous to steel I-beams. Each pyroxene I-beam is made up of pairs of downward-pointing tetrahedra joined across an octahedral layer (Fig. 15.5). I-beams are packed in staggered fashion, and the broad backs of these are joined by Ca^{2+} ions of the CN VIII polyhedra (M-2 sites). The Mg—O octahedral layer bond is considerably stronger than the Ca—O bond, and there is more empty space between Ca^{2+} ions than between Mg^{2+} ions. Because of this, cleavage follows along the Ca—Ca lineations across the backs of I-beams, and is connected in zigzag fashion to form cleavage planes that are roughly 90° to one another (89° and 93°, Fig. 15.6). Illustrations in many textbooks erroneously show the cleavages as breaking across the I-beams at the centers of the MgO (M-1) octahedra. The steel rods bent around the I-beams in the ball model of diopside (Fig. 15.5) show

Figure 15.5. Packing model (upper) and polyhedral drawing (lower) representing I-beam structure of monoclinic *C*2/*c* pyroxenes projected along the *c* axis. I-beams are outlined by bent rods of packing model. Translation of chains is up toward viewer.

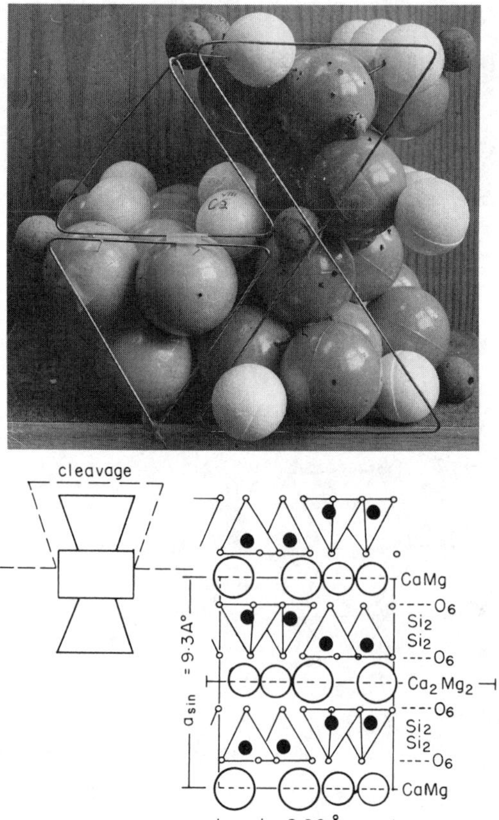

how cleavage breaks across Ca—Ca lines at the backs of I-beams. I-beams are thus not broken by cleavage paths.

Exsolution

Many pyroxenes crystallized at high temperatures (~1100–1200°C) in silicate magmas or melts will consist of a chemically complex solid solution. If this is *slowly* cooled, one chemical phase may separate from another in a postsolidification or subsolidus state. The most abundant phase then becomes the host; and the separated phase, the exsolution lamella. Depending on the

Figure 15.6. I-beam structures and cleavages {110} in pyroxenes and amphiboles. Cleavage breaks along weakly bonded backs of I-beams and not through the octahedral layers at I-beam centers (see Fig. 15.5). I-beam widths control cleavage angles.

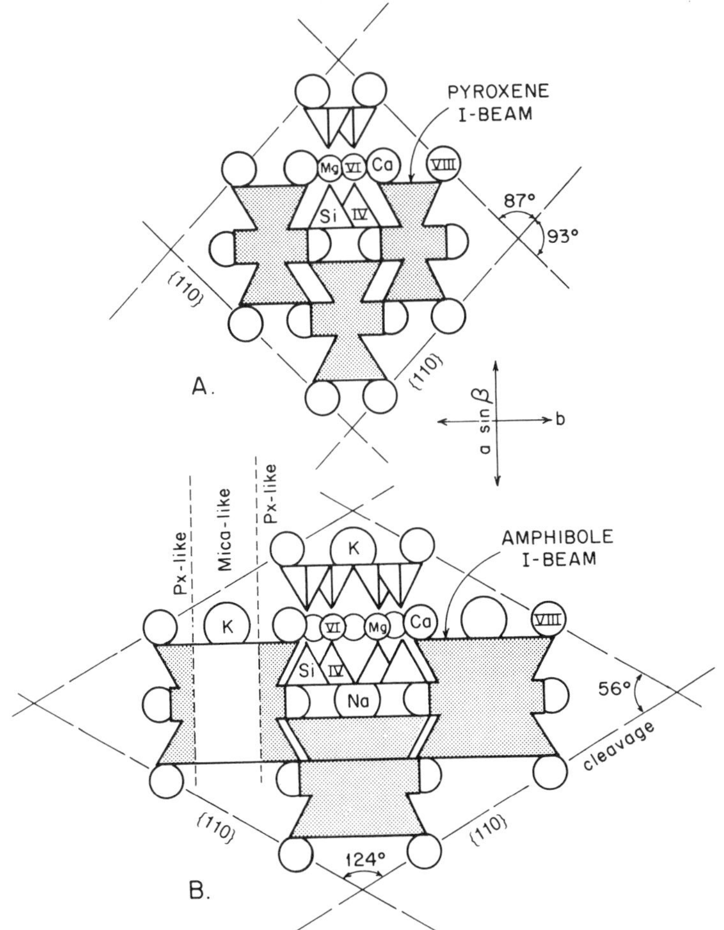

amount of initial solid solution and the cooling history, one or more sets of lamellae of different pyroxenes may exsolve within a single host grain (Figs. 15.7, 15.8). Lamellae exsolved at high temperatures (~ 1000°C) will be coarse (~ 5–10 μm thick), and those exsolved at about 600° will be fine (~ 0.*X* μm). Rocks of diverse origin and chemistry that were metamorphosed in the earth at temperatures of about 700–800°C and pressure of 5–10 kbar (equivalent to 17.5–35 km depth), if slowly cooled to about 600°C, may contain pyroxenes that exsolve thin exsolution lamellae. Thus, thickness of exsolution lamellae in host pyroxenes may be used as a yardstick for the thermal crystallization and cooling history of the rock in which they occur. Because the exsolution processes are associated with chemical changes, both compositional and lattice dimensions of lamellae and host will change over time. The relation of the dimension of each lattice will be recorded inside each such complex grain. It has been observed that angles between exsolution lamellae and the *c* axis of the host grains are related to lattice misfits of lamellae and host. These lattice misfits may be correlated with composition. Magnesium-rich monoclinic pyroxene hosts contain exsolution lamellae with larger angles than those seen in iron-rich hosts, because lattice misfits are larger (Robinson et al. 1971; Jaffe et al. 1975).

In high temperature magmatic rocks, (010) plates of host augite may show

Figure 15.7. Diagram relating the ($Fe^{2+} + Fe^{3+} + Mn$)/($Fe^{2+} + Fe^{3+} + Mn + Mg$) ratio with angle between "001" pigeonite exsolution lamellae and *c* axis of host augite. Both low temperature (metamorphic) and higher temperature exsolution patterns are shown. Reprinted with permission from Jaffe, Jaffe, Ollila, and Hall (1983), Univ. Mass., Dept. Geol. and Geog. Contrib. No. 46, p. 16.

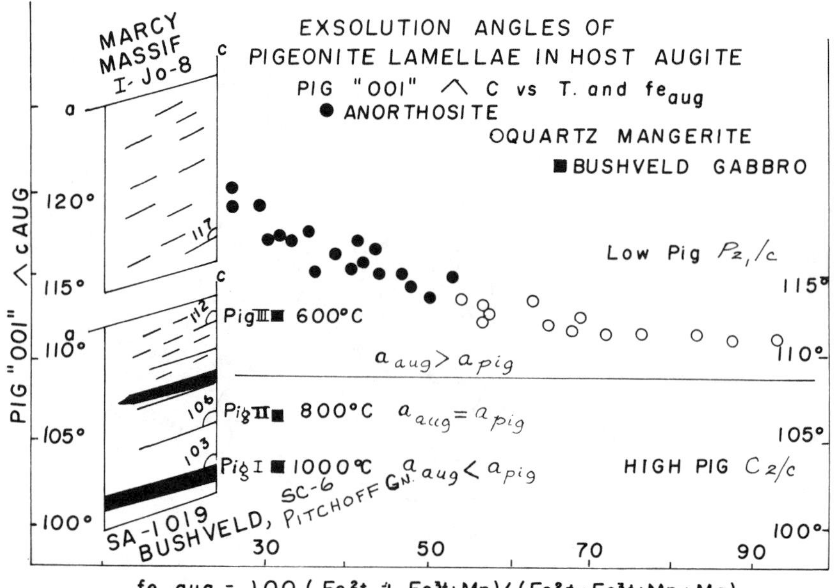

coarse (5–15 μm) exsolution lamellae of pigeonite oriented into the obtuse angle $a \wedge c$ (Fig. 15.9). Host pigeonite may also exsolve comparable coarse lamellae of augite. On slow cooling, host pigeonite will undergo phase inversion to hypersthene while still retaining the coarse augite exsolution lamellae. This texture is called *inverted pigeonite* (Fig. 15.9).

Exsolution features that took place at lower temperatures, ~600 °C, such as in intensely metamorphosed rocks, will show on (010) plates of host augites fine exsolution lamellae of hypersthene parallel to the *c* axis of the host, and yet finer exsolution lamellae of pigeonite oriented, here, in acute angle β ($a \wedge c$ acute) of the host augite. The angles that these lamellae make with the *c* axis of the host augite are controlled by chemistry and lattice misfit between *a* host versus *a* lamella and *c* host versus *c* lamella. Such pigeonite lamellae will exsolve at crystallographically irrational angles in the augite host and will thus virtually never lie parallel to (001) or (100) directions or planes in the host. Angles between these pigeonite exsolution lamellae and the *c* axis of the augite host are largest (122° and 22°) for hosts with high Mg/(Mg + Fe) ratio and decrease to small values (110° and 0°) for hosts with high Fe/(Fe + Mg) ratio (Figs. 15.7,

Figure 15.8. Diagram relating the ($Fe^{2+} + Fe^{3+}$ + Mn)/($Fe^{2+} + Fe^{3+}$ + Mn + Mg) ratio with angle of intersection of two sets of pigeonite exsolution lamellae ("001" $\wedge$ "100") in host augite. Exsolution angles are for a 600° (metamorphic) isotherm. Reprinted with permission from Jaffe et al. (1983), Univ. Mass., Dept. Geol. and Geog. Contrib. No. 46, p. 17.

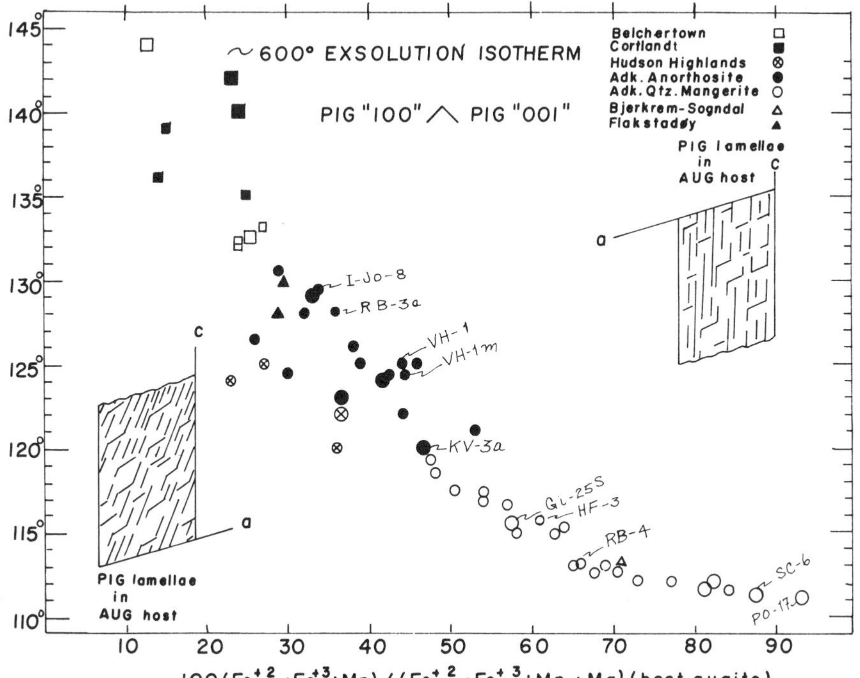

15.8). Host hypersthenes in these lower temperature rocks will contain very fine lamellae of augite parallel to the *c* axis of the host; these are never the inverted pigeonites of high temperature origin (Fig. 15.9). Thus the thickness and angles of exsolution of lamellae in pyroxenes may record much of the very complex igneous and metamorphic history of a region in the earth's crust and serve as a geologic thermometer for such processes. On rare occasions, giant megacrysts of aluminum-rich orthopyroxenes, occurring as exotic fragments of the earth's mantle, are found in anorthosites (plagioclase–feldspar-dominant rocks). Many such pyroxene megacrysts contain exsolution lamellae of both calcium-rich plagioclase and Fe–Mg-rich garnet (Figs. 15.10, 15.11; Jaffe and Schumacher 1985).

Control over the orientation of the exsolution lamellae inside host grains is a function of both the relative site of the β angle of host and lamellae and the relative size (or misfit) of the a_{host}-a_{lam} and c_{host}-c_{lam}. If the host has the smaller β angle – for example, augite $\beta = 106°$ – and the lamellae, pigeonite, $\beta = 109°$, then the *a-a* and *c-c* misfits will control the orientation of two sets of exsolution

Figure 15.9. Typical exsolution textures in pyroxenes of high temperature magmatic and lower temperature granulite facies metamorphic rocks. Both coarseness of exsolution lamellae and their orientation in host pyroxenes are geothermometers.

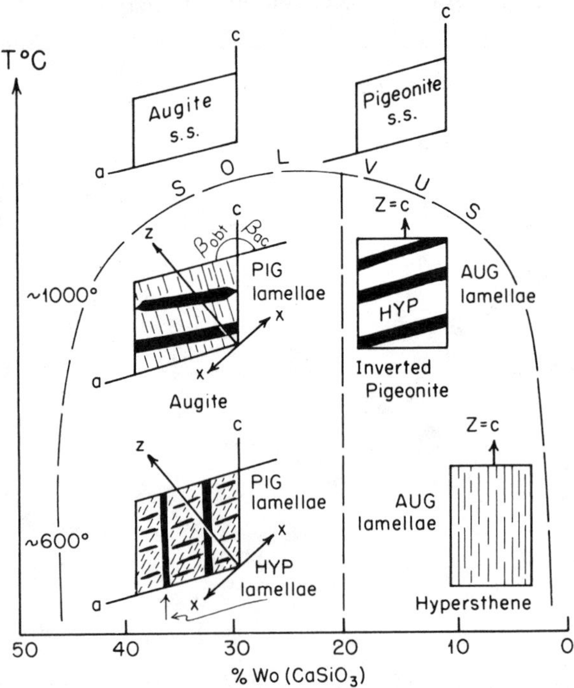

lamellae, designated "001" and "100" (see Figs. 15.7–15.9):

$a_{aug} > a_{pig} \longrightarrow$ results in exsolution of "001" lamellae of pigeonite in acute β augite host

$c_{aug} > c_{pig} \longrightarrow$ results in exsolution of "100" lamellae of pigeonite in acute β augite host

These relations are found for exsolution in pyroxenes crystallized at lower metamorphic temperatures. In magmatic rocks these relations are reversed, because the pigeonite phase expands rapidly for temperatures greater than about 700°C that exist in most Mg–Fe-rich silicate magmas.

Although relative β angles of pigeonite remain larger than that of the host augite, the a and c dimensions of pigeonite exceed those of augite. Thus, for magmatic rocks, a large degree of solid solution plus very slow cooling may result in high temperature exsolution at about 1000°C, where

$a_{aug} < a_{pig} \longrightarrow$ results in exsolution of "001" lamellae of pigeonite in obtuse β augite host

$c_{aug} < c_{pig} \longrightarrow$ results in exsolution of "100" lamellae of pigeonite in obtuse β augite host

Thus temperature of crystallization and subsequent exsolution exercise con-

Figure 15.10. Schematic illustration showing exsolution lamellae of garnet, calcic plagioclase, ilmenite, and clinopyroxene oriented parallel to (100) of host Al-rich orthopyroxene megacrysts PR-2 and I-Jo-12 from the Mt. Marcy quadrangle. Indices of refraction and optical orientation are shown. Reprinted with permission from H. W. Jaffe and J. C. Schumacher, *Can. Mineral.*, 23: 464.

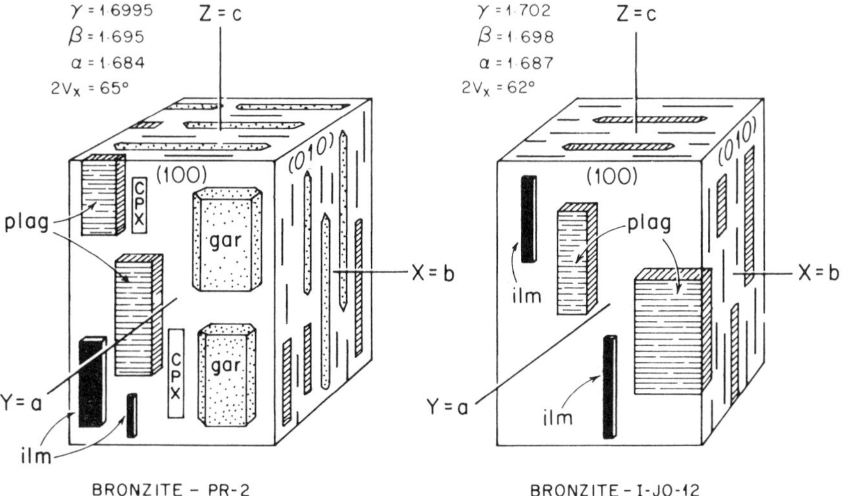

trol over the dimensions of each lattice and thus the orientation of exsolution lamellae in a host grain (Figs. 15.7–15.9, and Robinson et al. 1971).

Relationship of optical properties of inosilicates and phyllosilicates

Most single chain pyroxenes are optically positive, with the slow vibration direction **Z** making a 40° angle with the *c* axis and oriented in obtuse angle β. In diopside and jadeite, $C2/c$ pyroxenes, this orientation of the slow velocity wave appears to correlate with the alignment of strips of underbonded O-2 oxygens, each surrounded by Mg^{2+}/VI, Ca^{2+}/VIII, and Si^{4+}/IV (ΣEBS = +1.58 instead of +2.00). Because cations draw closer to the underbonded O-2 oxygens (see Chapter 5), electron density will increase along this lineup, resulting in an increase in polarizability, retardation of the electric vector of visible light, and a maximum index of refraction (Fig. 15.12) at about 40° to the *c* axis.

It has already been shown (see Chapter 11) that planar structures, such as phyllosilicates, are most often optically negative because the electric vector of light is markedly retarded parallel to the sheets and their increased electron density.

Figure 15.11. Core of orthopyroxene megacryst PR-2 showing broken lamellae of garnet (black) and bytownite (white). Fractures that transgress garnet lamella do not transgress plagioclase lamella. Garnet and plagioclase lamellae are 10 μm thick. Note fine opaque ilmenite lamellae and extremely fine clinopyroxene lamellae (white) (crossed nicols). Reprinted with permission from H. W. Jaffe and J. C. Schumacher, *Can. Mineral.*, 23: 466.

Double chain amphibole minerals represent a compromise between single chain and sheet structures, because they possess "mica-like" strips oriented parallel to the *b-c* plane (100), which contains the intermediate and close to the maximum index of refraction. These interrupted amphibole sheets are offset from one another because of monoclinicity. This offset parallel to the (100) planes results in an angle of $\mathbf{Z} \wedge c \simeq 10°–20°$.

Details of orientation of the maximum retardation velocity vector **Z** in pyroxenes and amphibole need much more study. In acmite, *C*2/*c* pyroxene, $NaFe[Si_2O_6]$, for example, the slow velocity wave **Z** moves from about 40° to the *c* axis to about 85°–88° to the *c* axis, and this change undoubtedly relates to orientation of a covalent $Fe^{3+}—O_2$ bond.

Double chain silicates: the amphiboles

Some of the relationships between single chain silicates (pyroxenes), phyllosilicates (clays and micas), and double chain silicates (amphiboles) were illustrated in Figures 15.1 and 15.6 and discussed under pyroxenes.

The structures of pyroxenes and amphiboles have much in common, and both frequently occur in the same rocks. Because amphiboles contain essential

Figure 15.12. Lineup of staggered rows of electrostatically underbonded oxygens, O-2, EBS = +1.58 (see Fig. 5.6) is roughly parallel to the slow vibration velocity vector **Z** of diopsidic and augitic pyroxenes, (010) projection.

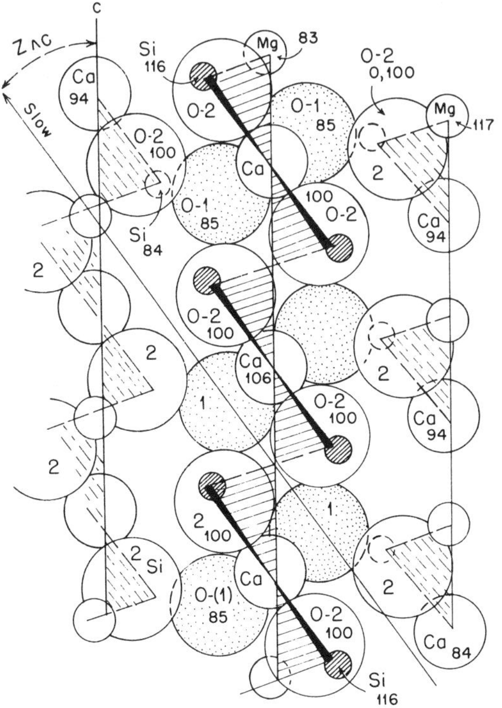

water in their formulae, they are abundant in metamorphic rocks crystallized at intermediate temperatures and pressures, and tend to give way to anhydrous pyroxenes at the temperatures and pressures associated with the highest grades of metamorphism and magmatism. Amphiboles are abundant in Mg- and Fe-rich rocks crystallized in a range of about 500°–700°C, and pyroxenes are abundant in similar metamorphic rocks crystallized at 700°–800°C. Pyroxenes also occur in magmatic rocks crystallized at 800°–1200°C, where amphiboles are absent.

If we compare diopside, $CaMgSi_2O_6$, with tremolite, $Ca_2Mg_5Si_8O_{22}(OH)_2$, and enstatite, $MgSi_2O_6$, with anthophyllite, $(Mg,Fe)_7Si_8O_{22}(OH)_2$: all of these may be related to the unit cell of diopside: $a = 9.75$ Å, $b = 9.00$ Å, and $c =$ 5.25 Å, by a doubling of a or b unit cell dimensions, with the c dimension remaining constant, as shown in Table 15.2. Thus tremolite is related to diopside by a doubling of the b dimension, enstatite to diopside by a doubling of the a dimension, and anthophyllite to diopside by a doubling of both the a and b dimensions of the unit cell, with the volumes of the unit cell closely approximating multiples of the 439-Å^3 volume for the twenty-four O^{2-} unit cell of diopside. The orthorhombic symmetry of the enstatite is created by twinning back-to-back unit cells of diopside parallel to (100) while reintegrating the resulting unit cell to orthorhombic symmetry.

In Figure 15.3, it was shown that SiO_4 tetrahedra of single chain pyroxenes are related by a c axial glide (010) $c/2$. If two such single chains are joined as in the lower part of Figures 15.13 and 15.14, two of each three oxygens labeled O-2 in diopside fuse into one oxygen now labeled O-7. The third O-2 oxygen of diopside is now labeled O-4 in amphibole terminology. Similarly, the pairs of O-3 bridging oxygens of diopside are now renumbered as bridging oxygens O-5 and O-6 that alternate along the c axis. Fusion of the two single pyroxene chains also creates a mirror plane m parallel to (010) in the amphibole double chain structure (Figs. 15.13, 15.14).

Table 15.2. *Comparative unit cell parameters of representative pyroxenes and amphiboles*

	Diopside *C*2/*c*	Tremolite *I*2*m*	Enstatite *Pbca*	Anthophyllite *Pnma*
a (Å)	9.75	9.92	18.22	18.56
b (Å)	9.00	18.05	8.82	18.08
c (Å)	5.25	5.27	5.18	5.28
β (°)	105.6	106.6	—	—
V_{uc} (Å^3)	439	904	832	1772
O^{2-}/uc	24	48	48	96

Figure 15.13. Packing (left) and polyhedral representation (right) of complex monoclinic amphibole $I2/m$ (= $C2/m$) structure.

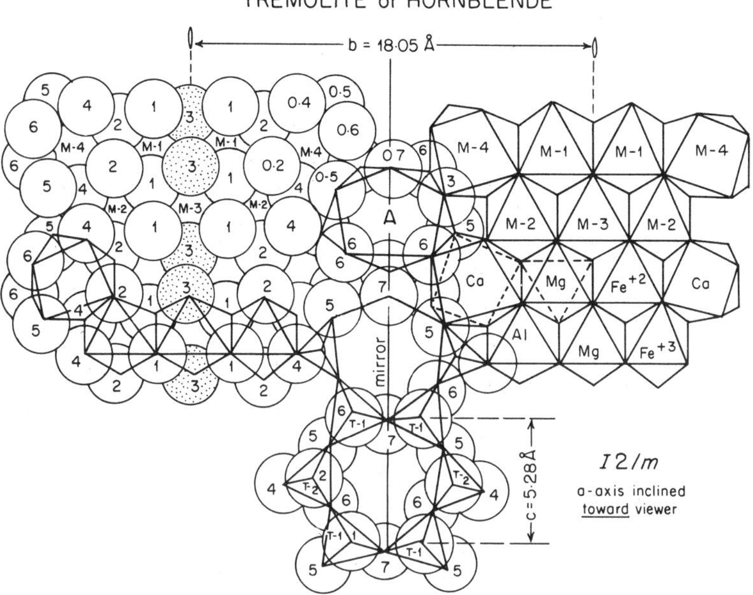

Figure 15.14. Three-dimensional perspective view of packing model of hornblende with front face of model parallel to (100) of Figure 15.13. Bent rod encloses an amphibole I-beam of Figure 15.6.

Now there are two types of silicon tetrahedron: T-1 formed of O-1, O-5, O-6, and O-7 around Si, and T-2 formed of O-2, O-4, O-5, and O-6 around Si of the amphibole structure (Fig. 15.13). The mirror plane bisects the O-1 oxygens of T-1, and both T-1 and T-2 reflect symmetrically across this (010) mirror plane. The oxygen labeled O-3 is an (OH) centered between apical O-1 and O-2 oxygens of tetrahedral rings. Thus the *c* glide symbol of diopside is replaced by the *m* plane symbol of tremolite. All common amphiboles, both monoclinic and orthorhombic, contain a mirror plane *m* parallel to (010). Thus the space groups are $I2/m$ for common monoclinic amphiboles (= $C2/m$ of most other authors) and *Pnma* for orthorhombic amphiboles. A rare amphibole, manganoan cummingtonite or tirodite, crystallizes in space group $P2_1/m$ and is an exception.

Where the simple structure of diopside contains but two coordination polyhedra other than Si^{IV} (i.e., $M\text{-}2^{VIII}$ and $M\text{-}1^{VI}$) amphiboles, by combining elements of both the phyllosilicates and pyroxenes, now have a complicated array of polyhedra (Figs. 15.13, 15.14). $M\text{-}2^{VIII}$ of diopside becomes $M\text{-}4^{VIII}$ of tremolite, $M\text{-}1^{VI}$ of diopside becomes $M\text{-}2^{VI}$ of tremolite, and $M\text{-}1^{VI}$ and $M\text{-}3^{VI}$ of tremolite are elements of the mica- or talc-like segment of the tremolite structure. Thus tremolite and the other $I2/m$ amphiboles have one CN VIII site, M-4, similar in shape to M-2 of diopside (a distorted square antiprism), and three different octahedral sites, M-2, M-1, and M-3, all distorted in different ways (Figs. 15.13, 15.14). The square antiprismatic sites are enclosed by the following oxygen types:

$M\text{-}4^{VIII}$	O-2, O-4, O-5, O-6	(two of each)
$M\text{-}2^{VI}$	O-1, O-2, O-4	(two of each)
$M\text{-}1^{VI}$	O-1, O-2, O-3	(two of each)
$M\text{-}3^{VI}$	O-1 (×4), O-3 (×2)	

Table 15.3. *Crystal chemistry, unit cell, and optical properties of selected amphiboles*

	$2/A^X_{0\text{-}1}$	$(M\text{-}4^{VIII})_2$	$(M\text{-}2^{VI})_2$	$(M\text{-}1^{VI})_2$	$(M\text{-}3^{VI})_1$	$[(T\text{-}1^{IV})_4$	$(T\text{-}2^{IV})_4$	$O_{22}]$	$(OH,F)_2$	*a* (Å)
I2/m										
Tremolite	□	Ca	Mg	Mg	Mg	Si	Si	O	OH,F	9.98
Cummingtonite	□	(Fe,Mg)	(Fe,Mg)	Mg	Mg	Si	Si	O	OH,F	9.89
Grunerite	□	Fe	Fe	Fe	Fe	Si	Si	O	OH,F	9.94
Glaucophane	□	Na	Al	Mg	Mg	Si	Si	O	OH,F	9.76
Riebeckite	□	Na	Fe^{3+}	Mg	Mg	Si	Si	O	OH,F	9.96
Hornblende	$Na_{0.5}$	Ca	$(Al,Fe_{0.5},Mg_{0.5})$	(Fe,Mg)	(Fe,Mg)	(Si_3Al)	$(Si_{3.5}Al_{0.5})$	O	OH,F	9.90
K-richterite	K	Ca,Na	$(Ti_{0.5}Fe_{1.5})$	Mg	Mg	Si	(Si_3Al)	O	OH,F	10.05
$P2_1/m$										
Mn-cummingtonite	□	Mn	(Mn,Mg,Fe)	Mg	Mg	Si	Si	O	$(OH_{1.6}O_{0.4})$	9.88
Pnma										
Anthophyllite	4/□	Mg	Mg	Mg	Mg	Si	Si	O	OH,F	18.56
Gedrite	$4/Na_{0.5}$	(Fe,Mg)	$(Al_{1.5}Mg_{0.5})$	(Mg,Fe)	(Mg,Fe)	(Si_3Al)	(Si_3Al)	O	OH,F	18.60

Note that the amphibole formulae contain two each of M-4, M-2, and M-1, but only one M-3 site. Careful perusal of Figure 15.13 will reveal the octahedral edge sharing common to phyllosilicates or layer-lattice minerals, and the M-4 and M-2 parts of the structure that are "pyroxene-like."

An additional polyhedral site, the A site, now appears in amphiboles in the position of the interlayer K^{XII} site (a hexagonal prism) of muscovite. In tremolite and similar monoclinic amphiboles, the A site has CN X rather than XII because the monoclinic β angle ($\sim 106°$) offsets the $[Si_8O_{22}](OH)_2$ chains, appreciably, parallel to (100), and also because the rings enclosing the A sites are not true hexagons (see Figs. 15.13, 15.14). This places two sets of five oxygens around each A site cation, rather than two sets of six. In many amphiboles this A site is vacant in the same manner as in the phyllosilicate talc, where the CN XII site is void. Amphiboles may contain from zero to one atom or ion per formula in these A sites. Table 15.3 shows that the A site is void in those amphiboles where the T-sites are fully occupied by Si, and partially to completely occupied where Al partially replaces Si in T-sites. Na^+ in A sites thus electrically compensates charge deficiency induced by $Al^{3+} \leftrightarrows Si^{4+}$ exchange. A-site occupancy also results from substitution of $Na^+ \leftrightarrows Ca^{2+}$ in M-4 sites; thus the substitution scheme as in richterites is $Na^{+X}Na^{+VIII} \leftrightarrows \square^{X}Ca^{2+VIII}$.

Another, more common, type of substitution occurs as a consequence of extensive tetrahedral replacement of Si, as shown in gedrite (Table 15.3). Here, the complicated substitution and consequent electric charge balance is effected as follows:

$$\underset{A}{Na_{.5}}\,\underset{M\text{-}2}{Al_{1.5}}\,\underset{T}{Al_2} \rightleftarrows \underset{A}{\square}\,\underset{M\text{-}2}{Mg_{1.5}}\,\underset{T}{Si_2}$$

Space does not permit listing the many known amphiboles in Table 15.3, and the interested reader is referred to the Mineralogical Society of America's *Short Course Volume* on amphiboles (1981) and to Hawthorne (1983) for many additional details on the chemistry of this complex group of minerals.

Table 15.3. *(cont.)*

b (Å)	c (Å)	β (°)	V_{uc} (Å^3)	(g/cm^3)	γ	β	α	2V°	sign	$Z \wedge c$ (°)	k	α_G (Å^3)
18.05	5.27	106.6	904.2	2.99	1.602	1.593	1.581	86	(−)	21	.198	63.89
18.19	5.33	109.9	902.1	3.21	1.657	1.644	1.634	86	(+)	15	.199	68.70
18.39	5.34	109.8	919.0	3.54	1.728	1.709	1.685	86	(−)	10	.200	77.59
17.74	5.29	108.2	870.9	3.13	1.642	1.637	1.620	51	(−)	5	.202	65.80
18.06	5.33	108.1	913.0	3.40	1.718	1.716	1.709	80	(−)	−3[a]	.210	77.85
18.06	5.31	105.9	912.7	3.23	1.692	1.687	1.670	60	(−)	17	.211	74.41
18.04	5.29	105.6	922.7	3.08	1.632	1.627	1.616	70	(−)	26	.203	68.83
18.10	5.33	109.6	897.9	3.19	1.650	1.642	1.628	74	(−)	22	.201	68.59
18.08	5.28	—	1771.8	2.93	1.613	1.605	1.593	65	(−)	—	.206	63.76
17.84	5.28	—	1775.3	3.14	1.682	1.671	1.661	85	(+)	—	.214	70.21

[a] X $\wedge$ c.

C2/m versus I2/m

Most scientific papers and textbooks assign monoclinic amphiboles to space group *C*2/*m*, but this author remains adamant in opting for the equivalent space group *I*2/*m* as preferable. It can be seen that either cell may be outlined from the points shown in Figure 15.15, and one may be transposed to the other as shown in Figure 15.15, where the short diagonal of the *C* cell = *a* of the *I* cell, and the *β* angle of the *I* cell is larger than that of the *C* cell. The *b* and *c* axes and the unit cell volume are not affected by this transposition. The choice of the unit cell as *C*2/*m* or *I*2/*m* and the reasons offered in favor of each have been discussed by Jaffe, Robinson, and Klein (1968). The principal argument in favor of the *I* cell is shown in Figure 15.16. In *C*2/*c* pyroxenes and *I*2/*m* amphiboles, which contain similar crystal chemical units, the *a* axes point in the same direction, and all octahedra orient or cant in a common direction. If the *C* cell is chosen for amphibole, pyroxene and amphibole orient their *a* axes and octahedra in opposite directions: this choice makes no sense. The *C* cell is chosen by crystallographers who look at spots on X-ray patterns, whereas the *I* cell is favored by optical mineralogists who must orient plates of the mineral under the microscope.

Exsolution in amphiboles

Because most amphiboles crystallize at temperatures of ~ 500–600°C in metamorphic rocks, lower than that for magmatic rocks, the solid solution

Figure 15.15. The *C*2/*m* and *I*2/*m* cells outlined from the same set of lattice points. The *I* cell, not the *C* cell, corresponds to the *C* cell of monoclinic pyroxene.

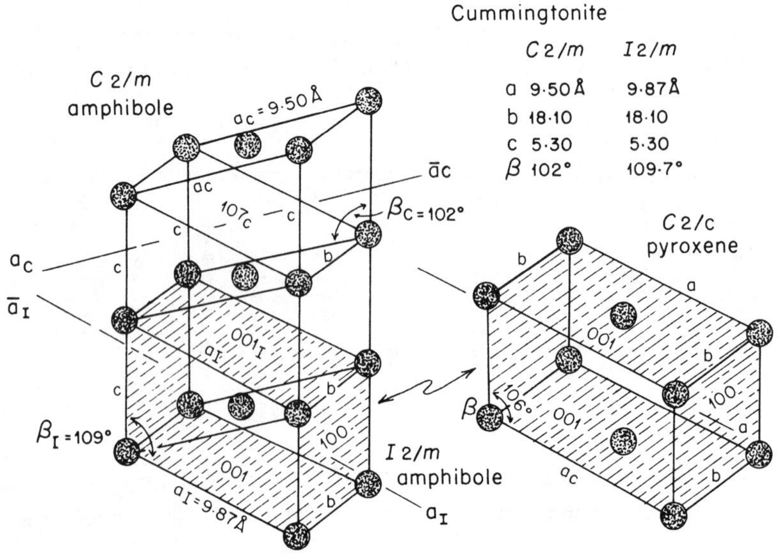

will be more limited. Lamellae will exsolve at a lower temperature and be much thinner and more difficult to identify than those found in pyroxenes. Nevertheless, they have been identified in amphiboles of very diverse composition. Many of them have been discussed by Ross, Papike, and Shaw (1969), Jaffe et al. (1968), and Robinson et al. (1971) from a number of American locations.

Figure 15.16. Optic orientation and alternative sets of cells for cummingtonite (CUMM), hornblende (HNBLD), and diopside. Diagram to the right shows cummingtonite host (white) and hornblende host (black) in contact with both phases showing exsolution lamellae. Exsolution lamellae parallel to (100) and traces of (110) cleavages (not shown) meet at 178° angle at contact between hosts. Alternate cells with *Y* vibration directions and exsolution planes are shown to left, and are compared with an oriented portion of the internal structure that is similar in all three minerals. The drawing is adapted from the structure of diopside, and the octahedra shown are M-1. For amphiboles the corresponding octahedra are M-1 and M-2. Reprinted with permission from H. W. Jaffe, P. Robinson, and C. Klein. Exsolution lamellae and optic orientation of clinoamphiboles, *Science,* 160: 777–8. Copyright © 1968 by the American Association for the Advancement of Science.

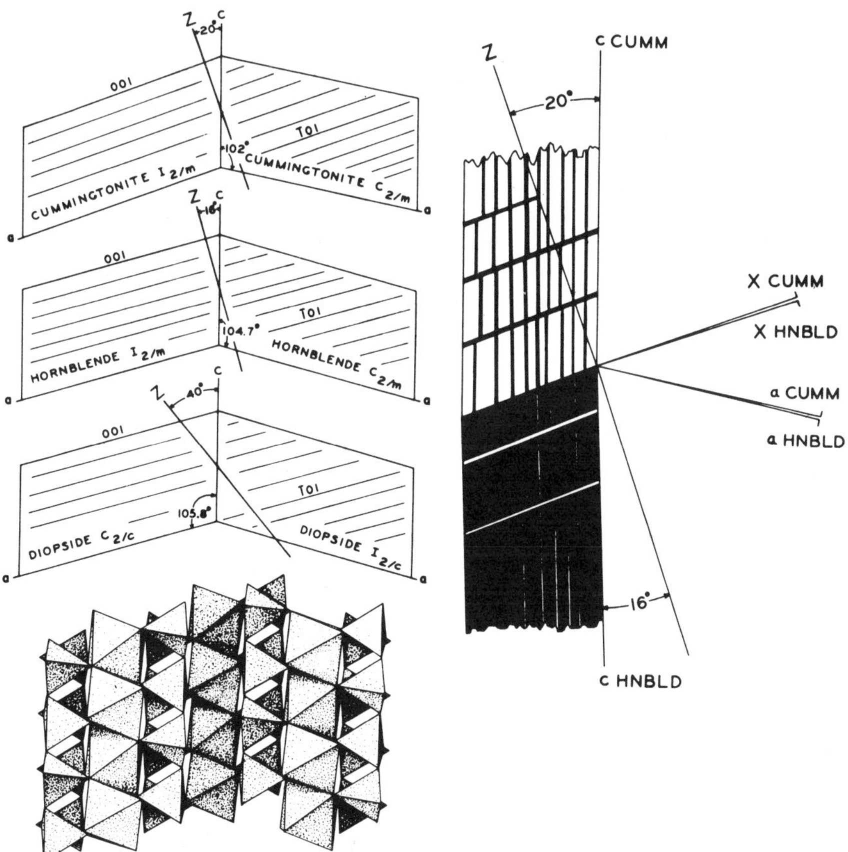

Lamellae exsolve from host amphiboles at similar angles and for the same reasons of lattice misfit as described for pyroxenes, but because of their fine size they are often overlooked in microscopic work. The following are some of the kinds of exsolution lamellae that have been identified in amphiboles:

Host hornblende	lamellae cummingtonite
Host cummingtonite	lamellae hornblende
Host tremolite	lamellae *P* cummingtonite
Host riebeckite	lamellae magnesioarfvedsonite
Host gedrite	lamellae anthophyllite

The **Z** vibration direction of slowest velocity of light, readily identifiable in the microscope, has been found to be oriented nearly perpendicular to the most common set of amphibole exsolution lamellae, those exsolved on irrational crystallographic planes near but not equal to (001) of the $I2/m$ cell. This helps to fix the orientation of mineral grains in the microscope. Illustrations of some amphibole exsolution lamellae are given in Figure 15.16.

In the discussion of exsolution in pyroxenes, the importance of lattice misfit a_{host}/a_{lam} and c_{host}/c_{lam} in establishing exsolution angles and their directions was underscored. In applying the same theory to amphiboles, one must be careful to use unit cell dimensions of either the *I* cell or the *C* cell consistently in comparing exsolution angles of host and lamellae.

The substitution of various ions of different radius may systematically affect some lattice dimension of amphiboles in a useful way; for example, the *b* dimension of amphiboles is controlled largely by the radius of the cation occupying the M-2 site, the smallest of the three octahedral sites designated as M-2, M-1, and M-3. Thus, from Table 15.3, glaucophane with small Al (0.57 Å) in M-2 has a small $b = 17.74$, and gedrite has $b = 17.84$; riebeckite with Fe^{3+} (0.64 Å) in M-2 has $b = 18.06$; tremolite with Mg^{2+} (0.72 Å) has $b = 18.05$ Å; and grunerite with the larger Fe^{2+} (0.77 Å) in M-2 has $b = 18.39$ Å.

In monoclinic amphiboles, $I2/m$, the *a* and *c* lattice dimensions do not show any clear pattern of change with cation occupancy. Their β angle ($a \wedge c$) does show a tendency to decrease with increasing radius of the cation in the $M\text{-}4^{VIII}$ site. Thus

	Ion in M-4	Radius (Å)	β (°) *I* cell
Cummingtonite	Mg^{2+}	0.89	109.9
Grunerite	Fe^{2+}	0.91	109.8
Glaucophane	Na^{+}	1.16	108.2
Riebeckite	Na^{+}	1.16	108.1
Tremolite	Ca^{2+}	1.12	106.6

By careful measurement of X-ray powder diffraction patterns, one can learn to identify or index each line with an interplanar spacing (*hkl*) and then use the relations given earlier for variation in the unit cell dimensions with occupancy of certain sites by cations of different radii.

Phyllosilicates, clays, and micas

Almost everyone is familiar with the white, fine-grained, soft clays; pale green extremely soft talc; green serpentines; lustrous, dark red-brown flakes of biotite; lustrous, silvery flakes of muscovite; and pearly green flakes of chlorite. All of these are phyllosilicates = sheet silicates = layer-lattice silicates.

All these minerals have their SiO_4 tetrahedra polymerized into infinite two-dimensional sheets of composition $[SiO_{2.5}]_n$, and all contain hydroxyl (OH) ions coplanar with six oxygens that form the upward- and downward-pointing apices of hexagonal rings of tetrahedra (see Figs. 12.1, 15.1, 15.17). Each tetrahedron of a phyllosilicate shares all three of its basal corner oxygens with those of adjoining tetrahedra.

The simplest example of a phyllosilicate is kaolinite, $Al_2[Si_2O_5](OH)_4$ (Fig. 15.17). Kaolinite crystallizes in the triclinic system in space group $C\bar{1}$. The C-centered lattice is shown by each void in the basal oxygen tetrahedral layer 1, and each (OH) in the basal octahedral layer 2A (Fig. 15.17). Thus, a basal oxygen sheet of hexagonal rings capped by hexagonally disposed silicon atoms (one above each triple junction of oxygens) has the composition $[Si_4O_6]$. One oxygen placed above each silicon forms a tetrahedron, and these apical oxygens form a ringlike sheet of composition O_4.

Now layers 1 and 2 (Fig. 15.17) form a tetrahedral layer, symbolized T or T-layer. If hydroxyl ions are placed in each O^{2-} ring of layer 2, coplanar with each O^{2-} ion, they will lie directly above the voids of layer 1. Now the layer marked 2A becomes the base of an octahedral layer but also remains the top of the tetrahedral layer. Addition of the hydroxyl in the O^{2-} rings now makes available a sheet of triangular cation sites, with each three anions, two O^{2-} and one $(OH)^-$, forming the lower triangle of an octahedron. When this is covered by a top sheet of anions, all $(OH)^-$ in kaolinite, each Al^{3+} lies inside the triangular antiprism we call an octahedron.

It is important to see that all of these sites, usually cited as three of three, may be filled by divalent cations, giving a trioctahedral layer; or some of them, most commonly two of three, may be occupied by trivalent cations, yielding a dioctahedral layer. All phyllosilicates may then be subdivided into dioctahedral and trioctahedral types (Fig. 15.18). Most phyllosilicates are monoclinic; some are triclinic; rare examples are orthorhombic. The monoclinic symmetry results solely from the placement of the upper layer over the voids in the lower anion sheets of the octahedral layer. The upper layer may be placed or staggered in three different 120° directions or vectors on the lower layer. We may refer to these as N-, SE-, and SW-stagger directions (Fig. 15.18). In kaolinite, the hydroxyl layer 3 (Fig. 15.18) is staggered N, and this completes the stacking of the simple layer as one T-layer overlain by one O-layer followed by a repetition: this

Figure 15.17. Layers of anions and cations in phyllosilicate or sheet structure minerals. Kaolinite is illustrated. It is dioctahedral, having two of three octahedral sites occupied. Photos reprinted with permission from J. Wear, J. E. Steckel, V. M. Fried, and J. L. White, Clay mineral models, *Soil Sci.*, 66(2): 113. Copyright © by Williams & Wilkins, 1948.

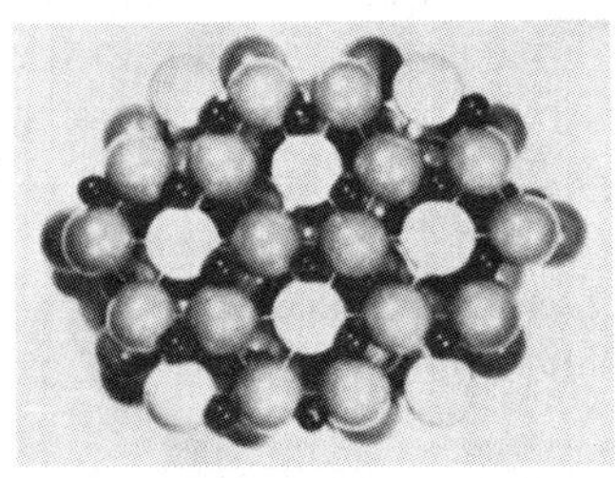

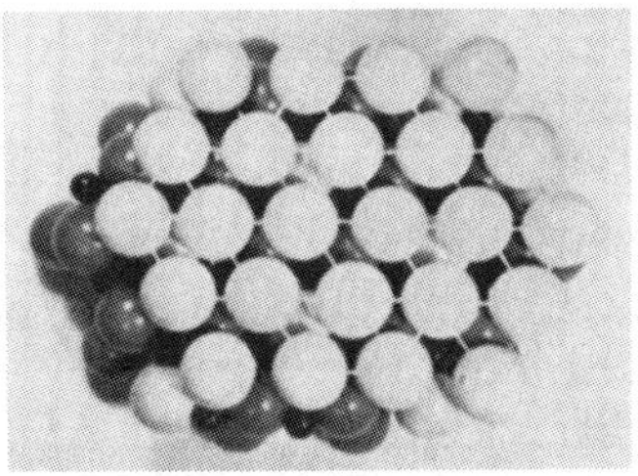

A

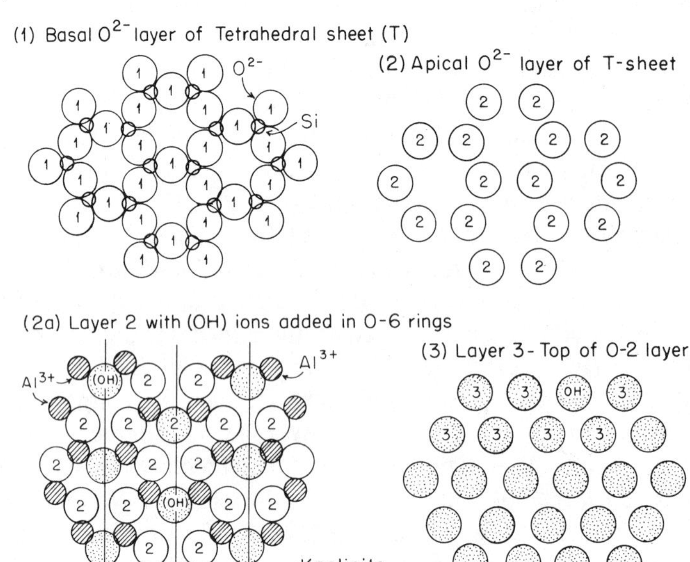

B

is called a two-layer sheet silicate, analogous to an open-faced sandwich (Fig. 15.19A). Now the single positive charge on each H^+ ion of each $(OH)^-$ anion will be oriented upward, bonding the top of the octahedral (O) layer to the basal oxygens of the succeeding tetrahedral (T) layer of the next unit cell. This is a hydrogen bond (see Fig. 12.17). Every upper sheet of oxygen is staggered N, in the same sense, in each successive unit cell.

Talc is an example of a trioctahedral three-layer phyllosilicate, $\square Mg_3[Si_4O_{10}](OH)_2$ (Fig. 15.19B), where an octahedral layer with all three of three sites occupied by Mg^{2+} ions is sandwiched between opposite-pointing tetrahedral layers, giving a stacking along the *c* axis of the T-O-T repeat. Again, all octahedral sheet stagger is unidirectional, N, in each unit.

Note that the six oxygen ions in rings of each basal tetrahedral layer will be back to back with and directly below the six in each succeeding repeat layer. This creates the double-six ring that forms the hexagonal prismatic coordination site, CN XII, of the micas. In talc, this site is empty, □, and oxygens of each basal layer are held together solely by weak van der Waals chemical bonds, which accounts for the softness of talc and the great ease of slippage of one layer over another. This property gives rise to its utility in face powders and as a carrier medium for dispersing insecticide powders. In chlorite (Fig. 15.19C), a

Figure 15.18. Directions of stagger of upper over lower anion sheet in the octahedral layer are limited to three, at 120°. Stagger may be unidirectional or polydirectional in consecutive layers, leading to polymorphism of micas. Only the lower sheet is shown.

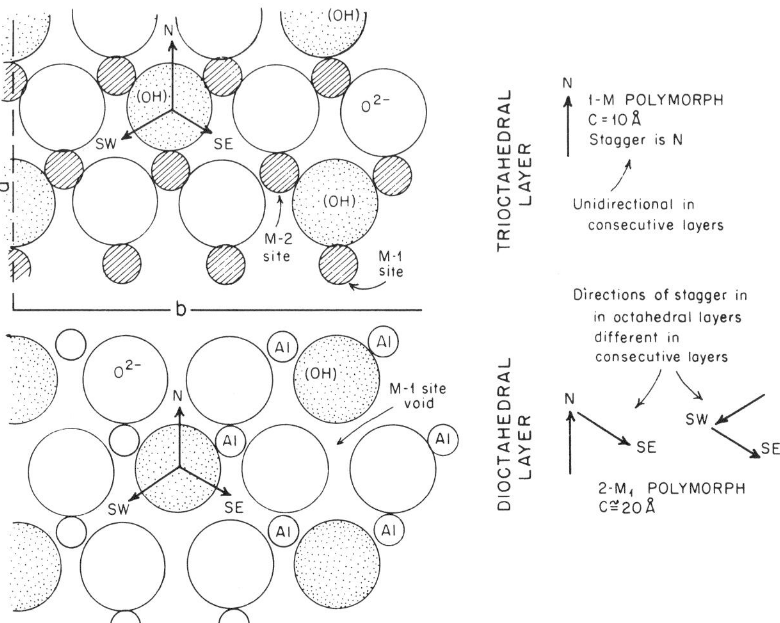

second octahedral layer is placed over this talc-like unit before a repeat takes place.

In a trioctahedral mica such as phlogopite (Fig. 15.19D), one Si^{4+} of each four is replaced by an Al^{3+} in each tetrahedral layer, undercharging the sheet. Electrical compensation is achieved by placing a K^+ ion into a CN XII site. Thus

$\square Mg_3Si_4O_{10}(OH)_2$	Talc
$KMg_3Si_3AlO_{10}(OH)_2$	Phlogopite
$K^+Al^{3+} \rightleftharpoons \square Si^{4+}$	

which demonstrates a now familiar coupled substitution involving a vacant lattice site. Each O-layer has a unidirectional N-pointing stagger, and phlogopite repeats itself above the K^+ layer; we have here a 1-M trioctahedral sheet silicate polymorph. Note that the basal layers are now held together by K^+ ions, which produces an ionic bond much stronger than the weak van der Waals or H bonds of talc and kaolinite, respectively. Mica sheets do not slip by one another as do talc sheets, but have to be peeled off one another in groups of sheets. Differences in the plasticity or "peelability" of sheets will vary with the radius and electric charge of the interlayer cation.

In example E (Fig. 15.19) a doubled mica cell is built like a double-decker sandwich. The example shown is muscovite, a dioctahedral, double-layer mica designated a 2-M polymorph (see Fig. 15.1). Such differences in staggering of successive layers and somewhat similar composition of micas may result in bizarre, "skyscraper"-like edifices known as mixed-layer lattices, with repeat distance of 20 or 30 Å.

Figure 15.19. Modified Pauling (1930) layer thicknesses for building phyllosilicate minerals.

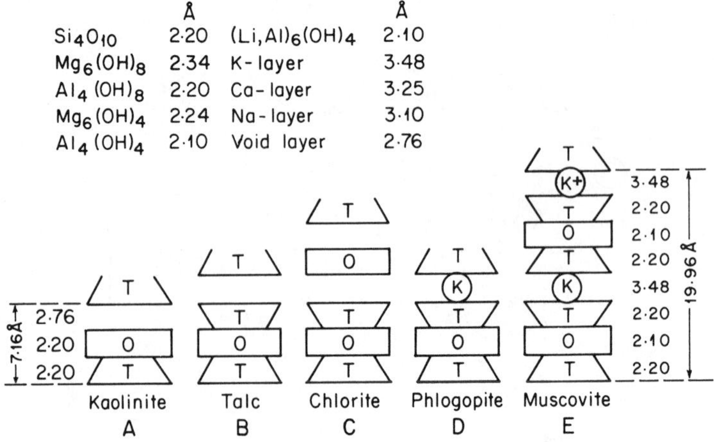

The large CN XII site in micas such as biotite becomes the repository for a number of large-radius ions that cannot fit into other mineral lattices. Rb^+ and Cs^+, the very large radius rare alkalies, concentrate heavily in the CN XII or K^+ site in biotite and muscovite, particularly those that grow during late magmatic or vapor-rich pegmatitic stages. Other large ions that concentrate in this site are Ba^{2+}, Sr^{2+}, Pb^{2+}, Tl^+, and $(NH_4)^+$. Of particular interest to geologists is the concentration of ^{40}Ar captured from the decay of radioactive ^{40}K, which is present in small amounts in all potassic minerals rich in nonradioactive ^{39}K, and useful in determining ages of minerals and rocks.

Li^+, unlike its large-radius radioactive neighbors (K^+,Rb^+) of periodic group IA, is a relatively small ion, comparable in radius to Fe^{2+} and Mg^{2+}, and almost always occupies octahedral CN VI sites in micas, such as in flesh-pink lepidolite common in the pink tourmaline-rich pegmatites of Pala, California.

Schematic T-O-T diagrams of the most common phyllosilicates are given in Figure 15.19. From each of these, it is possible to add up the chemistry of each layer to obtain the formula. Pauling (1930, 1960), after studying the fit of layers of oxygen in sheet silicate stacks, derived a value of 2.20 Å for the thickness of both tetrahedral and octahedral layers along *c*, and fairly successfully reconstructed calculated repeat distances that matched reasonably well with values from X-ray data. This author has modified Pauling's 2.20-Å value for layers of different composition and derived the values of Figure 15.19, which, when added for approximate composition, will match measured repeat distances along *c* (001) even more closely.

X-ray powder diffraction data for phyllosilicates

Perhaps more X-ray powder diffraction data have been obtained for phyllosilicates than for any other silicate group. Because of the layered nature of this group and the expandability of the layers, both when heated and when treated with ethylene glycol, measurement of basal spacing of repeat layers serves to identify most species. In addition, the X-ray spacing 060, the sixth order of reflection of the (010) planes of micas, is present in all X-ray powder diffraction patterns. Its measurement permits a determination of the dioctahedral or trioctahedral nature of any phyllosilicate mineral. Because Al^{3+} (r^{VI} = 0.53 Å) is considerably smaller than either Mg^{2+} or Fe^{2+} (r^{VI} = 0.66, 0.72 Å), and because octahedral sites are so numerous, the octahedra contract or expand in the *b* lattice dimension, giving values of

d_{060} = 1.487–1.504 Å   dioctahedral
d_{060} = 1.520–1.541 Å   trioctahedral

By using values for the basal spacing, d_{00l}, along with those of d_{060} and measurements of expanded lattices, we can easily identify phyllosilicates and their unit cell dimensions (Fig. 15.20 and Table 15.4). An example of the calculation of unit cell dimension from limited X-ray data for muscovite is given in Table 15.5.

Optical properties of phyllosilicates

Examples of the slow velocity of the electric vector associated with the high electron density parallel to layers in phyllosilicates and other layered minerals has been emphasized throughout this book (see Chapters 11, 12) and need not be further explored here. Summing up these data, it is evident that most layer-lattice minerals are optically negative with a high birefringence because of the strong contrast between the velocity of light vibrating parallel to the layers (high index of refraction) and perpendicular to the layers (low index of refraction). The *c* axis of most such minerals is perpendicular to the layers and parallel to the fast velocity (direction of the low index of refraction). When the electric vector vibrates parallel to the layers, high absorption and a high index of refraction are associated with it, aided by the overlapping of atomic electron orbitals in the plane of layering.

Figure 15.20. Classification of common phyllosilicate minerals on the basis of the identification of the repeat distance in Å of the first basal spacing measured by X-ray diffraction.

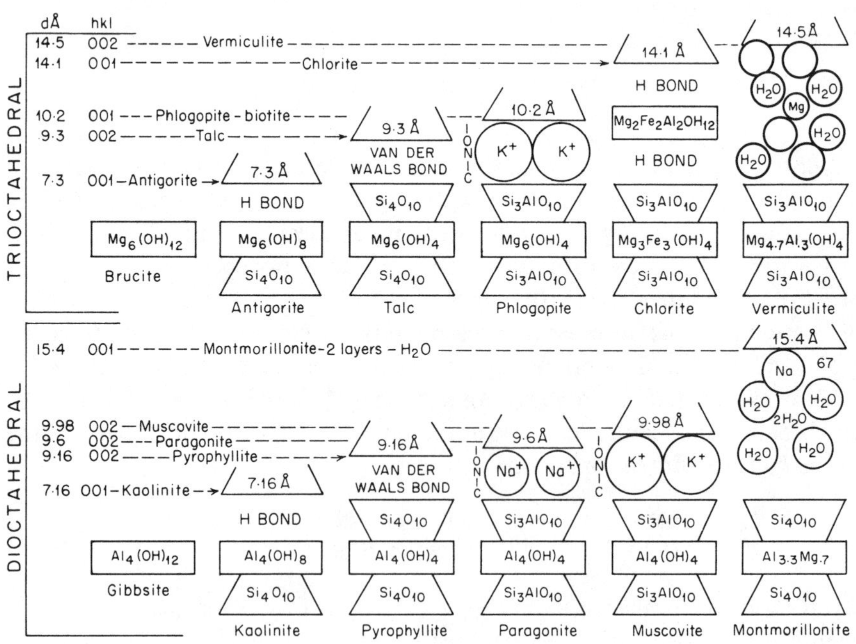

Table 15.4. *Diagnostic X-ray d-spacings of phyllosilicates*

	hkl			Octahedral
	001	002	060	layer
Montmorillonite	15.4		1.500	Di
Vermiculite	—	14.4	1.537	Tri
Chlorite	14.3	7.18	1.539	Tri
Halloysite	10.1		1.487	Di
Muscovite	—	9.98	1.504	Di
Biotite	10.1		1.527	Tri
Phlogopite	10.2		1.528	Tri
Talc	—	9.30	1.520	Tri
Pyrophyllite	—	9.16	1.488	Di
Dehydrated montmorillonite	9.5		1.500	Di
Kaolinite	7.16		1.489	Di
Metahalloysite	7.41		1.480	Di
Antigorite	7.30		1.536	Tri
Clinochrysotile	—	7.36	1.541	Tri

Table 15.5. *Important relations between X-ray diffraction data and unit cell dimensions of phyllosilicates*

Example: muscovite, polymorph 2M-1

$\beta = 95°$, $\sin\beta = 0.9962$, space group $C2/c$

$d_{060}Å \times 6 = b_0$ $1.504 \times 6 = 9.024\ Å = b_0$

$b_0/\sqrt{3} = a_0$ $9.024/\sqrt{3} = 9.024/1.732 = 5.21\ Å = a_0$

$d_{001} = c \sin\beta/\sin\beta = c_0$

$d_{002} = 9.98 \times 2 = 19.96 = c \sin\beta = 19.96/0.9962 = 20.04\ Å = c_0$

Bibliography

Bailey, S. W. (1980). Structures of layer silicates. In *Crystal structures of clay minerals and their identification.* Mineral. Soc. Monograph 5, pp 1–124.

Brown, W. (1972). La symétrie et les solutions solides des clinopyroxènes. *Bull. Soc. France Minéral. Cristallog.,* 95: 574–82.

Burnham, C. W. (1965). Ferrosilite. Carnegie Inst. Yearbook 65, Geophys. Lab., Washington, D.C., pp. 285–90.

Burnham, C. W., Clark, J. R., Papike, J. J., and Prewitt, C. I. (1967). A proposed crystallographic nomenclature for clinopyroxene structures. *Z. Krist.,* 125: 1–6.

Cameron, M., and Papike, J. J. (1980). Crystal chemistry of silicate pyroxenes. *Mineral. Soc. Am. Rev. Mineral.,* 2: 5–92.

Clark, J. R., Appelman, D. E., and Papike, J. J. (1969). Crystal chemical characterization of clinopyroxenes based on eight new structure refinements. *Mineral. Soc. Am. Spec. Pap.,* 2: 31–50.

Hawthorne, F. C. (1983). The crystal chemistry of the amphiboles. *Can. Mineral.,* 21: 173–480.

Jackson, W. W., and West, J. (1930). The crystal structure of muscovite – $KAl_2(AlSi_3O_{10})(OH)_2$. *Z. Krist., Kristallogeom.,* 76: 211–27; 85: 160–4.

Jaffe, H. W., Robinson, P., and Klein, C., Jr. (1968). Exsolution lamellae and optic orientation of clinoamphiboles. *Science,* 160: 776–80.

Jaffe, H. W., Robinson, P., and Tracy, R. J. (1978). Orthoferrosilite and other iron-rich pyroxenes in microperthite gneiss of the Mount Marcy area, Adirondack Mountains. *Am. Mineral.,* 63: 1116–36.

Jaffe, H. W., Robinson, P., Tracy, R. J., and Ross, M. (1975). Orientation of pigeonite exsolution lamellae in metamorphic augite: correlation with composition and calculated optimal phase boundaries. *Am. Mineral.,* 60: 9–28.

Jaffe, H. W., and Schumacher, J. S. (1985). Garnet and plagioclase exsolved from aluminum-rich orthopyroxene in the Marcy anorthosite, northeastern Adirondacks, New York. *Can. Mineral.,* 23: 457–78.

Klein, C., Jr. (1964). Cummingtonite-grunerite series: a chemical, optical, and X-ray study. *Am. Mineral.,* 49: 963–82.

Kroll, H., Bambauer, H. V., and Schirmer, U. (1980). The high albite-monalbite and analbite-monalbite transitions. *Am. Mineral.,* 65: 1192–1211.

Liebau, F. (1962). Die Systematik der Silikate. *Naturwissenschaften,* 49: 481–91.

(1985). *Structural chemistry of silicates.* Springer-Verlag, New York.

Matsumoto, T. (1974). Possible structure types derived from *Pbca* orthopyroxenes. *Min. Mag.,* 7: 374–83.

Morimoto, N., and Koto, K. (1969). The crystal structure of orthoenstatite. *Z. Krist.,* 129: 65–83.

Morimoto, N., and Güven, N. (1970). Refinement of the crystal structure of pigeonite. *Am. Mineral.,* 55: 1195–1209.

Papike, J. J., and Clark, J. R. (1968). The crystal structure and cation distribution of glaucophane. *Am. Mineral.,* 53: 1156–73.

Papike, J. J., Ross, M., and Clark, J. R. (1969). Crystal-chemical characterization of clinoamphibole based on five new structure refinements. *Mineral. Soc. Am. Spec. Pap.,* 2: 117–36.

Pauling, L. (1930). The structures of the micas and related minerals. *Proc. Natl. Acad. Sci. U.S.A.,* 16: 123–9.

(1960). *The nature of the chemical bond,* 3d ed., Cornell Univ. Press, 558–9.

Prewitt, C. T., and Burnham, C. W. (1966). The crystal structure of jadeite. *Am. Mineral.,* 51: 956–75.

Robinson, P. (1980). The composition space of terrestrial pyroxenes – internal and external space limits. *Mineral. Soc. Am. Rev. Mineral.,* 2: 419–94.

Robinson, P., Jaffe, H. W., Ross, M., and Klein, C., Jr. (1971). Orientation of exsolution lamellae in clinopyroxenes and clinoamphiboles: consideration of optimal phase boundaries. *Am. Mineral.,* 56: 909–39.

Robinson, P., Ross, M., Nord, G. L., Jr., Smyth, J. R., and Jaffe, H. W. (1977). Exsolution lamellae in augite and pigeonite: fossil indicators of lattice parameters at high temperature and pressure. *Am. Mineral.,* 62: 857–73.

Ross, M., Papike, J. J., and Shaw, K. W. (1969). Exsolution textures in amphiboles as indicators of subsolidus thermal histories. *Mineral. Soc. Am. Spec. Pap.,* 2: 275–99.

Smith, J. V., and Yoder, H. S. (1956). Experimental and theoretical studies of the mica polymorphs. *Min. Mag.,* 31: 209–35.

Smyth, J. R. (1974). The high temperature crystal chemistry of clinohypersthene. *Am. Mineral.,* 59: 1069–82.

Sueno, S., Cameron, M., and Prewitt, C. T. (1976). Orthoferrosilite: high temperature crystal chemistry. *Am. Mineral.,* 61: 38–53.

Wear, J. I., Steckel, J. E., Fried, M., and White, J. L. (1948). Clay mineral models: construction and implications. *Soil Sci.,* 66(2): 111–7.

16 Tektosilicates

Tektosilicates (Greek τεκτο = framework, carpenter's lattice) are true three-dimensional frameworks in which each tetrahedron shares all four of its corners with other tetrahedra. The ratio of tetrahedral cations to oxygen is thus always 1 : 2, and it must be emphasized that tetrahedral Al^{3+} must be added to Si^{4+} to obtain the proper ratio. The following groups of tektosilicate minerals are all extremely important to the petrology of silicate rocks: feldspars, feldspathoids, and zeolites.

Because of this extensive corner sharing of Si and Al tetrahedra, tektosilicates form open structures with low density, dominated by atoms with low coordination numbers. It was noted earlier that minerals built largely from regular polyhedra tend to be isotropic or to have low birefringence, and this feature is common to quartz, feldspars, feldspathoids, and zeolites. Many tektosilicates thus build large "birdcage" structures, in which Si + Al tetrahedra enclose large void spaces or cavities. This is true of the feldspars, but much more so of the zeolites. In many such minerals, relatively small cations occupy large sites that are by no means true coordination sites. Such cations may occupy only a very small volume of these large voids; they are located in these sites to balance the charge on the negatively charged tetrahedral framework. Because they are relatively loosely bound, they can easily be chemically exchanged by treatment with solutions rich in other cations. Such birdcage zeolite structures have the ability to trap large molecules that are passed through their structure. They are true molecular sieves and are so used in the cracking or refining processes in the petroleum industry, as well as in other industrial uses. Because of this property of natural zeolites, several industrial research laboratories have synthesized zeolites with void sites that are tailor-made for a particular industrial use.

Quartz and SiO_2 polymorphs

Silica or silicon oxide crystallized from its melt at atmospheric pressure will, with an increase in temperature, undergo a series of polymorphic transformations before melting at 1713°C (Fig. 16.1). Trigonal α-quartz, stable up to 573°C, will invert to hexagonal β- or high quartz by a mere twisting of its links. Such a change in which no bonds are broken is classed as a *displacive transformation.* When the temperature is further raised to 870°C, a major reorganization of tetrahedra takes place, with breaking and restructuring of Si—O tetrahedral bonds. This is classed as a *reconstructive phase transformation.*

Hexagonal tridymite formed at this temperature remains stable up to 1470°C, where it gives way, also by reconstructive phase transformation, to cristobalite, which becomes isometric at its highest temperature. Cristobalite is stable to 1713°C, where it melts to liquid SiO_2. The temperature of the displacive transformation of low or α-quartz to high or β-quartz is reduced by about 25°C/kbar of increasing pressure (Fig. 16.1). Such polymorphic transitions depend only on the increase of total pressure (at similar temperature), and not on the presence of a vapor phase.

Figure 16.1. Polymorphism in SiO_2: the *P-T* diagram for the system SiO_2:H_2O with the melting curve of granite for reference.

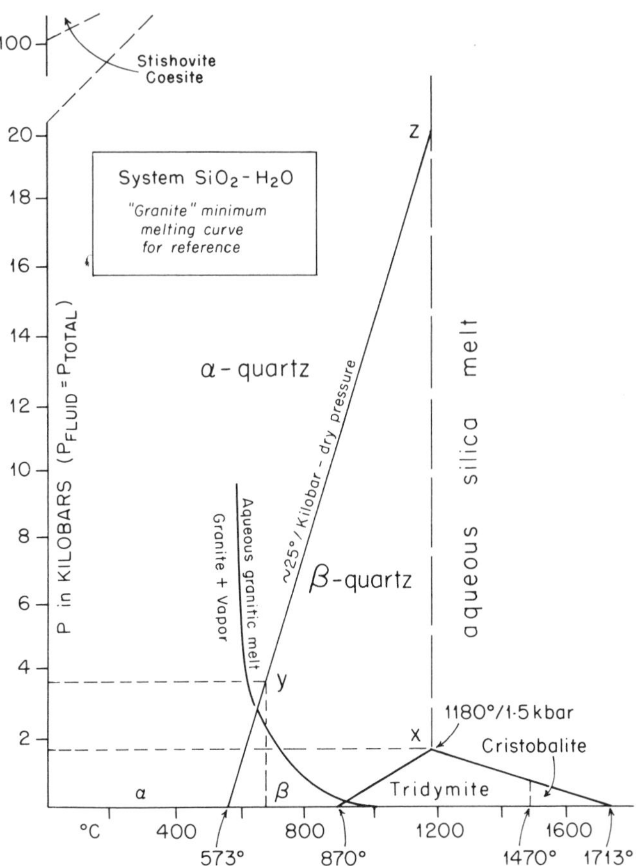

The polymorphic forms of SiO_2 just described remain tektosilicates, with all Si atoms residing in the tetrahedra. This is also true of an additional monoclinic polymorph, coesite, which crystallizes at high pressure (Fig. 16.1, Table 16.1). At extreme pressures, such as are induced by the explosion and vaporization of a meteorite weighing many hundreds of tons, the common α-quartz of sandstone may be converted to stishovite, the highest pressure form of SiO_2. This phase was, incidentally, synthesized in the laboratory before it was found in nature in sandstones of Meteor Crater, Arizona. The enormous pressures operative in its formation place six rather than four oxygens around each Si ion. Stishovite is thus no longer a tektosilicate. This change is accompanied by an increase in the radius of Si^{4+} from 0.26 to 0.42 Å and by a corresponding decrease in the radii of the enclosing oxygen ions. The extremely small interatomic distance of 2.15–2.29 Å has been measured by two different investigators across the centers of the two oxygens that form a common edge to two SiO_4 tetrahedra; in rutile, isostructural with stishovite, the analogous O—O distance is 2.50 Å (see Figs. 11.7, 18.7, 18.8). In stishovite, SiO_2, as in rutile, TiO_2, there are three metal atoms around each oxygen; it could thus be grouped with the oxides, as could the other SiO_2 polymorphs discussed earlier. However, all will be grouped here with the tektosilicates.

When a melt is saturated with water ($P_{fluid} = P_{total}$), the operative pressure,

Table 16.1. *Polymorphs of SiO_2 and their properties*

	Polymorph					
	α-quartz	β-quartz	Tridymite[a]	Cristobalite[a]	Coesite	Stishovite
Formula	$3/Si^{IV}O_2^{II}$	$3/Si^{IV}O_2^{II}$	$64/Si^{IV}O_2^{II}$	$4/Si^{IV}O_2^{II}$	$16/Si^{IV}O_2^{II}$	$2/Si^{VI}O_2^{III}$
Crystal System	Trigonal	Hexagonal	Orthorh.	Tetrag.	Monoclinic	Tetragonal
Space Group	$P3_221$	$P6_422$	$P222$	$P4_32_12$	$C2/c$	$P4_2mnm$
a (Å)	4.913	5.00	9.88	4.97	7.17	4.18
b (Å)	—	—	17.1	—	12.38	—
c (Å)	5.405	5.46	16.3	6.93	7.17	2.66
(g/cm³)	2.65	2.53	2.32	2.34	2.92	4.28
ω	1.544	—		1.487		1.800
ϵ	1.553	—		1.484		1.845
γ			1.473		1.599	
β			1.470		1.595	
α			1.469		1.594	
k	0.206	—	0.203	0.208	0.204	0.187
α_G ($Å^3$)	4.906	—	4.83	4.95	4.86	4.46

[a] β or high forms of tridymite and cristobalite also exist.

written P_{H_2O}, is transmitted via the very compressible fluid phase. This results in the lowering of all melting temperatures. Note that an increase of only 1.5 kbar as P_{H_2O} lowers the melting point of β-quartz about 533°C, from 1713°C to about 1180°C (point x, Fig. 16.1). At about 3.6 kbar, an aqueous granitic melt will crystallize α-quartz directly, bypassing the crystallization of β-quartz (point y, Fig. 16.1). Accordingly, β- or high quartz tends to occur principally in lavas or bodies of melt crystallized at very shallow crustal levels, whereas α-quartz will crystallize directly from granitic melts solidified at deeper crustal levels. Because the inversion of high or β-quartz to low or α-quartz is reversible but nonquenchable, β-quartz will always change to α-quartz on cooling, regardless of the rate of cooling. That quartz in an igneous rock crystallized originally as β-quartz can only be verified by crystal morphology. A flattened hexagonal bipyramid, rather than the familiar trigonal prism, verifies the original crystallization as β-quartz.

Crystal structure of α-quartz

The crystal structure of α-quartz is illustrated in both packing and polyhedral drawings (Fig. 16.2). All of the polyhedra shown are related by 3_1 (right-handed) screw triad axes parallel to the c axis, and hexagonal axes a_1, a_2, and a_3 are twofold axes of symmetry. The last entry in the space group symbol, 1, lacking a superscript bar, emphasizes the absence of a center of symmetry. There are no mirror planes of symmetry.

Right-handed α-quartz, $P3_12\ 1$, transforms on heating to right-handed β-quartz, based on the inheritance of right-handed 3_1 screw triad axes in the transformation. Although the screw axis at the center of the drawing of low quartz, also 3_1, changes to 6_4 (left-handed) in high or β-quartz, it is the 3_1 axes that determine the handedness.

Feldspars

Crystal chemistry

Feldspars are the most important of all naturally occurring solids (minerals), in that they make up more than 60% of the crust of the earth. Feldspars are difficult to illustrate or visualize, partly because they build complex tektosilicate frameworks, and partly because of their low symmetry. All feldspars crystallize in either the monoclinic or the triclinic crystal system.

Because feldspars are tektosilicates, they build open structures as full of void space as Emmenthal (Swiss) cheese. Built of elements with relatively low atomic number and low mass (O, Si, Al, Na, K, and Ca), their open framework leads to structures with large volume, low density (2.56–2.76 g/cm^3), low polarizability, low index of refraction, and low birefringence (Table 16.2). Structures such as these, made up essentially of [Si + AlO_4] tetrahedra, cannot build up a chain or sheet of electron density much greater in one direction of the crystal than in any other. A condition of isotropy or low birefringence thus prevails.

The simplest and best approach to visualization of the basic feldspar crystal structure is to think of it as being built of fourfold rings of upward- and downward-pointing tetrahedra of composition $[AlSi_3O_8]$ or of $[Al_2Si_2O_8]$. Every oxygen forms a corner of two such tetrahedra, because all tetrahedral corner oxygens are shared with those of other tetrahedra. Now, if these rings are joined, they form chainlike elements parallel to the *a* crystal axis (Fig. 16.3), and these elements, in turn, are joined laterally to form sheetlike elements parallel to (010) and (201) planes in the crystal (Fig. 16.3). When so joined together, the tetrahedral oxygens form large voids, each bounded by ten oxygens, with a few bounded by eight oxygens. Because the tetrahedral framework carries a negative charge $[AlSi_3O_8]^{1-}$ and $[Al_2Si_2O_8]^{2-}$ per formula unit, the large-radius alkali ions K^+ and Na^+ balance the charge in the 1− framework, and Ca^{2+} or

Figure 16.2. Packing model (upper) and polyhedral drawing (lower) of α-quartz, SiO_2.

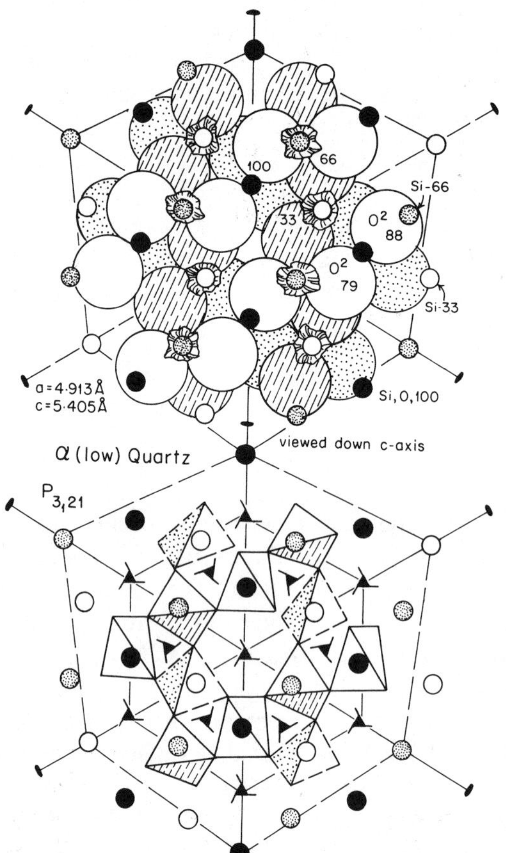

Table 16.2. *Unit cell dimensions, space group, density, indices of refraction, refractivity, and polarizability of feldspars*

Feldspar	Formula	a (Å)	b (Å)	c (Å)	α °	β °	γ °	Space group	γ	β	α	ρ	k	α_G (Å^3)
Sanidine	$4/KAlSi_3O_8$	8.56	13.03	7.17	90	116	90	$C2/m$	1.524	1.523	1.518	2.56	0.204	22.51
Orthoclase	$4/KAlSi_3O_8$	8.56	12.99	7.19	90	116	90	$C2/m$	1.522	1.521	1.518	2.56	0.203	22.40
Microcline	$4/KAlSi_3O_8$	8.56	12.97	7.22	90.7	116	87.7	$C\bar{1}$	1.521	1.518	1.514	2.56	0.202	22.28
Monalbite (high)	$4/NaAlSi_3O_8$	8.15	12.88	7.11	93.3	116.3	90	$C2/m$	1.534	1.532	1.527	2.60	0.204	21.65
Albite (low)	$4/NaAlSi_3O_8$(ab)	8.14	12.79	7.16	94.3	116.6	87.6	$C\bar{1}$	1.538	1.531	1.527	2.62	0.203	21.54
Labradorite	An 50	8.18	12.86	7.11	93.5	116	90	$C\bar{1}$	1.562	1.558	1.555	2.69	0.207	22.40
Bytownite	An 80	8.17	12.87	14.18	93.3	116	90.5	$I\bar{1}$	1.575	1.570	1.566	2.72	0.2045	22.93
Anorthite	$8/CaAl_2Si_2O_8$(an)	8.18	12.88	14.17	93.2	116	91.2	$P\bar{1}$	1.589	1.584	1.575	2.76	0.211	23.27
Celsian	$8/BaAl_2Si_2O_8$	8.63	13.05	14.41	90	115	90	$I2/c$	1.608	1.599	1.593	3.35	0.179	26.64
Reedmergnerite	$4/NaBSi_3O_8$	7.83	12.36	6.80	93	116	92	$C\bar{1}$	1.573	1.565	1.554	2.78	0.203	19.80
Buddingtonite	$4/(NH_4)AlSi_3O_8 \cdot H_2O$	8.59	13.04	7.18	—	113	—	$P2_1$	1.534	1.531	1.530	2.38	0.2235	22.74

Ba^{2+} in the 2– framework, to build all of the common feldspars (Table 16.2). These relatively large-radius cations occupy some, but not all, of the volume of the tenfold void sites, designated M-sites, built by the tetrahedral oxygens of the framework. Further, none of these cations in M-sites has CN X, because the framework oxygens tend to pack as closely as they can around each alkali or alkaline earth cation. To a first approximation, the radius ratio holds well, and the ions with the largest radii, K^+ ($r = 1.55$ Å) and Ba^{2+} ($r = 1.47$ Å), have CN IX, whereas Na^+ ($r = 1.13$ Å) and Ca^{2+} ($r = 1.07$ Å) occupy irregular CN sites with CN either VI or VII. Evidently the tetrahedral oxygen framework collapses only slightly around K^+ and Ba^{2+}, but considerably around the smaller Na^+ and Ca^{2+} ions. In the rare feldspar buddingtonite, $(NH_4)^+$ occupies the large M-site, substituting for K^+. In the equally rare feldspar reedmergnerite, B proxies Al in the tetrahedral sites of albite. Sanidine, $KAlSi_3O_8$, and monalbite the simplest of the feldspar structures, have the highest symmetry, crystallizing in the monoclinic space group $C2/m$, whereas albite, $NaAlSi_3O_8$, and anorthite, $CaAl_2Si_2O_8$, crystallize in the triclinic space groups $C\bar{1}$ and $P\bar{1}$, respectively. Evidently the collapse of the tetrahedral framework around the smaller cation results in a lowering of the symmetry, accompanied by reduction in volume and increase in both density and indices of refraction (Table 16.2).

Figure 16.3. Tetrahedral model of the basic feldspar structure, sanidine, $KAlSi_3O_8$, (010) projection.

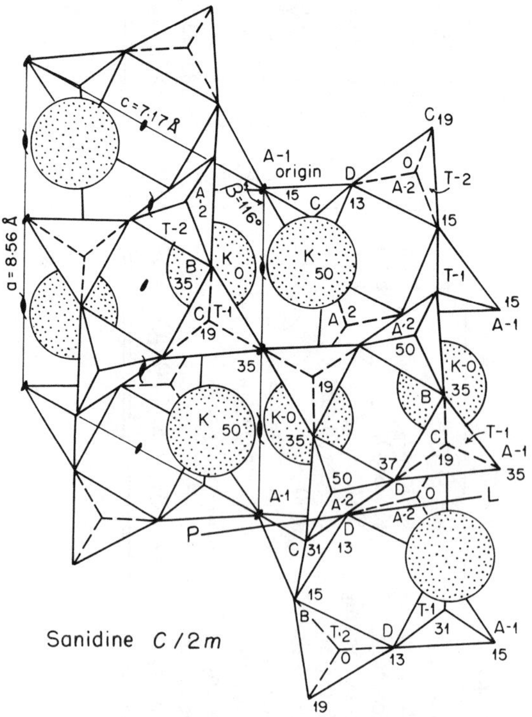

Some feldspar species exhibit complete solid solution at high temperature, as in the alkali feldspar group $KAlSi_3O_8$ and $NaAlSi_3O_8$ (usually symbolized *Or* and *Ab*, respectively), and in the plagioclase feldspar group $NaAlSi_3O_8$ and $CaAl_2Si_2O_8$ (symbolized *An*). These relations are illustrated in the ternary diagram (Fig. 16.4). Note that K^+ and Na^+ have substantially different radii but carry the same positive charge. Conversely, Na^+ and Ca^{2+} have very similar ionic radii; both these pairs show complete solid solution. Between K^+ and Ca^{2+}, hardly any solid solution occurs, because both the radii and the positive charges are different. Note that as the temperature of crystallization decreases so does solid solution, and complicated breaks in the solid solution series appear (Fig. 16.4). Let us state this in another way. In lavas and igneous melts crystallized at temperatures above about 800°C, extensive to complete solid solution series occur, whereas in igneous melts crystallized at lower temperatures and/or cooled very slowly over geologic time, interruptions in the solid solution series are observed.

Al—Si ordering and polymorphism

Closely allied to the phenomena of complete solid solution and restricted solid solution is the ordering, or its lack, of Si^{4+} and Al^{3+} in the tetrahedral site of each feldspar. It is the concentration, or its lack, of Si and Al atoms or ions in tetrahedral sites that characterizes the different polymorphs of the feldspar family. Thus, sanidine, orthoclase, and microcline represent the high, intermediate, and low temperature polymorphs, respectively, of $KAlSi_3O_8$ (Table 16.2). The high temperature polymorph sanidine, with sixteen tetrahedral sites in one unit cell, has its twelve Si + four Al atoms totally disordered in all sites, whereas the low temperature polymorph microcline has virtually all Al

Figure 16.4. Ternary diagram showing degree of solid solution between K, Na, and Ca feldspars crystallized at high and low magmatic temperatures. Solid solution fields are dotted.

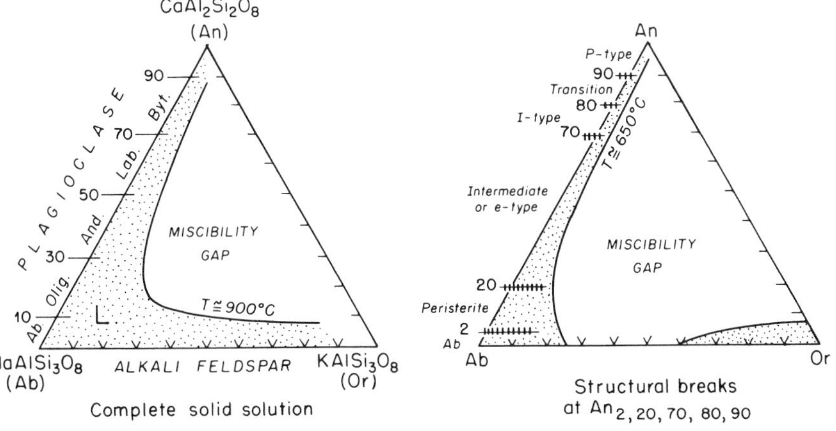

Table 16.3. *Tetrahedral–oxygen (T—O) distances*[a] *for selected feldspars: T—O = Si—O or Al—O*

Formula	Feldspar	Space group	T-site designation and T—O distance (Å)		Degree of ordering	T-sites/ unit cell
4/[$KAlSi_3O_8$]			T_1	T_2		
	Sanidine	*C*2/*m*	1.642	1.642	Total disorder	
	Orthoclase	*C*2/*m*	1.652	1.633		
			T_1o T_1m	T_2o T_2m		12 Si + 4 Al
	Microcline	$C\bar{1}$	1.741 1.614	1.611 1.612	Complete order	
4/[$NaAlSi_3O_8$]						
	Monalbite	*C*2/*m*	1.646 1.639	1.642 1.643	Disorder, near total	
	Albite	$C\bar{1}$	1.742 1.608	1.614 1.616	Order	
8/[$CaAl_2Si_2O_8$]						
			T_1ozo T_1ooo T_1mzo T_1moo	T_2ozo T_2ooo T_2mzo T_2moo		
	Anorthite	$P\bar{1}$	1.758 1.613 1.608 1.752	1.613 1.746 1.608 1.752	Total order	16 Si + 16 Al
			T_1ozi T_1ooi T_1mzi T_1moi	T_2ozi T_2ooi T_2mzi T_2moi		
			1.746 1.616 1.626 1.741	1.610 1.753 1.752 1.628		

[a] T—O distance (Å) indicates degree of ordering of Al and Si into specific tetrahedral sites: 1.744 = Al site; 1.610 = Si site; 1.642 = disordered Si + Al site.

atoms ordered into a site designated T_1O, and its twelve Si atoms all reside in the three remaining sites T_1m, T_2O, T_2m.

Because Al (atomic number 13) and Si (atomic number 14) have nearly identical atomic structure, it is virtually impossible to distinguish one from another on a single crystal X-ray photograph. They diffract X-radiation with equal efficiency. Occupancy of tetrahedral sites, or Si—Al ordering, must be determined by measuring T–O interatomic distance (Table 16.3) or O—O distances on X-ray film. Large distances indicate large tetrahedra containing Al, and small distances indicate tetrahedra containing Si. Thus, a T–O distance of 1.744 Å indicates complete ordering or occupancy of a tetrahedral site by Al, whereas a T–O distance of 1.611 Å indicates occupancy by Si alone. Thus, in reedmergnerite, where the small B atom proxies for Al in site T_1O, the T–O distance for the three remaining T-sites verifies the value for Si—O occupancy alone in albite. The value for T–O = 1.642 for both tetrahedral sites T_1 and T_2 in sanidine (Table 16.3) indicates random occupancy by Al and Si equivalent to total disorder for the polymorph crystallized at high temperature.

As temperature falls, feldspar polymorphs are expected to show increased cation ordering of Al into sites designated T_1O, reduced symmetry, increased density associated with reduction in volume, and, in some examples, an increase in cation coordination numbers. Because alkali feldspars and albite have twelve Si + four Al atoms in their sixteen tetrahedral sites of one unit cell, a spectrum of ordering is observed that results in the polymorphism shown (see Table 16.2). Anorthite, however, has a doubled *c* dimension (see Table 16.2) and has sixteen Al + sixteen Si in its thirty-two tetrahedral sites of one unit cell. Although there are sixteen different tetrahedral sites, the equal amount of Si and Al atoms present results in the perfect ordering of Si and Al, as seen in Table 16.3. Sixteen sites have the large, almost equal T–O distances appropriate to an Al—O tetrahedron, and the other sites, the small, almost equal T–O distance appropriate to Si—O occupancy. Thus, the two major factors that determine what feldspar will form are the Al—Si ordering in the tetrahedra and the radius of the cation that occupies the large CN VI–IX metal or M-site.

One further important restriction on Si—Al ordering in feldspars applies in all tektosilicates and probably in other silicate groups as well. This is the *aluminum avoidance principle,* which states that no two adjoining and linked tetrahedra may both be occupied by Al, because the linkage Al—O—Al across a bridging oxygen would severely undercharge this oxygen with respect to Pauling's Rule 2 of electrostatic bond strength (see Chapters 4 and 5). This rule obviously places severe limitations on possible patterns of Al—Si tetrahedral ordering in feldspars.

Careful X-ray studies made of feldspars show that the large cations K, Na, and Ca are disc-shaped and flattened rather than round. Because classical EBS solutions are difficult for feldspars, the discoid shape of the large alkali cations may represent the attempt of these cations to synchronize their occupancy of the large metal sites (CN IX in sanidine, Fig. 16.5) with partial, sequential occupancy of other large voids of CN VIII–X (Fig. 16.6).

Feldspar structure as typified by sanidine

The structure of sanidine (Figs. 16.3, 16.5, 16.6, 16.7) is representative of the feldspars in general and is the simplest to illustrate, because of its higher symmetry and the presence of a mirror or reflection plane parallel to (010). Whereas many mineralogists prefer to illustrate the sanidine crystal in the $(20\bar{1})$ projection (plane P-L, Figs. 16.3, 16.5), this author finds the (010) projection perpendicular to the *b* axis easier to represent and easier to show in packing models. The *b* axis or (010) projection (Figs. 16.3, 16.5, 16.6) shows twofold rotation axes parallel to *b* passing through those oxygen atoms designated A-1 and also through the centers of TO_4 rings. Screw diad axes, 2_1, also parallel to *b* (Figs. 16.3, 16.5) are located at $\frac{1}{4}$ and $\frac{3}{4}$ of the distance along the *z* axes and inside the unit cell midway between them.

Note (Figs. 16.3, 16.5) that oxygens in sanidine are labeled A-1, A-2, B, C, and D. There are two types of tetrahedra: T-1, consisting of oxygens A-1, B, C, and D; and T-2, made up of A-2, B, C, and D. Each fourfold tetrahedral ring contains two each of T-1 and T-2 tetrahedra, which point alternately up and down the *b* axis. Atoms designated A-2 and K^+ are located on the sole reflection plane present (010) and are the only atoms on the reflection or mirror plane. They are thus located at elevations 0, 50, and 100 up the *b* axis. A packing or ball

Figure 16.5. Sanidine, in ion-packing drawing, projected along *b*, (010) projection. Rings of $[TO_4]_4$ form warped oxygen sheets parallel to (010), shown at elevation near 65.

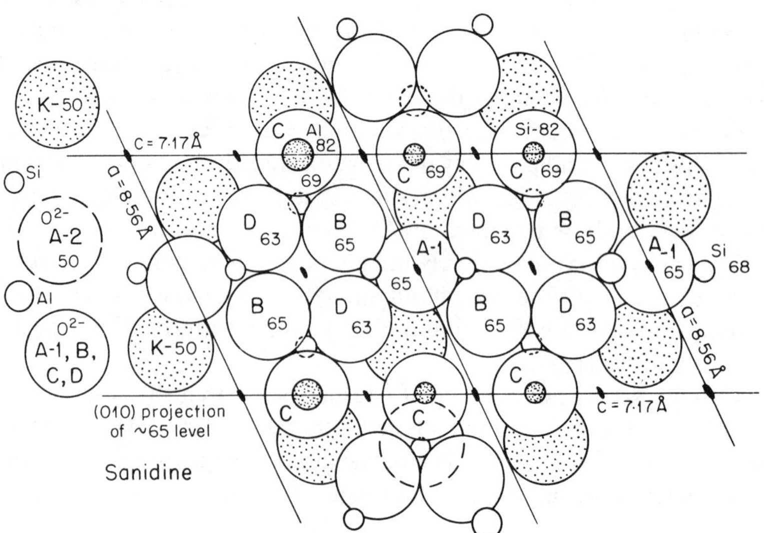

model projected on (100) along $a \sin \beta$ (Fig. 16.7) shows the mirror plane very well and reveals the presence of large tunnels of void sites running parallel to $a \sin \beta$ and located on the mirror plane, where they intersect the large tunnels parallel to the b axis (Fig. 16.6) as well as tunnels parallel to the c axis. The

Figure 16.6. Idealized polyhedral drawing and overlay of packing model of Figure 16.5. Large voids form tunnels parallel to both c and $a \sin \beta$. Tunnels down b alternate as CN X–VIII–X–VIII.

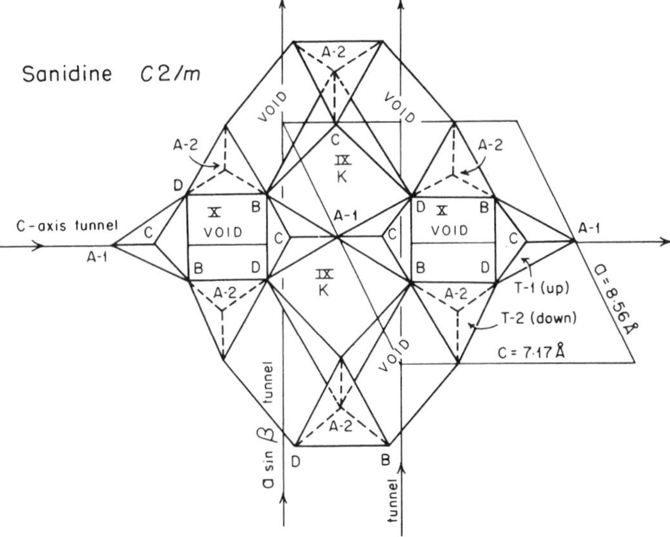

Figure 16.7. Packing model of sanidine projected along $a \sin \beta$, (100) projection, showing K^+ and O—A^2 ions on mirror plane (010) and tunnels along $a \sin \beta$.

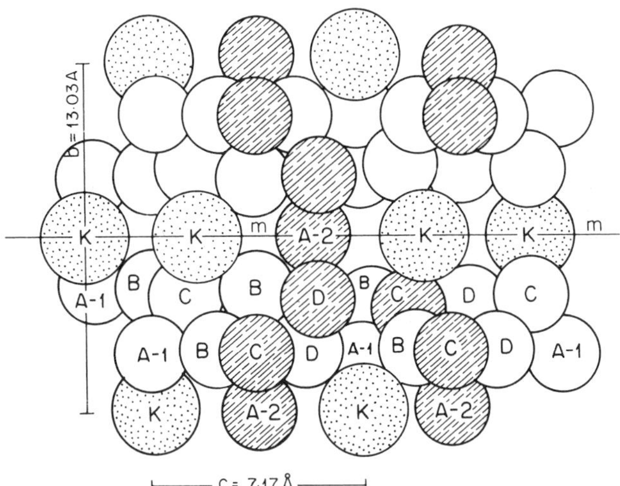

tunnels parallel to *b* that pass through the centers of tetrahedral rings (Figs. 16.3, 16.6) consist of a stacking of alternating CN X–CN VIII–CN X voids. These large voids could be partially and synchronously occupied by the large alkali metals in sanidine. Similar voids in other feldspars would be smaller in diameter.

None of the large metal cations that occupy the CN IX–CN VI–VII sites, designated the M-sites, are centered inside these large oxygen cages, nor are the small Si and Al atoms centered in the tetrahedral sites. These departures from centrosymmetric location of cations in anion coordination polyhedra of feldspars are evidently caused by local imbalance of electric charge, such as has been demonstrated for pyroxene (see Chapters 5 and 7). Cations move away from overcharged or overbonded oxygens and move closer to undercharged or underbonded oxygens. This gives interatomic distances (M–O) that are larger or smaller respectively, than the sum of the ionic radii. In all feldspars, the distance M–$O_{A\text{-}2}$ is smaller than all the others, implying an underbonding of $O_{A\text{-}2}$ that makes contact with only one M ion and joins two pairs of tetrahedra across mirror planes (Fig. 16.5). Oxygen A-1, which makes contact with two M ions, should reduce its EBS by preferentially incorporating more Al than Si into tetrahedral T-sites.

Exsolution, ordering, and iridescence

In the alkali feldspar series, both orthoclase and albite or their polymorphs have the same Si : Al ratio of 1 : 3, very similar unit cell dimensions, and separation of one phase from the other by an exsolution process involves only the exchange of equally charged M cations, K^+ and Na^+, across areas of identical Al—Si ordering. Overall volume reduction, however, must occur, because albite exsolution lamellae in host orthoclase will occupy a smaller volume than they did in solid solution in an unexsolved "homogeneous" alkali feldspar.

Virtually all alkali feldspars crystallized from a magma and slowly cooled in plutonic or deep levels of the earth's crust show fine blebs, rods, or strips of albite exsolved as lamellae inside host grains of orthoclase or microcline. These exsolution lamellae may be submicroscopic in thickness ($< 0.1\ \mu m$), are commonly microscopic ($0.5–5.0\ \mu m$), and are often macroscopic and visible in the hand specimen under a hand lens or even with the unaided eye. Marked differences in index of refraction, and often in color as well, enhance their recognition.

These kinds of submicroscopic, microscopic, and megascopic exsolutions of albite lamellae in host orthoclase are designated cryptoperthite, microperthite, and perthite, respectively (Fig. 16.8). They represent an attempt of the chemically inhomogeneous host crystal of alkali feldspar to reequilibrate or minimize its free energy and overall space or volume requirements at lowered temperature by separating a large-volume disordered atomic array into two ordered arrays of smaller volume, each more economical of space. Exsolution is thus an

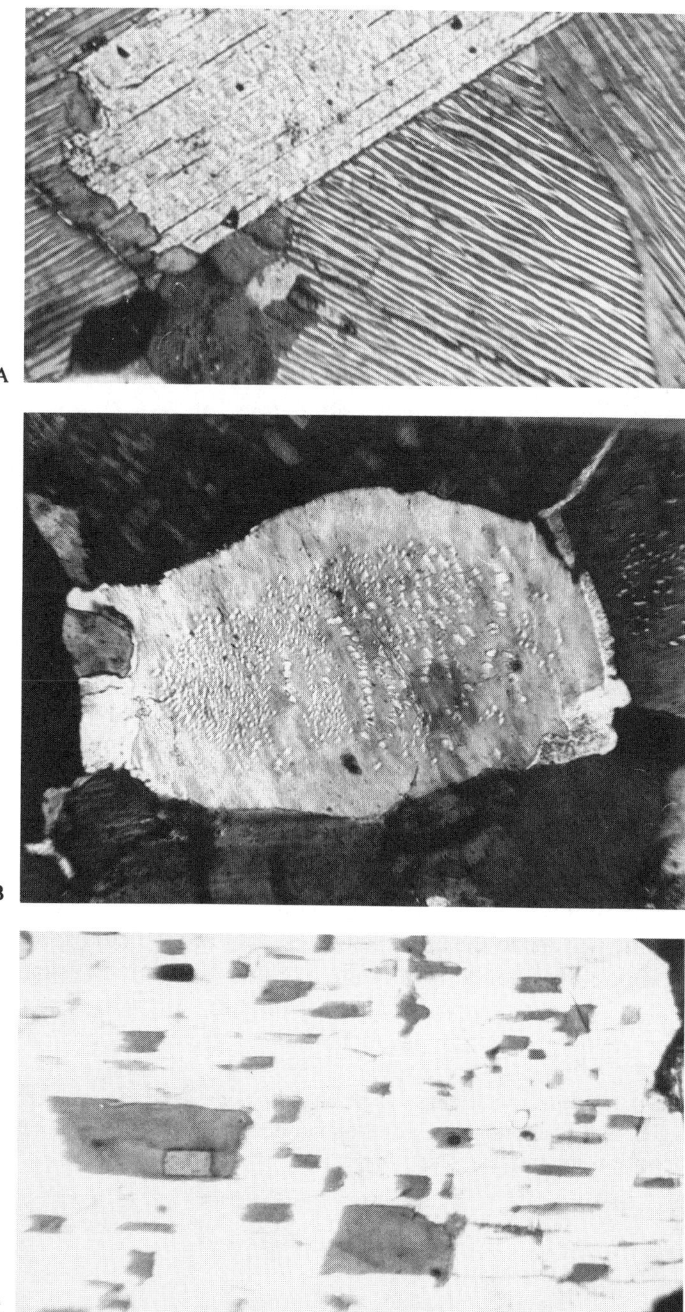

Figure 16.8. Microscopic textures of mesoperthite (A), microperthite (B), and antiperthite (C). Reprinted with permission from H. W. Jaffe, Petrology of the Precambrian crystalline and associated Paleozoic rocks of the Monroe area, Hudson Highlands, New York, United States of America. Dr. ès Sc. Thèse #1597, 1972, Université de Genève, Switzerland, p. 110.

ordering process that requires initial solid solution at high temperature and, above all, very slow cooling to lower temperature to enhance the kinetic exchange of K^+ and Na^+ ions within the structure. Exsolution is a subsolidus phenomenon and involves an ion exchange in a solidified crystal. No melt is involved (also see Chapter 15). The chemical limits of solid solution and exsolution are the subject of petrology, and are best explored in the many fine texts on this subject, among them the summaries of Ribbe (1975b), Yund (1975), and Smith (1974, 1975).

Exsolution within the plagioclase feldspar series is a much more complicated process than within the alkali feldspar series or in pyroxenes (see Chapter 15). Because, in anorthite, $c = 14.17$ Å and in albite, $c = 7.16$ Å, the anorthite unit cell is doubled in size with respect to that of albite (see Table 16.2). Anorthite and albite also differ markedly in their Al—Si ordering, which is perfect in anorthite, where Al : Si $= 1$, and very different in albite, where Al : Si $= 0.333$. To state this differently, in anorthite Al and Si are perfectly ordered with sixteen ions of each occupying the thirty-two tetrahedral sites, so each AlO_4 tetrahedron is surrounded by four SiO_4 tetrahedra, and vice versa. This alternation is the reason for the doubling of the length of the *c* axis. In albite, Si orders into sites T_1O. Each such site is surrounded by four SiO_4 tetrahedra while two-thirds of the SiO_4 tetrahedra are surrounded by one Al and two Si neighbors in the tetrahedra. It is thus indeed surprising that albite and anorthite form a solid solution series at high temperature of about 900 °C or greater. At lower temperatures of about 650 °C, this plagioclase series is broken up by several discontinuities into smaller ranges of solid solution, each controlled by the degree of structural similarity. Within these ranges (Figs. 16.4, 16.9), exsolution of orthoclase in plagioclase hosts and even of one plagioclase in another do occur.

Because the structurally different units of albite and anorthite cannot fit together in the same way as do the alkali feldspars, they are joined together in submicroscopic modulated domains of albite-like and anorthite-like slabs of each within ranges of bulk composition An_{0-2} (albite), An_{2-16} (peristerite), An_{16-70} (intermediate or 'e' structure plagioclase), An_{70-80} (transitional anorth-

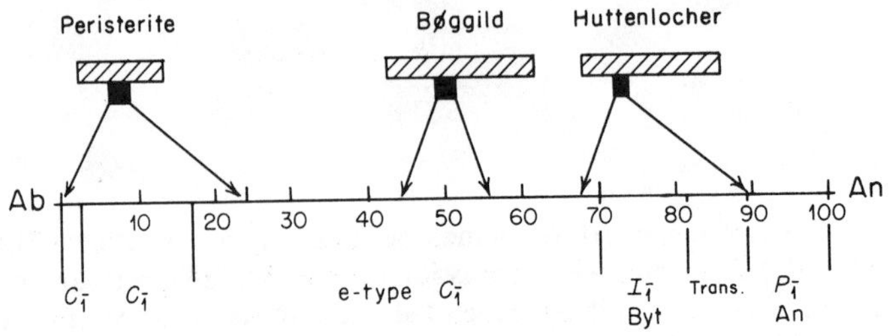

Figure 16.9. Compositional ranges of plagioclase feldspars that form peristerite, Bøggild, and Huttenlocher exsolution lamellar integrowths. Arrows point to compositions of lamellae pairs.

ite), and An_{93-100} (primitive anorthite) (Fig. 16.4). All of these are microscopically homogeneous, and microscopic measurements of their indices of refraction give reliable estimates of their bulk composition while revealing no information on their very fine or submicroscopic structural inhomogeneity.

When large bodies of deep level magmas or intensely metamorphosed rocks containing these plagioclases are very slowly cooled to about 500–600°C and held at these temperatures for millions of years, reequilibration or ordering in the solid state takes place, and one feldspar exsolves as lamellae inside the host grain representing the chemically more abundant constituent. When microscopic exsolution lamellae of orthoclase occur inside plagioclase, the new crystals are classed as antiperthite (Fig. 16.8C): this phenomenon is most common in the intermediate or 'e' structure type (Fig. 16.4, Table 16.4). In the peristerite range, feldspars of bulk composition An_{2-16} may unmix or exsolve into submicroscopic peristerites made up of alternating thicker lamellae of pure albite (An_0), and thinner lamellae of sodic plagioclase (An_{25}). The individual lamellae are submicroscopic, but hand specimens of these crystals show a beautiful blue iridescence.

Within the range of bulk composition An_{42-61} in the intermediate series (Figs. 16.4, 16.9), many specimens, slowly cooled as we have just described, may separate into two sets of submicroscopic lamellae differing in composition by about 12% anorthite; for example, slabs of intermediate plagioclase composition An_{56} (labradorite) alternating with slabs of composition An_{44} (andesine). These have been designated as Bøggild lamellae (Table 16.4).

Where the bulk composition is determined to have been An_{50}, the lamellar An_{56} slabs (designated d_a) and the An_{44} slabs (designated d_b) are of equal thickness, each about 675 Å, and the repeat distance $d_{a+b} = 1350$ Å, although submicroscopic, is capable of diffracting visible light with a strong maximum at 4100 Å of the visible spectrum. Specimens of this composition held at the proper angle to a source of sunlight show a spectacular blue iridescence (interference color), which is commonly seen in the labradorites from Labrador, Norway, and the Adirondacks. The color observed is a function of the sum of the repeat distance of the lamellae and of the mean refractive indices of components $a + b$ in air, expressed as a function of Bragg's law of diffraction (see Fig. 6.6) $N\lambda = 2d \sin \theta$. Thus, the wavelength $N\lambda$ of the iridescent labradorite, 4100 Å, is expressed as

$$2(n_a d_a \sin \theta_a + n_b d_b \sin \theta_b)$$

where n_a and n_b are mean indices of refraction in air of slabs $a(An_{56})$ and $b(An_{44})$, d_a and d_b are slab thicknesses, 675 Å each, and θ is the angle of incidence. (A simpler equation,

$$N\lambda = 2nd_{a+b} \sin \theta$$

is also useful.) For more calcic bulk compositions near An_{53-55}, the slab richer in anorthite is thicker ($d_a = 1800$ Å) than the slab poorer in anorthite

Table 16.4. *Classification, bulk and lamellae composition, thickness d_{a+b} and iridescent properties of finer-sized feldspar exsolution intergrowths*

	Composition		Lamel. thick.	Iridescent	
Classification	Bulk	Lamel.	d_{a+b} Å	color	λ (Å)
Peristerite	An_{2-13}	$An_{0,25}$	1500	Light blue	~4400
Bøggild	An_{42-61}				
	An_{50}	$An_{44,56}$	1350	Deep blue	~4100
	↓	↓	↓	Green	↓
				Yellow	
	An_{53}	$An_{47,59}$	2150	Red	~6400
Huttenlocher	An_{65-85}	$An_{67,90}$	(μm)	Noniridesc.	
Larvik type	$Or_{20}Ab_{70}An_{10}$	Or_{90}, An_{16}[a]	1500	Blue	~4400

[a]Alternating Or_{90} and An_{16} lamellae.

($d_b = 850$ Å), and the repeat distance, $d_{a+b} = 2650$ Å, is consistent with an iridescent maximum of $N\lambda = 6500$ Å, in the red range of the visible electromagnetic spectrum (see Chapter 2, Table 16.4). Intermediate values yield a green iridescent interference color. Large plagioclase crystals from rocks known as anorthosites, from Labrador, the Adirondacks, and southern Norway, may show in sunlight a zonation of iridescent colors, with a core of red succeeded by a zone of green, succeeded in turn by a thicker, dominant zone of blue. This is consistent with small yet significant chemical zonation of the crystal during growth from a core of An_{55} to An_{52}, succeeded by a dominant zone of An_{50}, where blue iridescence dominates.

For bulk compositions in the range An_{67-90} (bytownite), the separation of thin lamellae of composition An_{90} (anorthite) inside thicker slabs of composition An_{67} (labradorite of intermediate structure type) occurs at a thickness visible with the polarizing microscope. Intergrowths of these compositions and thicknesses are called Huttenlocher intergrowths or Huttenlocher lamellae (see Table 16.4, Fig. 16.9). They are found exclusively in plagioclase feldspars of rocks slowly cooled for geologically long periods. These are too thick to produce the iridescent colors associated with the Bøggild lamellae (Nissen 1974).

A chemically complex solid solution of anorthite-bearing alkali feldspars crystallized at high temperatures (>900°C) is often designated anorthoclase. A typical composition is $Ab_{70}An_{10}Or_{20}$ (point L of Fig. 16.4). Submicroscopic exsolution in these may result in separation by exsolution into lamellae of orthoclase, $Or_{90}Ab_{10}$, interlayered with those of plagioclase of composition $Ab_{85}An_{15}$ (oligoclase). When they are about 500–1000 Å thick and repeat layers of combined thickness d_{a+b} are about 1500 Å, they also may show spectacular, brilliant pale blue iridescence attributable to the same theory as described for Bøggild exsolution lamellae of labradorites. Coarser exsolution may produce mesoperthite (Fig. 16.8A). This iridescent anorthite-bearing alkali feldspar, or anorthoclase, is the major mineral phase in the gray-black quartz-bearing syenites of Larvik, southern Norway, locally named larvikites. When this rock is cut and polished, the remarkable blue iridescence endows the slabs with a distinctively beautiful twinkling blue on black mottling that has resulted in its worldwide use as tombstones and as decorative external slabs on banks and other buildings, including English pubs (hence it has at times been irreverently known as "pubite").

In summary, exsolution is a process of ordering whereby an initial high temperature solid solution, on slow cooling, orders itself into two phases separated in the solid state, with each phase more economical of space or volume. The driving force or kinetic impetus for the exsolutions is, in feldspars, Al—Si tetrahedral ordering in plagioclases, and Na^+—K^+ or M-cation ordering in alkali feldspars. Exsolution into two pure end members can never be achieved, because low temperature kinetic barriers arrest the process before such complete separation can occur; ions simply stop moving.

Bibliography

Jaffe, H. W. (1972). Petrology of the Precambrian crystalline and associated Paleozoic rocks of the Monroe area, Hudson Highlands, New York, United States of America. Dr.ès Sc. Thèse # 1597, Université de Genève, Switzerland.

Kroll, H., Bambauer, H. V., and Schirmer, U. (1980). The high albite-monalbite and analbite-monalbite transitions. *Am. Mineral.,* 65: 1192–1211.

McConnell, J. D. C. (1974). Electron-optical study of the fine structure of a Schiller labradorite. In *The feldspars,* eds. W. S. McKenzie and J. Zussman. Manchester Univ. Press, Manchester, pp. 478–90.

Megaw, H. D. (1974). The architecture of the feldspars. In *The feldspars,* eds. W. S. McKenzie and J. Zussman. Manchester Univ. Press, Manchester, pp. 2–24.

Morse, S. A. (1969). Feldspars. Carnegie Inst. Yearbook 67, Geophys. Lab., Washington, D.C., pp. 120–6.

Nissen, H. -U. (1974). Exsolution phenomena in bytownite plagioclases. In *The feldspars,* eds. W. S. McKenzie and J. Zussman. Manchester Univ. Press, Manchester, pp. 491–521.

Phillips, M. W., Colville, A. A. and Ribbe, P. H. (1971). The crystal structures of two oligoclases: a comparison of low and high albite. *Z. Krist.,* 133: 43–65.

Phillips, M. W., and Ribbe, P. H. (1973). The variation of tetrahedral bond lengths in sodic plagioclase feldspars. *Contrib. Mineral. Petrol.,* 39: 327–39.

Ribbe, P. H. (1975a). The chemistry, structure, and nomenclature of feldspars. *Mineral. Soc. Am. Rev. Mineral.,* 2: R-1–R-51.

(1975b). Exsolution textures and interference colors in feldspars. *Mineral. Soc. Am. Rev. Mineral.,* 2: R-73–R-96.

Smith, J. V. (1974). *Feldspar minerals,* Vols. 1, 2. Springer-Verlag, Heidelberg.

(1975). Some chemical properties of feldspars. *Mineral. Soc. Am. Rev. Mineral.,* 2: Sm-18–Sm-29.

Smith, J. V., and Ribbe, P. H. (1969). Atomic movements in plagioclase feldspars. *Contrib. Mineral. Petrol.,* 21: 157–202.

Taylor, W. H. (1933). The structure of sanidine and other feldspars. *Z. Krist.,* 85: 425–42.

Yund, R. A. (1975). Microstructure, kinetics, and mechanisms of alkali feldspar exsolution. *Mineral. Soc. Am. Rev. Mineral.,* 2: Y-29–Y-57.

17 The crystal chemistry of the borate minerals

General crystal chemistry

Because of their limited natural occurrence, borate minerals are less familiar to mineralogists and students than their much more numerous silicate counterparts. Nonetheless, the crystal chemistry of borate minerals is more varied, more complex, and thus more fascinating than that of the silicates.

Boron, atomic number 5, $1s^2 2s^2 2p^1$, can in the excited state readily place its three valence electrons in a hybrid bond sp^2 orbital configuration as follows:

$$ {}_5B_{sp^2} \quad 1s\ (\uparrow\downarrow)\ \overbrace{[(\uparrow)\ (\uparrow)\ (\uparrow)]}^{sp^2} $$

In so doing, it directs its three hybrid orbital lobes of high electron density at 120° angles toward the corners of an equilateral triangle. It would thus be anticipated that the boron atom, B, would normally adopt CN III with bonding elements such as oxygen, O. The B atom is most commonly found in triangular coordination with three oxygens, but it also occurs in tetrahedral coordination with four oxygens in some minerals. Further complicating the situation is the common occurrence of borate polyions, in which the B atom occurs in both CN III and CN IV polyhedra joined together in the same mineral.

Many careful measurements obtained in borates by X-ray methods show that the interatomic distances are

$$ \text{B—O (triangle)} = 1.35\text{–}1.38\ \text{Å} $$

and

$$ \text{B—O (tetrahedron)} = 1.44\text{–}1.49\ \text{Å} $$

Subtracting the ionic radius of the oxygen anion, $O^{2-} = 1.36$ Å, from these B—O interatomic distances leaves a radius of only 0.02 Å for a B^{3+} cation centered in a triangle, and only 0.12 Å inside a tetrahedron bounded by O^{2-}

Table 17.1. *Crystal chemical, unit cell, and optical data for selected borates*

Formula	Space group	a (Å)	b (Å)	c (Å)	β (°)	ρ	γ (ω)	β	α (ϵ)	k	α_G (Å^3)
Isolated [BO_3] triangles											
Nordenskioldine 3/$Ca^{VI}Sn^{VI}[B^{III}O_3^{III}]_2$	$R\bar{3}$	4.85		15.92		4.20	1.778		1.660	.176	14.42
Sassolite 4/$B^{III}(OH^{I})_3$ or $H_3^{II}[B^{III}O_3^{III}]$	$P\bar{1}$	7.04	7.05	6.58	101	1.498	1.459	1.456	1.340	.279	6.85
Hambergite 8/$Be_2^{IV}[B^{III}O_3](OH)$	*Pbca*	9.73	12.18	4.43		2.37	1.617	1.580	1.543	.245	9.11
Ludwigite 4/$Fe^{3+VI}Mg_2^{VI}O_2[BO_3]$	*Pbam*	9.14	12.05	3.05		4.0	1.985	1.865	1.850	.225	17.42
Synthetic 4/$Na_4^{VI}[B_2^{III}O_5]$	*C*2/*c*	10.62	8.62	6.29	110.1	2.56	—	(1.516)[a]	—	(.202)	(15.50)
Synthetic 4/$Eu^{VIII}[B_2^{III}O_4]$	*Pnca*	6.59	12.06	4.34		4.61	—	(1.733)	—	(.159)	(14.97)
Synthetic 18/$Ba^{VI}Ba_2^{IX}[B_3O_6]_2$	$R\bar{3}$	7.23		39.19		3.74	—	(1.602)	—	(.161)	(14.23)
Isolated BO_4 tetrahedra											
Sinhalite 4/$Mg^{VI}Al^{VI}[B^{IV}O_4]$	*Pbnm*	4.33	9.88	5.68		3.50	1.705	1.698	1.669	.197	9.85
Cahnite 2/$Ca_2^{VIII}[B^{IV}(OH)_4][As^{IV}O_4]$	$I\bar{4}$	7.09		6.19		3.06	1.662		1.655	.215	28.12
BO_3 + BO_4 polyions (triangles + tetrahedra)											
Kernite 4/$Na_2^{VI}[B_2^{III}B_2^{IV}O_6(OH)_2]\cdot 3H_2O$	$P2_1/c$	7.02	9.16	15.68	108.9	1.901	1.488	1.473	1.454	.248	26.87
Inyoite 4/$Ca^{VIII}[B^{III}B_2^{IV}O_3(OH)_5]\cdot 4H_2O$	$P2_1/a$	10.63	12.06	8.40	114	2.13	1.517	1.505	1.492	.237	25.98
Colemanite 4/$Ca^{VII}[B^{III}B_2^{IV}O_4(OH)_3]\cdot H_2O$	$P2_1/a$	8.74	11.26	6.10	110	2.42	1.614	1.592	1.586	.247	20.13
Roweite 4/$Ca_2^{VIII}Mn_2^{VI}(OH)_4[B_2^{III}B_2^{IV}O_7(OH)_2]$	*Pbam*	9.06	13.36	8.29		2.93	1.660	1.658	1.646	.223	35.03
Kaliborite 4/$H^{II}K^{VIII}Mg_2^{VI}[B_6^{III}B_6^{IV}O_{16}(OH)_{10}]\cdot 4H_2O$	*C*2/*c*	18.53	8.43	14.67	100.1	2.11	1.550	1.527	1.508	.250	71.02

[a] Parenthesis indicate values were calculated using the *k* values of Table 11.1.

anions. These cation and anion values yield radius ratios that exceed geometric limits of 0.155–0.212 imposed for CN III, and 0.212–0.414 imposed for CN IV (see Chapter 4). The B^{3+} cation would be much too small to make tangent contact with either three or four O^{2-} anions, and would "rattle" inside these oxygen polyhedral cages. From the foregoing, it should be evident that B—O bonds in most minerals are largely of a covalent nature, with, of course, some partial ionic character. This would be predicted from the electronegativity value of boron, $\Delta\chi = 2.0$, in combination with oxygen, $\chi = 3.5$ ($\Delta\chi = 1.5$, equivalent to about 56% covalent, 44% ionic bonding). This is further supported by electron density contour maps published by Cooper et al. (1973) for the complex hydrated borate kernite (Table 17.1; see also Fig. 17.10). These investigators show a high electron density for B—O bonds and a low electron density for Na—O bonds in kernite, and conclude that these represent predominantly covalent and ionic bonds, respectively.

Note that the void coplanar with three spheres or oxygen atoms in an equilateral triangle is smaller than the void centered above three spheres capped by a fourth, as in a tetrahedron. If an apical oxygen is removed from a tetrahedron and the cation compressed into the triangular void remaining, one can compare SiO_4 tetrahedra with BO_3 triangles and examine any similarities in polymerization. It turns out that most of the simpler silicate polymerization classes are indeed mimicked by borates (Fig. 17.1, Table 17.1). Thus, nesosilicates, cyclosilicates, and inosilicates have their boron–oxygen triangular polyhedral analogues (Table 17.1).

Boron and oxygen also polymerize into complex sheets and framework structures that mimic those of phyllosilicates and tektosilicates.

Borates based upon isolated $[BO_3]$ triangles

The pseudohexagonal or pseudotrigonal structure of sassolite, triclinic boric acid $H_3[BO_3]$ or $B(OH)_3$, is the simplest of these. The sheets of boric acid consist of $[BO_3]$ triangles linked by hydrogen or hydroxyl bonds (Fig. 17.2). Shortened interatomic distances of O—O = 2.36 Å in triangles (2.80 Å = normal) and B—O distance = 1.36 Å suggest that sp^2 hybridization of the B atom has also induced O—O orbital overlap in $[BO_3]$ triangles (Fig. 17.2). Note that distances for O—O (between two different triangles) are 2.73 Å, close to the sum of spheres of two O^{2-} anions of radius 1.37 Å. It is important to see that H atoms or H^+ ions (thus protons) are not centered between two O atoms but lie at a distance, H—O = ± 1.0 Å (0.85–1.05 Å), and thus are not centered within an O atom as in an ideal spherical $(OH)^{-1}$ anion.

The way in which the asymmetrically disposed H atom on each O atom effectively bonds all $[BO_3]$ triangles together is readily seen in Fig. 17.2. Note that sassolite is triclinic, so the *c* crystal axis is not perpendicular to the drawing (Fig. 17.2) but is inclined at 101° ($\beta, a \wedge c = 101°$) in a northwesterly direction. Thus, consecutive or superposed sheets of $H_3[BO_3]$ stacked along *c* will shift northwesterly, building a triclinic rather than a hexagonal lattice. In this type of stacking, B atoms of the upper sheet will not lie directly above B atoms of the lower sheet; instead, the shift will place O atoms above B atoms, thus avoiding the complication of placing residually charged cations too close to one another.

Although B is covalently bonded, sp^2, it will still retain a residual positive charge in keeping with its 44% ionic character. This latter is emphasized by the perfect solubility of boric acid in water, a polar solvent.

The marked planarity of the structure, parallel to the *a-b* plane containing the sp^2 plane triangular orbitals as well as O—O overlap, results in an associated high electron density and a marked reduction in the velocity of light vibrating parallel to this (001) plane (indicates a high index of refraction). Conversely, light vibrating parallel to the *c* crystal axis will meet with a much lower electron density, because the sheets are not closely stacked ($c = 6.58$ Å). This marked contrast in atomic packing results in the large double refraction or high birefringence and optically negative sign found in sassolite and in most of the common carbonate and nitrate minerals as well (see Chapter 11). Because all chemical bonds, B—O, sp^2, and O—H—O, lie in an only slightly warped plane, superposed sheets can be linked together only by weak van der Waals bonds.

Figure 17.1. $[BO_3]$ polymerization group analogues of $[SiO_4]$ polymerization groups.

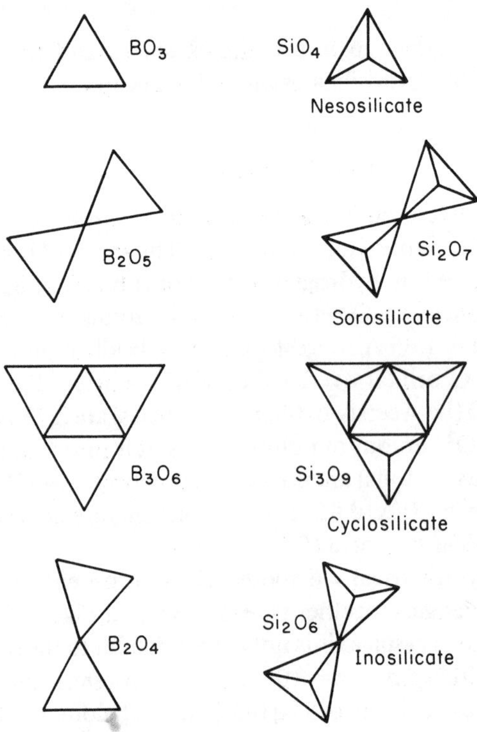

The rare borate mineral nordenskioldine is interesting in that it represents the isostructural tin–boron analogue of the common carbonate, dolomite:

$$CaSn[BO_3]_2 \quad \text{and} \quad CaMg[CO_3]_2$$

In both minerals, perfect cation ordering places Ca ions in one plane that alternates with planes of Mg ions in dolomite and with Sn ions in nordenskioldine. Because the $[CO_3]$ and $[BO_3]$ anions lie in the (0001) plane perpendicular to the *c* axis, both minerals are, predictably, optically negative with high birefringence. Note that trigonal nordenskioldine has much higher indices of refraction than pseudohexagonal or pseudotrigonal sassolite (Table 17.1); both are optically negative with about the same birefringence, $\omega - \epsilon = 0.118$, for nordenskioldine, and $\omega - \epsilon = 0.119$, sassolite. The large difference in magnitude of indices of refraction is explained by the large mass of the Sn atom and the anhydrous nature of nordenskioldine, when contrasted with the low mass of the hydrated sassolite structure, $H_3[BO_3] \cdot H—O$. In combination as water, H_2O (OH), or O—H—O, always results in a marked lowering of the indices of refraction of any mineral. The Larsen and Berman specific refractivity value,

Figure 17.2. Packing drawing of sassolite, boric acid, $4/H_3[BO_3]$.

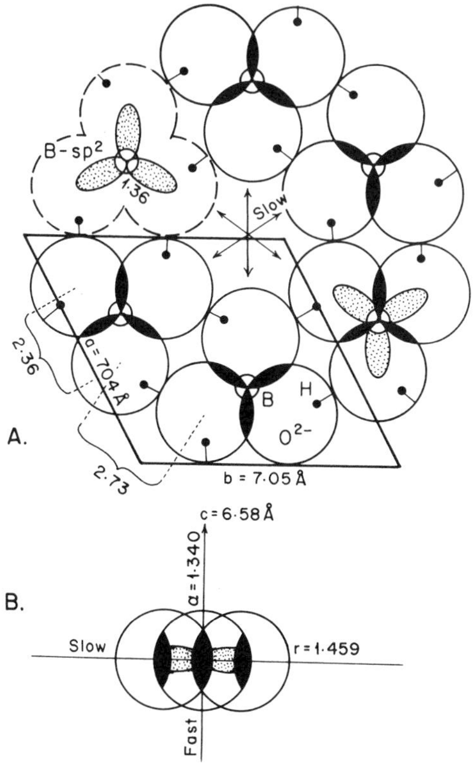

$k_{H_2O} = 0.34$, adopted herein (see Tables 11.2, 16.2) appears to work equally well for all combinations of H and O in minerals.

The structure of the borate mineral ludwigite, $4/Fe^{3+}Mg_2O_2[BO_3]$ (Table 17.1), will now be analyzed in detail because it illustrates many principles of crystal chemistry. The unit cell packing model and polyhedral representation of ludwigite are shown in Figure 17.3. The following features are important:

1. Ludwigite is orthorhombic, *Pbam,* and interrupted sheets of atoms containing $[BO_3]$ triangles stack one above the other up the *c* axis. $[BO_3]$ triangles in one layer would lie too close to those in the next layer were these not spaced further apart by octahedral chains of Fe^{3+}—O and Mg—O that run in staggered lines or rows up the *c* axis. Effectively, O atoms at elevation 50 (Fig. 17.3–17.5) increase the separation of (BO_3) planes in which all atoms lie at elevation 0 and 100. Thus, the *c* dimension of 3.05 Å is appreciably larger than the ideal O—O distance of 2.80 Å for two layers in superposed contact.

Figure 17.3. Packing model (upper) and polyhedral model (lower) of ludwigite, $4/FeMg_2O_2[BO_3]$, (001) projection.

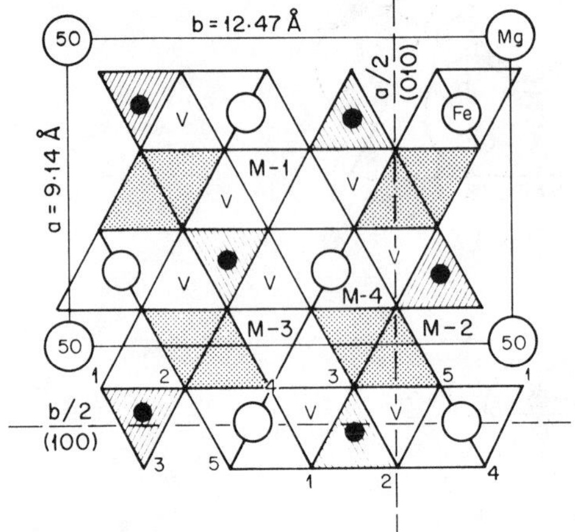

2. The space group *Pbam* indicates a primitive lattice, which is readily discernible in Figures 17.3, 17.4, and axial glide planes parallel to (010) and (100). Axial *a* glide planes run parallel to (010) and bisect all O-2 atoms; note reflection and translation *a*/2 of all numbered atoms and polyhedral elements (Fig. 17.3). Axial *b* glide planes lie parallel to (100) and bisect all Fe atoms along this plane; note, again, reflection and translation of all numbered atoms and polyhedral elements along this plane. Because ludwigite is orthorhombic, with $c = 3.05$ Å, it is easy to see from Figures 17.3 and 17.4 that a mirror plane will lie parallel to the plane of projection (001). A twofold axis of symmetry, not recorded in the space group symbol, emerges from the Mg ion (elevation 50) at the center of the unit cell (Fig. 17.3).

3. There are five different oxygen atom environments in ludwigite, numbered O-1, O-2, O-3, O-4, O-5, as indicated in Figure 17.4. Note that $[BO_3]$ groups are planar with O-1, O-2, O-3, and B lying directly over one another at elevations 0 and 100. These planes are isolated from one another, and cross-connected by the Fe^{3+}- and Mg-O octahedra in which oxygens O-4 and O-5 lie at elevation 50. This atomic array results in the disposition of four different octahedral sites, here numbered M-4, M-2, M-1, and M-3 in a manner analogous to the four polyhedra in amphiboles (see Chapter 15). The four octahedral

Figure 17.4. Unit cell projection (001) of ludwigite, showing numbering of anions, O-(1)–O-(5) (upper), and EBS distribution around each anion (lower).

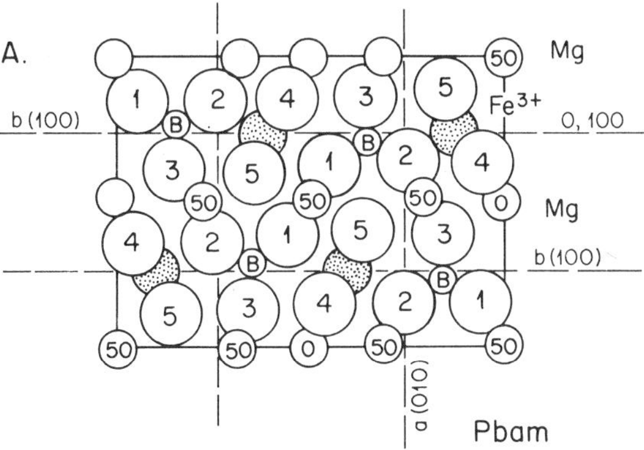

sites differ in that cations or metal atoms have the following oxygen neighbors (see Fig. 17.3):

M-4: 4, 4, 5, 5, 1, 2 (all elevation 0, 100)
M-2: 2, 2, 3, 3, 4, 5 (all elevation 50)
M-1: 1, 1, 1, 1, 5, 5 (all elevation 50)
M-3: 4, 4, 4, 4, 3, 3 (all elevation 0, 100)

In the (120) projection of Figure 17.5, the octahedral keels readily characterize each polyhedron as follows:

M-4:	0-4, 0-5 keel	M-2:	0-2, 0-3 keel
M-1:	0-1, 0-1 keel	M-3:	0-4, 0-4 keel

There are four each of M-4 and M-2 atoms and two each of M-1 and M-3 per unit cell, which contains $Fe^{3+}{}_4Mg_8B_4O_{20}$ (Fig. 17.4).

4. The EBS shows overbonded and underbonded oxygens, but the summation over all five O^{2-} anions is exactly 10+, satisfying Pauling's Rule 2C (see Chapter 5). The EBS solution is diagrammed in Figure 17.4. One noteworthy feature is the unusual occurrence of a five-coordinated oxygen, O-4, which has as neighboring cations two Fe^{3+}/VI and three Mg^{2+}/VI types, summing exactly to an EBS of +2.00. Close study of Figures 17.3 and 17.4 will show that these five loci of positive charge are opposite a void site. Thus combined, these simulate an octahedral point charge distribution with one point equal to 0.

Figure 17.5. (120) projection of ludwigite, showing octahedral chains or strips of Fe—O (M-4) and Mg—O (M-1,-2,-3), all staggered parallel to *c* and cross-connected by B atoms in BO_3 triangles.

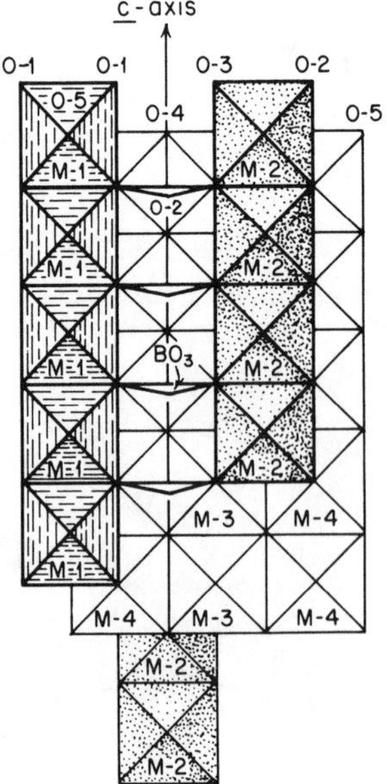

5. The formula $4/Fe^{3+}Mg_2O_2[BO_3]$ can be expanded to show more of the features of the ludwigite structure, as follows:

$$\begin{array}{llllll} (M\text{-}4)_2 & (M\text{-}2)_2 & (M\text{-}1)_1 & (M\text{-}3)_1 & ④_2\,⑤_2 & [B\,①\,②\,③]_2 \\ 2/Fe^{3+}{}_2 & Mg_2 & Mg & Mg & ④_2\,⑤_2 & [B\,①\,②\,③]_2 \end{array}$$

6. As indicated under 1, there are no true $[BO_3]$ sheets, because staggered endless chains or strips of Fe^{3+} and Mg edge sharing octahedra run up the *c* axis, separating the triangles and preventing any linkage of these one to another. It is the presence of these staggered, edge sharing strips of octahedra running parallel to *c* (Fig. 17.5), particularly the Fe^{3+} octahedra, that are responsible for the high electron density and consequent very high index of refraction of ludwigite (Table 17.1). Light vibrating parallel to the rows of atoms staggered parallel to the *c* axis results in optically positive sign and high birefringence (see Chapter 11).

The planar arrays with relatively low electron density lie parallel to (001) and the *a* and *b* crystal axes and contain vibration velocities **X** (fast) almost equal to **Y** (intermediate), both strongly contrasting with **Z** (slow) perpendicular to these planes and parallel to the *c* axis. Because **X** and **Y** vibration directions are nearly equal, the α- and β-indices of refraction are close to one another and much lower than the γ-index of refraction parallel to *c* (Table 17.1). This combination of refractive indices $\alpha \cong \beta \ll \gamma$ is compatible with a small, optically positive axial angle ($2V = 20°\,+$) and a prolate triaxial ellipsoid model that approaches the uniaxial prolate spheroid of revolution (see Chapter 11). Thus, from an indicatrix viewpoint, ludwigite mimics rutile (prolate spheroid of revolution, optically positive), whereas sassolite and nordenskioldine model the trigonal carbonates and pseudohexagonal micas (oblate spheroids of revolution, optically negative). Once again, for emphasis, the velocity of light is retarded parallel to staggered rows or planes, increasing index of refraction in these directions, because of the high electron density resulting from overlapped orbitals.

Borates built of isolated $[BO_4]$ tetrahedra

Although examples of borate minerals containing $[BO_4]$ tetrahedra alone are less common, sinhalite, $MgAl[BO_4]$ provides a fine example. It was originally misidentified as forsterite (Mg–olivine) in the British Museum collection. Careful detective work resulted in the identification of sinhalite as a borate mineral new to science. Sinhalite resembles forsterite not only in the hand specimen but in atomic structure as well. The minerals are isostructural, with the Mg and Si of forsterite proxied by Al and B in sinhalite. The presence of the Al and B ions, both of smaller radius, results in a contraction of the unit cell dimensions of sinhalite with respect to those of forsterite. A packing model and its polyhedral representation (Fig. 17.6) is applicable to either of these minerals.

Electrostatic bond strength is satisfied by an alternation of Mg^2/VI, Mg^2/VI,

Al^3/VI, B^3/IV and Mg, Al, Al, B around oxygen ions in an ordered sequence, or by 3 × (Mg,Al) + B in a disordered sequence.

Borates built of complex polyions

Perhaps the most fascinating aspect of borate crystal chemistry is the occurrence of polyions made up of various combinations of boron in triangular and tetrahedral coordination (Figs. 17.7, 17.8) polymerized by both oxygen and hydroxyl ions. A few of the many minerals that contain such polyions are listed in Table 17.1. Many of these borate polyions and the minerals they form were worked out by Morimoto (1956) and by Christ, Clark, and Evans of the U.S. Geological Survey (1958), and since then have been studied by many

Figure 17.6. Packing model (upper) and polyhedral drawing (lower) of sinhalite, $MgAl[BO_4]$, isostructural with forsterite, $Mg_2[SiO_4]$. Packing model shows well the distorted HCP array.

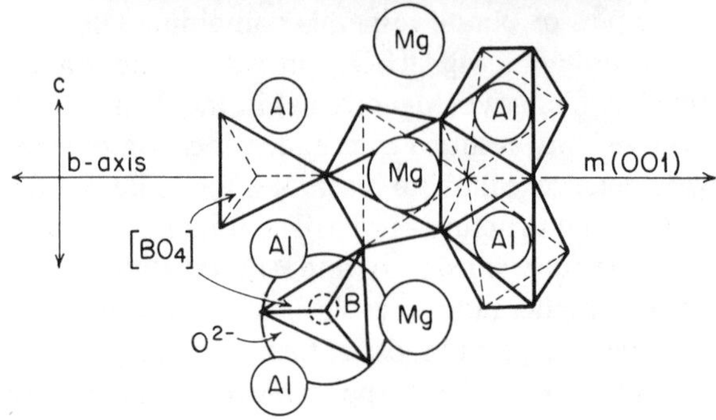

Figure 17.7. Polyhedral and packing drawings of some complex borate polyions built of (BO_3) and (BO_4) groups linked in various ways.

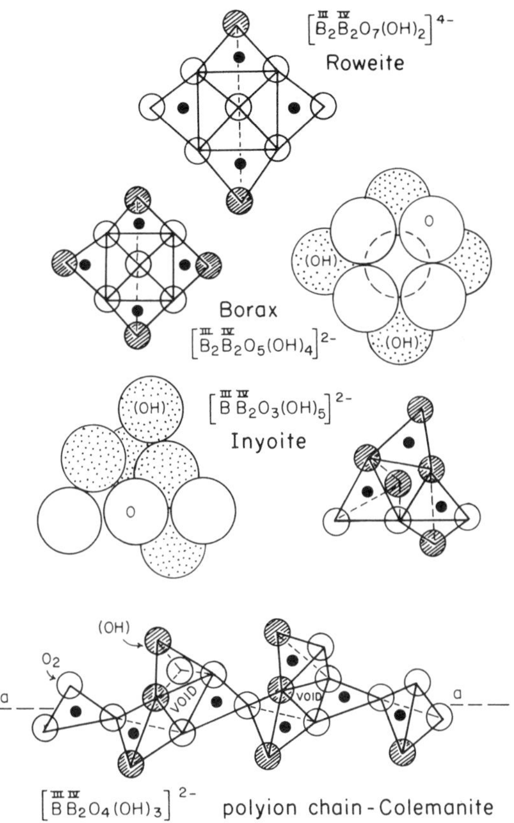

Figure 17.8. Packing models of borate polyions of Figure 17.7.

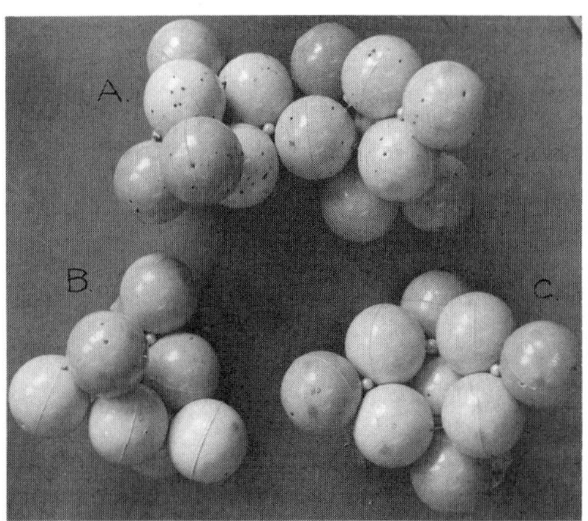

others. The best known of these minerals is probably borax, for which the formula is still given in some books as $Na_2B_4O_7 \cdot 10H_2O$. When the crystal chemistry of borax was studied by Morimoto (1956), it became apparent that the mineral is built of borate polyions of composition $[B_4O_5(OH)_4]^{2-}$ and chains or strips of edge sharing octahedra of composition $Na_2 \cdot 8H_2O$. Borate polyions link to one another by O—H—O bonds and form linear arrays parallel to the *c* axis and to the chains of Na—H_2O octahedra (Fig. 17.9). The structural formula of borax is thus

$$4/Na_2^{VI}[B_2^{III}B_2^{IV}O_5(OH)_4] \cdot 8H_2O$$

Borax thus contains oxygen, hydroxyl, and water molecules as well, all essential components of the crystal structure. This plus the low mass and large volume enclosed by structures based on these complex polyions leads to the very low densities and indices of refraction of the hydrated borate minerals. Thus, high refractivity K but low density give a low mean index of refraction $n - 1$. Similarly, high polarizability α_G (Å^3) divided by very high molar volume V_m(Å^3), also yield low indices of refraction $n - 1$ (see Chapter 11). Thus low indices of refraction are associated with low density and large molar volume.

Figure 17.9. Polyhedral representation of the structure of borax, $4/Na_2^{VI}[B_2^{III}B_2^{IV}O_5(OH)_4] \cdot 8H_2O$, (010) projection. Note octahedral Na—H_2O chains.

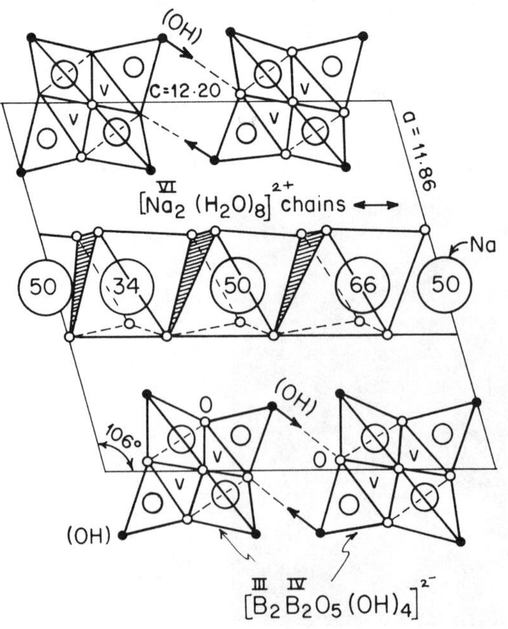

Whereas the polyions in borax pack together as $[B_4O_5(OH)_4]^{2-}$ packages, other polyions polymerize into endless chains, such as the $[B_3O_4(OH)_3]^{2-}$ chain of colemanite. This mineral is made up of one triangle and two tetrahedra translated infinitely along the *a* axis. Similar chains of polyions of composition $[B_4O_6(OH)_2]^{2-}$ run infinitely along the *b* axis of kernite (Table 17.1). Electron or difference density maps for kernite of Cooper et al. (1973) contrast the high electron density around the more covalent B—O triangles with the much lower electron density around more ionic Na—O bonds (Fig. 17.10). Borate mineralogy thus elegantly illustrates most of the principles of refractivity espoused in this book.

Figure 17.10. Electron or difference density maps of kernite. Contours are much more closely spaced (denser) along covalent B—O bonds than along ionic Na—O directions. Reprinted with permission from Cooper, Larsen, Coppens, and Giese, Electron population analyses . . . kernite. *Am. Mineral.*, 58: 29.

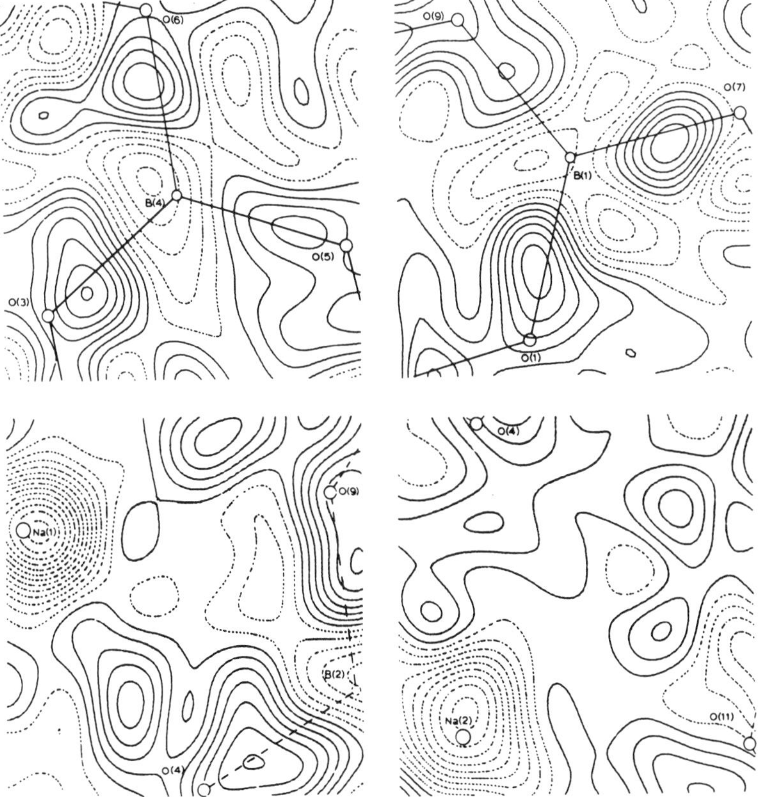

Difference density maps. Contours at 0.05 e/A^3; negative contours dotted. The sections in the upper half of the figure contain the three oxygen atoms, while those in the lower half are through one sodium and two oxygen atoms. The solid lines indicate bonds lying in the section; the dashed lines are projections of bonds to atoms not lying in the plane of the density map.

Bibliography

Christ, C. L. (1960). Crystal chemistry and systematic classification of hydrated borate minerals. *Am. Mineral.,* 45: 334–40.

Christ, C. L., Clark, J. R., and Evans, H. T. (1958). The crystal structure of colemanite $CaB_3O_4(OH)_3 \cdot H_2O$. *Acta Crystallogr.,* 11: *Pt. 11,* 761–70.

Clark, J. R. (1959). The crystal structures of inyoite, $Ca_3B_3O_3(OH)_5 \cdot 4H_2O$. *Acta Crystallogr.,* 12: *Pt. 2,* 162–70.

Clark, J. R., Appleman, D. E., and Christ, C. L. (1964). Crystal chemistry and structure refinement of five hydrated calcium borates. *J. Inorg. Nucl. Chem.* 26: 73–95.

Cooper, W. F., Larsen, F. K., Coppens, P., and Giese, R. F. (1973). Electron population analyses of accurate diffraction data. V. Structure and one-center charge refinement of the light atom mineral, kernite, $Na_2B_4O_6(OH)_2 \cdot 3H_2$. *Am. Mineral.,* 58: 21–31.

Corazza, E., and Sabelli, C. (1966). The crystal structure of kaliborite. *Atti Accad. Naz. Lincei, R. C. Cl. Sci. Fis. Mat. Nat.,* 41: 527–52.

Ehrenburg, W., and Ramdohr, P. (1935). The structure of nordenskioldite. *Neues Jahrb. Mineral. Geol.* 69A: 1.

Fang, J. H., and Newnham, R. E. (1965). The crystal structure of sinhalite. *Min. Mag.,* 35: 196–9.

König, H., Hoppe, R., and Jansen, M. (1979). The crystal structure of sodium borate, $Na_4B_2O_5$. *Z. Anorg. Chem.,* 449: 91–101.

Machida, K., Adachi, G., and Shiokawa, J. (1978). The crystal structure of europium metaborate, EuB_2O_4. *Acta Crystallogr. B,* 35: 149–51.

Mighell, A. D., Perloff, A., and Block, S. (1966). The crystal structure of the high temperature form of barium borate, $BaO \cdot B_2O_3$. *Acta Crystallogr.,* 20: *Pt. 6,* 819–23.

Moore, P. B., and Araki, T. (1974). Roweite, $Ca_2Mn_2(OH)_4[B_4O_7(OH)_2]$: its atomic arrangement. *Am. Mineral.,* 59: 60–65.

Morimoto, H. (1956). The crystal structure of borax. *Mineral. J. Jpn,* 2: 1–18.

Prewitt, C. T., and Buerger, M. J. (1961). The crystal structure of cahnite $Ca_2B[AsO_4](OH)_4$. *Am. Mineral.,* 46: 1077–85.

Takéuchi, Y., Watanabé, T., and Ito, H. (1950). The crystal structure of warwickite, ludwigite, and pinakiolite. *Acta Crystallogr.,* 3: *Pt. 2,* 98–107.

Zachariasen, W. H. (1931). The crystalline structure of hambergite, $Be_2BO_3(OH)$. *Z. Krist.,* 76: 289.

(1954). The precise structure of orthoboric acid. *Acta Crystallogr.,* 7: *Pt. 4,* 305.

Zachariasen, W. H., Plettinger, H. A., and Marezio, M. (1963). The structure and birefringence of hambergite, $Be_2BO_3(OH)$. *Acta Crystallogr.,* 16: *Pt. 11,* 1144–6.

18 The crystal chemistry of oxides and some fluoride minerals

Oxides, or compounds of oxygen with metals, represent close-packed atomic arrays showing high coordination numbers for both the metal and oxygen atoms, high density, and generally high indices of refraction. These properties have been interpreted by many to indicate that the chemical bonding in oxides is largely ionic, and that small metal cations occupy interstices inside hexagonal and cubic close-packed arrays of oxygen anions, O^{2-}. The high indices of refraction of oxides, many of the same magnitude as diamond, and their high refractivity values K, their high polarizability per molar volume α_G/V_m (Å^3), and the small electronegativity difference of their metal–oxygen bonds all suggest that covalent chemical bonds account for a large, if not major, component of these properties. As has been emphasized in this book, most metal–oxygen compounds, including silicates, owe their chemical, optical, and mechanical properties to mixed ionic–covalent bonds. These include especially the very important bonds that build pervasive molecular orbitals that act as conveyor belts to absorb and transport most of the electromagnetic spectrum of the visible light impinging on a crystal.

Detailed examination of the properties of oxide minerals shows that there is poor correlation between density, index of refraction, and percentage of covalent bonding calculated from electronegativity differences. The most important factor relating index of refraction, refractivity, and polarizability appears to be the *species of valence electron orbitals* used in bonding metal to oxygen. Refractivity, polarizability, and indices of refraction of oxides will be *lowest* where *s* and *p electrons* bond metal to oxygen, *extremely high* where *3d transition metal orbitals* are involved in the metal–oxygen bonds, and *intermediate* where either *4d, 5d,* or *4f valence electron orbitals* participate in the bonding (see Table 11.1).

Apparently oxides that use 3*d* transition metal *d* orbitals in bonding have metal atom (or ion) radii, coordination sites, and oxygen polyhedral volumes that enhance or optimize concentration of electron density, resulting in extremely high polarizability, refractivity, and index of refraction.

Corundum and hematite have virtually identical crystal structures, and Al^{3+VI} and Fe^{3+VI} have relatively small ionic radii for CN VI, 0.53 and 0.64 Å, respectively. Although the smaller Al^{3+} ion reduces the volume of the AlO_6 octahedron with respect to the FeO_6 octahedron of hematite, corundum has a mean n of 1.760, and hematite a mean n of 3.100. Al uses $3s^23p^1$ orbitals to bond to oxygen in corundum, whereas Fe^{3+} uses $4s^23d^1$ orbitals to bond to oxygen in hematite. The colorless, completely transparent nature of pure corundum, compared with the deep red, poorly transparent, almost metallic nature of hematite suggests that $3d$ orbitals of Fe^{3+} form π bonds with unused p orbitals of oxygen in hematite. Closeness of packing lowers energy and determines the structure, but chemical bonding determines the physical and chemical properties.

The occurrence of square or rectangular planar coordinations of oxygen around Pd and Pt in PdO and PtO, and around divalent Cu in tenorite, CuO, along with coordination number II for univalent Cu in cuprite, Cu_2O, can best be explained by the presence of hybrid covalent bond orbitals, dsp^2 in PdO and CuO, and s–p linear in Cu_2O (see Chapter 3). Square planar and linear coordinations are not to be expected in ionic oxide structures based upon close packing of oxygen ions. However, in some structures containing divalent Cu, square planar coordination may result from crystal field repulsion of d_{z^2} orbitals parallel to the c axis (see Chapter 7), allowing close approach of four O^{2-} anions to a divalent Cu ion in a plane perpendicular to c.

Most of the oxide minerals and the few fluoride minerals described in this chapter are based upon a close packing of anions containing interstitial metal atoms. From the previous discussion, however, it would be a mistake to assume that the closeness of packing results predominantly from ionic bonding.

Because most oxides and fluorides have simple formulae, compared with silicates, they will be divided, for purposes of discussion, into coordination groups such as VI–VI, IV–IV, VI–III, and so on.

VI–VI structures

Periclase (see Figure 18.1)

$4/Mg^{VI}O^{VI}$ Isometric $Fm3m$
$a = 4.21$ Å $\quad V_{uc} = 74.618$ Å^3
$M/\rho = V_m = 40.32/3.58 = 11.26\ cm^3$
Electronegativity: Mg—O = 2.3 (74% ionic)
Hardness: 5.5 Luster: vitreous
Color: white
A—X distance: Mg—O = 2.10 Å O—O = 2.8 Å
CN: Mg^{VI}–regular octahedron
O^{VI}–regular octahedron
Specific refractivity: $K = 0.2056$

Molecular polarizability: $\alpha_G = 3.285$ Å^3
Optical properties: $n = 1.736$ Isotropic
Color (transmitted light): colorless
Optical indicatrix model: sphere of $r = n$

Discussion. The example chosen as representative of this large group is periclase, MgO, which the reader will immediately recognize as having the halite structure, NaCl (Fig. 18.1). An extremely large number of compounds – oxides, halides, sulfides, and carbides – crystallize as VI–VI structures using the *Fm*3*m* space group of halite (Table 18.1 and see Chapters 4, 6, 8).

The structure is sometimes cited as XII–VI–VI (or VI–VI–XII) because the O^{2-} ion is itself cubic close packed, and accordingly each O is surrounded by twelve additional O ions or atoms (see Fig. 4.2). As the CN code indicates, each Mg is surrounded by six O, and each O by six Mg, at corners of geometrically regular octahedra. Thus, all six of six available octahedral sites around each oxygen are occupied; that is, each oxygen forms a corner common to six MgO_6 octahedra, and vice versa. Thus all twelve edges of each octahedron are shared, and there is no opportunity for shared edges to shorten and for unshared edges to lengthen. The eight tetrahedral voids or sites of the close-packed O arrays (see Figs. 4.2, 4.3) are empty.

VI–IV structures

Corundum (see Figs. 18.2, 18.3)

Trigonal: $R\bar{3}c$ $Al_2^{VI}O_3^{IV}$
uc_{rh} ($2/Al_2O_3$) $a_{rh} = 5.13$ Å $\alpha = 55.3°$ $V_{uc} = 84.65$ Å$^3_{rh}$
uc_{hex} ($6/Al_2O_3$) $a_{hex} = 4.76$ Å $c_{hex} = 13.00$ Å
$V_{uc} = 253.94$ Å$^3_{hex}$

Figure 18.1. Packing model of either periclase, MgO, or halite, NaCl.

$M/\rho = V_m = 101.96/4.00 = 25.49\ cm^3$
Electronegativity difference Al—O = 2.0 (63% ionic)
Hardness: 9.0 Luster: vitreous
Color: brown, white, red (ruby), blue (sapphire)
CN: Al^{VI} - distorted octahedron
O^{IV} - regular tetrahedron
A—X distances: Al—O = 1.86, 1.97 Å
O—O (shared edge) = 2.50 Å
O—O (unshared edge) = 2.80 Å
Fe^{3+}—O = 1.91, 2.06 Å
V^{3+}—O = 1.96, 2.06 Å

Table 18.1. *Crystal chemical data for VI–VI structures of the periclase–halite type, Fm3m*

Phase	Comp.	a (Å)	ρ	n
	LiF	4.02	2.64	1.392
Villiaumite	NaF	4.62	2.79	1.326
Carobbiite	KF	5.34	2.50	1.362
	RbF	5.64	3.86	1.394
	CsF	6.00	4.65	1.478
	LiCl	5.13	2.07	1.662
Halite	NaCl	5.64	2.17	1.544
Sylvite	KCl	6.29	2.00	1.490
	RbCl	6.54	2.80	1.494
(high T phase)	CsCl	7.02	3.23	1.534
	NaH	4.88	1.37	1.470
Periclase	MgO	4.21	3.58	1.736
Manganosite	MnO	4.44	5.36	2.17
Wüstite	FeO	4.30	6.00	2.32
Bunsenite	NiO	4.17	5.86	2.37_{Li}
	CoO	4.25	6.48	—
	CdO	4.70	8.21	2.49_{Li}
	SrO	5.15	5.04	1.856
	BaO	5.53	6.02	1.958
	MgS	5.20	2.84	2.272
Oldhamite	CaS	5.69	2.71	2.120
Alabandite	MnS	5.21	4.07	2.70_{Li}
	SrS	6.02	3.91	2.107
	BaS	6.39	4.38	2.155
	MgSe	5.45	4.27	2.48
	CaTe	6.34	4.87	2.605
	SrTe	6.47	5.22	2.408
	BaTe	6.99	7.59	2.520
Galena	PbS	5.95	7.5	3.89
	TiC	4.32	4.93	Opaque

Cr^{3+}—O = 1.97, 2.02 Å
Ti^{4+}—O = 2.01, 2.05 Å
Specific refractivity: $K = 0.1915$
Molecular polarizability: $\alpha_G = 7.74$ Å^3
Optical properties: $\omega = 1.769$ $\epsilon = 1.760$ $\omega - \epsilon =$ 0.009 uniaxial (−)
Color (transmitted light): colorless, pale blue, pale pink
Optical indicatrix model: oblate spheroid of revolution with maximum radius $r = n \perp c$

Discussion. In the corundum structure (Figs. 18.2, 18.3) the metal atom Al occupies four of the six available octahedral sites around each O atom in a hexagonal close-packed array, and the tetrahedral sites remain empty. Recall that in a hexagonal close-packed anion array, octahedra and tetrahedra are above and below one another. They share faces along a threefold axis (see Fig. 4.3).

To place Al^{3+} ions in four of the six octahedral sites around an oxygen, one may place three up and one down, or two up and two down, but in either case, placement of a fourth Al brings it geometrically closer to one of the other three

Figure 18.2. Polyhedral model of corundum, Al_2O_3, parallel to *c*, $(2\bar{1}\bar{1}0)$ projection (Newnham and deHahn 1962), and packing drawings of octahedra, (0001) projection perpendicular to *c*.

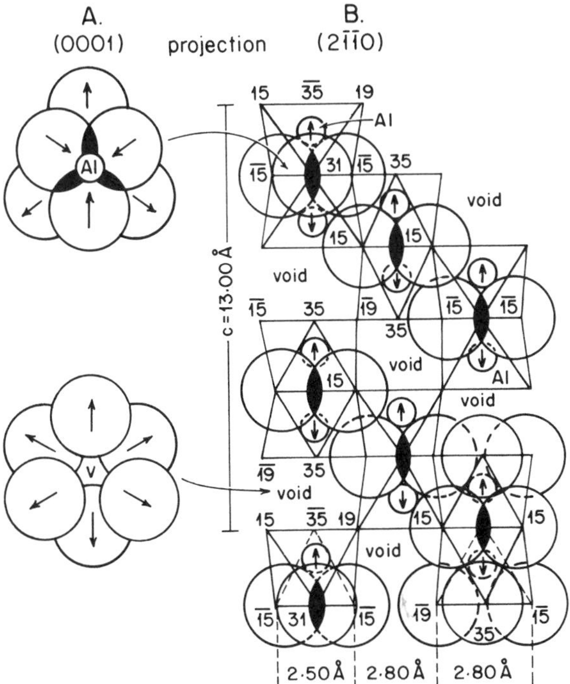

with which it might share a face. Cation repulsion should and does occur. The two Al atoms that approach too closely lie on opposite sides of a shared triangular octahedral face (Fig. 18.2), which should greatly reduce stability (see Pauling's rule 3, Chapter 5). Yet corundum is not only stable, but is the hardest of all minerals after diamond. Normal O—O distances for O^{2-} ions are about 2.80 Å ($r_{O^{2-}} = 1.40$, varying with CN; see Table 4.1). In corundum, the three edges bounding a shared octahedral face have O—O distances of only 2.50 Å. Shared-edge shortening has occurred, and all octahedra have become distorted as shown in Figures 8.3, 18.1, and 18.2.

This shared-edge shortening, or oxygen-pair overlap, coupled with cation–cation repulsion across a shared face prevents too close an approach of two positively charged Al ions. Meanwhile, the Al–Al repulsion is facilitated because an empty octahedral site lies above and below each shared-face pair of octahedra (Figs. 18.2, 18.3). The Al atoms can thus move in opposite directions away from the shared face, and, at equilibrium position, will be exactly equidistant from the other two. All four Al atoms now lie at corners of a tetrahedron. The trigonal symmetry $R\bar{3}c$ allows these distorted AlO_6 octahedra with contracted shared faces to run in spirals up the *c* axis, producing the corundum structure shown in Figure 18.2.

Hematite, Fe_2O_3, karelianite, V_2O_3, and eskolaite, Cr_2O_3 (Table 18.2), are isostructural with corundum. That the bonding in these minerals becomes more covalent, along with the use of 3*d* electrons, has been noted. All these minerals have very high indices of refraction and are strongly colored, which is typical of covalent compounds of first-period transition metals that use 3*d* electrons in bonding.

The titanates ilmenite, $FeTiO_3$, the rarer geikielite, $MgTiO_3$, and pyropha-

Figure 18.3. Packing model of corundum, (0001) projection.

nite, $MnTiO_3$, also crystallize with the VI–IV structure of corundum, but the (Fe, Mg, Mn) atoms of each of these are ordered into octahedral sites in planes that alternate along *c* with planes containing only TiO_6 octahedra. Thus, while preserving the basic structure type, cation ordering reduces the symmetry from $R\bar{3}c$ to $R\bar{3}$. The presence, and the ordering into layers of, two species of atoms rather than one eliminates the *c* glide plane of symmetry of corundum.

VIII–IV structures and derivatives: A:X

Fluorite (Fig. 18.4)

$4/Ca^{VIII}F_2^{IV}$ Isometric $Fm3m$
$a = 5.46$ Å $V_{uc} = 163.04$ Å^3
$M/\rho = V_m = 78.08/3.18 = 24.55$ cm^3
Electronegativity difference: Ca—F = 3.0 (89% ionic)
Hardness: 4.0 Luster: vitreous
Color: purple, green, white
CN: Ca^{VIII} – regular cube
 F^{IV} – regular tetrahedron
A—X distance: Ca—F = 2.36 Å
Specific refractivity: $K = 0.136$
Molecular polarizability: $\alpha_G = 4.22$ Å^3
Optical properties: $n = 1.433$ Isotropic
Color (transmitted light): colorless to very pale violet or green
Optical indicatrix model: sphere of $r = n$

Table 18.2. *Crystal chemical data for VI–IV structures of the corundum type, $R\bar{3}c$ and $R\bar{3}$*

Phase	Comp.	a (Å)	c (Å)	ρ	ω	ϵ
Corundum	Al_2O_3	4.76	13.00	4.0	1.769	1.760
Karelianite	V_2O_3	4.99	13.98	4.95	—	—
Eskolaite	Cr_2O_3	4.95	13.58	5.26	2.5_{Li}	
Hematite	Fe_2O_3	5.03	13.75	5.28	3.22	2.94
Ilmenite	$FeTiO_3$	5.08	14.03	4.82	Opaque	
Geikilite	$MgTiO_3$	5.07	13.95	3.85	2.33	1.96
Pyrophanite	$MnTiO_3$	5.14	14.36	4.57	2.48	2.21
	Ga_2O_3	4.98	13.49	6.44	—	—
	Rh_2O_3	5.11	13.82	8.09	—	—
	Co_2As_3	6.16	15.40	6.74	—	—

Uraninite

$4/U^{VIII}O_2^{IV}$ Isometric *Fm3m*
$a = 5.47$ Å $V_{uc} = 163.49$ Å^3
$M/\rho = V_m = 270.03/10.96 = 24.64$ cm^3
Electronegativity difference: U—O = 2.2 (70% ionic)
Hardness: 5.5 Luster: submetallic
Color: black
CN: U^{VIII} – cube
O^{IV} – tetrahedron
A–X distance: U—O = 2.37 Å
Specific refractivity: *K*: not applicable
Optical properties: opaque

Discussion. The structure of fluorite, CaF_2, will be familiar to most students of mineralogy. A calcium ion is enclosed within a cube of fluorine anions, each of which is, in turn, coordinated tetrahedrally by four calcium ions (Fig. 18.4). It may not be readily apparent that the cube in the center of Figure 18.4 is empty. To fill it with a Ca^{2+} ion would place five cations around an F anion, thereby overcharging it. Thus, in fluorite, every other cube translated along the three axial directions must be left empty, giving the checkerboard pattern of Figure 18.4.

Oxide minerals that adopt the fluorite structure are the radioactive species thorianite, ThO_2, and uraninite, UO_2. Uraninite is never found pure in nature, because radioactive decay to Pb and He accompanied by oxidation of U^{4+} to U^{6+} is always present.

Crystal chemical data for some of the many compounds that crystallize with the VIII–IV structure of fluorite are given in Table 18.3.

Figure 18.4. Packing model (left) of fluorite, CaF_2, and checkerboard pattern of alternating Ca filled and empty F_8 cubelets.

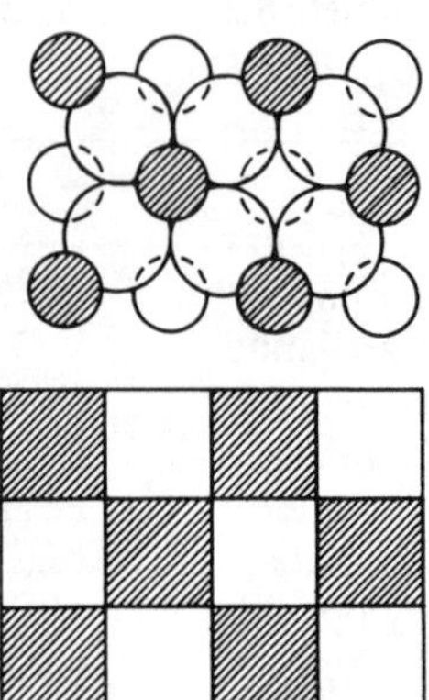

Bixbyite (Fig. 18.5)

$16/Mn_2^{VI}O_3^{IV}$ Isometric *Ia*3
$a = 9.419$ Å $V_{uc} = 835.63$ Å^3
$M/\rho = V_m = 157.876/5.02 = 31.45$ cm^3
Electronegativity difference: Mn—O = 2.0 (63% ionic)
Hardness: 6.0 Luster: submetallic
Color: black
CN: Mn^{VI} – a cube with opposite corners of one face missing gives a peculiar polyhedron that is ½ cube, ½ tetrahedron, and consists of seven faces: one square, four right triangular, two equilateral triangular
O^{IV} – tetrahedral
A—X distance: Mn—O = 2.00 Å O—O = 2.51 Å
Specific refractivity: K = 0.289
Molecular polarizability: $\alpha_G = 18.07$ Å^3
Optical properties: $n = 2.45$, isotropic, nearly opaque
Optical indicatrix model: sphere of $r = n$

Discussion. Bixbyite, $Mn_2^{VI}O_3^{IV}$, has a structure totally unlike that of corundum or hematite. It is a hybrid of the fluorite structure, in which opposite corners of one edge of a cube are missing (Fig. 18.5). The CN of the cation is thus reduced from VIII to VI. It may be regarded as a fluorite structure with three-fourths of the F anions replaced by O, necessitating sixteen rather than four formulae for a unit cell repeat. Numerous trivalent rare earth oxides

Table 18.3. *Crystal chemical data for VIII–IV structures of the fluorite type, Fm3m*

Phase	Comp.	*a* (Å)	*ρ*	*n*
Fluorite	CaF_2	5.46	3.18	1.433
Uraninite	UO_2	5.4	10.96	Opaque
Thorianite	ThO_2	5.60	9.7	2.20
	Rb_2O	6.74	—	—
	Rb_2S	7.65	—	—
	$SiMg_2$	6.39	—	—
	Li_2O	4.62	2.01	1.644
	Na_2O	6.53	—	—
	$AuAl_2$	6.00	—	—
	BaF_2	6.20	4.89	1.475
	Be_2C	4.33	—	—
	CdF_2	5.39	6.64	1.560
	$PtSn_2$	6.43	—	—
>1400°C	ZrO_2	5.07	—	—

crystallize with the bixbyite structure, for example Yb_2O_3, Nd_2O_3, Pr_2O_3, Y_2O_3, La_2O_3, and others (Table 18.4).

Although some investigators state that Mn^{3+} in bixbyite has only four near O neighbors, there must be six coordinating oxygens in order to satisfy Pauling's Rule 2 (see Chapter 5). The extremely unusual CN VI polyhedron: $\frac{1}{2}$ cube, $\frac{1}{2}$ tetrahedron, is found in the many simple rare earth oxides, A_2X_3, that crystallize with this same structure. Although Mn^{3+} is a small ion, $r = 0.65$ Å, ions as large as La^{3+}, Pr^{3+}, Nd^{3+}, and Y^{3+}, with $r = 1.06$, 1.01, 1.00, and 0.89 Å, respectively, for CN VI, should be much too large to occupy a CN IV site, and are never so found. Geller determined (1971) that pure Mn_2O_3 is orthorhombic and crystallizes in space group *Pbca*, but as little as 0.02 formula percent of Fe^{3+}, proxying for Mn^{3+}, stabilizes the isometric *Ia*3 structure of Figure 18.5.

Baddeleyite (Fig. 18.6)

4/Zr^{VII} ①^{III} ②^{IV} Monoclinic $P2_1/c$
$a = 5.15$ Å $b = 5.21$ Å $c = 5.31$ Å
$\beta = 99.4°$ $V_{uc} = 140.562$ Å^3
$M/\rho = V_m = 123.22/5.82 = 21.17$ cm^3

Figure 18.5. Packing drawing of bixbyite, Mn_2O_3, viewed as stacking of ⑧ fluorite-type units lacking one-fourth of their anions.

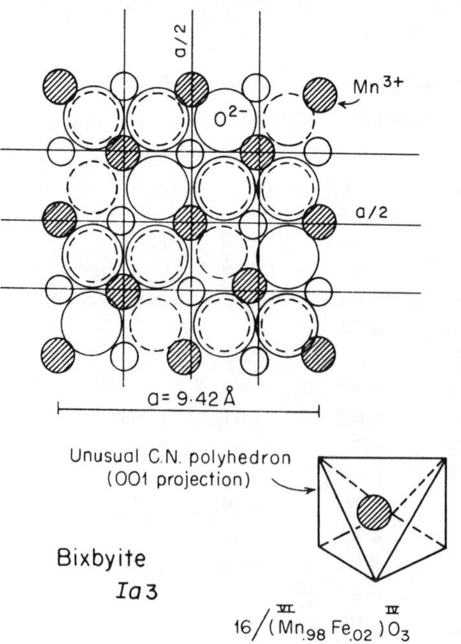

Electronegativity difference: Zr—O = 2.4 (67% ionic)
Hardness: 6.5 Luster: Adamantine
Color: brown
CN: Zr^{VII} - very distorted polyhedron
①^{III} - triangular
②^{IV} - tetrahedral
A—X distance: Zr-① = 2.04 Å Zr-② = 2.16–2.26 Å
O–O = 2.52 Å
Specific refractivity: $K = 0.2016$
Molecular polarizability: $\alpha_G = 9.85$ Å^3
Optical properties: $\gamma = 2.20$ $\beta = 2.19$ $\alpha = 2.13$ $\gamma - \alpha = .07$
biaxial (−) $\mathbf{X} \wedge c = 13°$ $\mathbf{Y} = b$ $2V = 30°$
Color (transmitted light): pale brown
Pleochroism: weak, yellowish-brown ‖**X**,
reddish-brown ‖**Y**, brown ‖**Z**; absorption **Z** > **Y** > **X**
Optical indicatrix model: oblate triaxial ellipsoid approaching an oblate spheroid of revolution

Discussion. Baddeleyite may, like bixbyite, also be regarded as a fluorite-derivative structure. The A^{VIII}, X^{IV} sites of fluorite are proxied by A^{VII}–X^{III} and X^{IV} sites of a very distorted fluorite lattice. Note that axial lengths for *a*, *b*, and *c* of monoclinic baddeleyite are almost equal in length, but distortion has produced a β angle, $a \wedge c = 99°$. In baddeleyite (Fig. 18.6), Zr^{4+}/VII makes three contacts with an undercharged or underbonded oxygen ①^{III}, and four contacts (as in fluorite) with an overbonded oxygen ②^{IV}. Electrostatic bond strength to ① is 1.714 and to ② is 2.286, which sum to +4 per two oxygens. Note (Fig.

Table 18.4. *Crystal chemical data for VI–IV structures of the bixbyite type, Ia3*

Phase	Comp.	*a* (Å)	ρ	*n*
Bixbyite	Mn_2O_3	9.42	5.02	2.45
	Sc_2O_3	9.85	—	—
	Y_2O_3	10.60	5.04	1.910
	La_2O_3	11.38	5.69	—
	Pr_2O_3	11.14	6.34	—
	Nd_2O_3	11.05	6.63	—
	Tl_2O_3	10.54	10.37	—
	Eu_2O_3	10.86	7.30	—
	Lu_2O_3	10.39	9.42	—
	Be_3N_2	8.13	2.22	—
	U_2N_3	10.68	11.30	—
	Be_3P_2	10.15	2.26	—

18.6) that ① oxygens lie along the (100) faces, and ② oxygens lie in a plane between these.

Because of its high mean index of refraction (2.173) and adamantine luster, colorless, synthetic "Cubic-ZrO_2" is now widely used in inexpensive jewelry as a diamond substitute, a role formerly held by YAG (see Chapter 9).

The TiO_2 polymorphs: rutile, anatase, and brookite

Rutile is by far the most common of the three polymorphs of TiO_2; anatase is less common, and brookite, rare. All three minerals have the same coordination pattern of VI–III, but the Ti—O_6 octahedra share different numbers of edges. Thus each Ti is in an octahedron of six oxygens, and each oxygen within a triangle of three Ti atoms. This is accomplished by the use of three different types of octahedral linkage (see Figs. 18.7 to 18.13).

In rutile, two of each twelve octahedral edges are shared with other octahedra;

Figure 18.6. Structure of baddeleyite, ZrO_2, featuring unusual CN VII polyhedron coordination of Zr^{4+}.

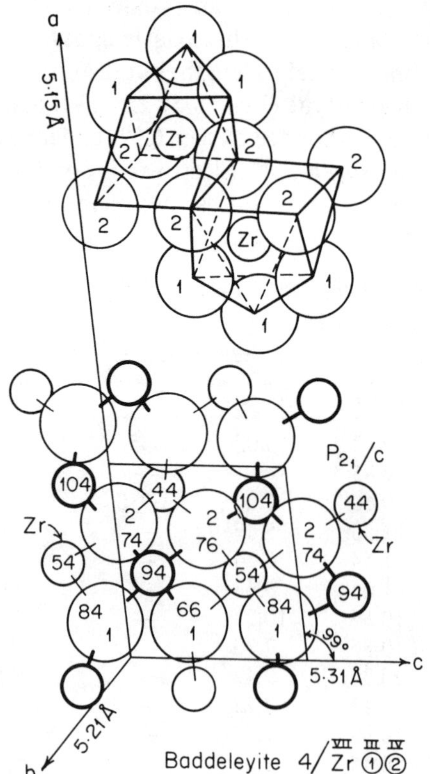

in anatase, four of each are shared; and in brookite, three of each are shared. Edge sharing is symmetrical and opposed in rutile and anatase, but, of necessity, lopsided in brookite. Although it might be expected that extensive polyhedral edge sharing would decrease the molar volume and increase the density and indices of refraction, exactly the reverse occurs (Figs. 18.7, 18.8).

Figure 18.7. Optical, density, and molar volume data for TiO_2 polymorphs; heavy lines in octahedral drawings represent shared edges.

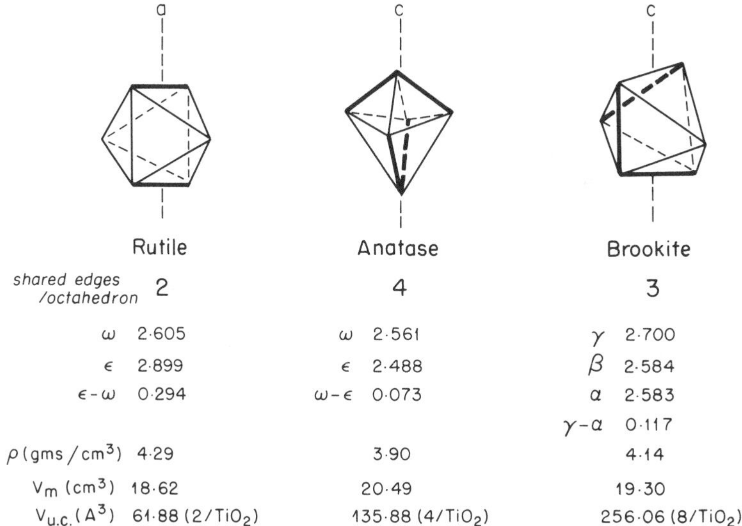

Figure 18.8. Packing models of one octahedron in rutile (A), anatase (B), and brookite (C). Small white Ti ions on octahedral edges show that rutile (A) shares two edges per octahedron, anatase (B), 4, and brookite (C), 3.

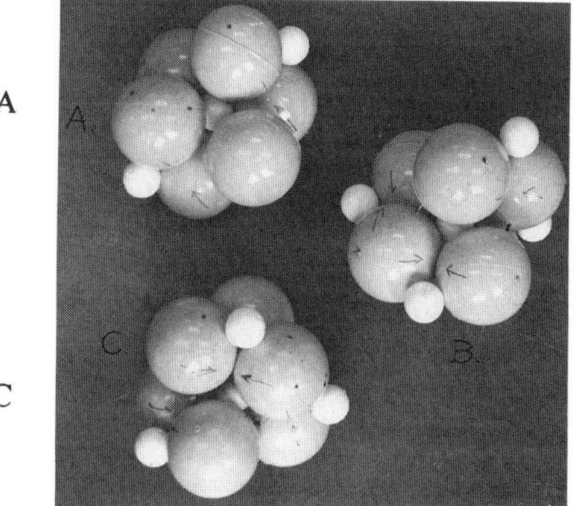

Rutile (see Figs. 18.9, 18.10)

$2/Ti^{VI}O_2^{III}$ Tetragonal $P4_2/m\ n\ m$ (c / (001)(100)(110))

$a = 4.58$ Å $c = 2.95$ Å $V_{uc} = 61.88$ Å^3
$M/\rho = V_m = 79.90/4.29 = 18.64$ cm^3
Electronegativity difference: Ti—O = 1.9 (59% ionic)
Hardness: 6.5 Luster: adamantine–submetallic
Color: deep red-brown
CN: Ti^{VI} – octahedral
O^{III} – equilateral triangle
A—X distance: Ti—O = 1.96 Å O—O (shared edge) = 2.50 Å
O—O (unshared) = 2.80 Å
Specific refractivity: $K = 0.398$
Molecular polarizability: $\alpha_G = 12.615$ Å^3
Optical properties: $\epsilon = 2.899$ $\omega = 2.605$ $\epsilon - \omega = 0.294$
uniaxial (+)
Color (transmitted light): orange, red-brown
Pleochroism: **E** > O, weak
Optical indicatrix model: prolate spheroid of revolution with maximum ellipsoidal $r = \epsilon$, parallel to c, and minimum r, ω equal to r the circular section perpendicular to c

Figure 18.9. (A) Polyhedral (octahedral) drawing of rutile, showing linkage of staggered rows of Ti octahedral chains parallel to c. (B) Polyhedral model of one unit cell of rutile is built around one octahedron.

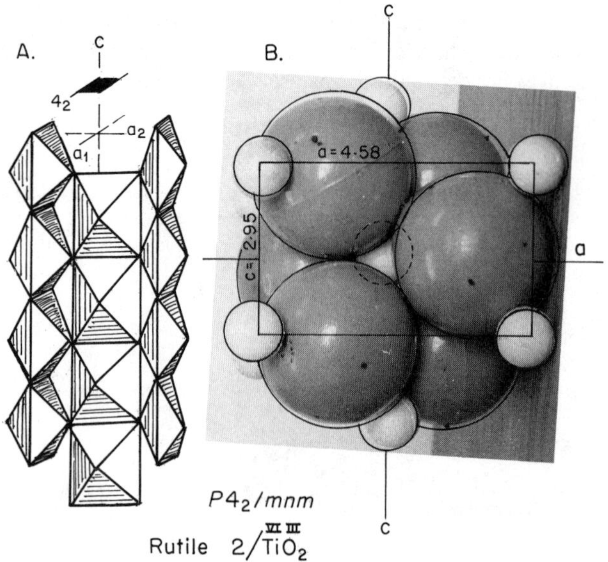

Discussion. Octahedral bands or chains run parallel to the *c* crystal axis and polymerize by sharing two opposed edges per octahedron (Fig. 18.9). Each such band is surrounded by and cross-linked to four identical octahedral bands turned at 90° to the first band. Adjoining bands are all staggered *c*/2 by the 4_2 screw axes that are oriented along tunnels parallel to *c* (Fig. 18.10). Shared O—O keels of each octahedron are shortened from an ideal O—O distance of 2.80 Å to 2.50 Å in accordance with Pauling's third rule for coordination structures. This shortening of shared edges of octahedra in staggered rows parallel to *c* increases the electron density and polarizability parallel to *c* relative to that parallel to *a* or perpendicular to *c*. Thus, $\epsilon \| c > \omega \perp c$, defining an optically positive uniaxial mineral (see Figs. 11.1, 18.9).

Isostructural with rutile are sellaite, MgF_2; cassiterite, SnO_2; pyrolusite, MnO_2; stishovite, SiO_2; and compounds GeO_2, PbO_2, VO_2, CrO_2, ZnF_2, PdF_2, and others (Table 18.5).

Figure 18.10. (left) Polyhedral model of rutile (001) projection showing *n* glides (100), mirror planes (110), and 4_2 screw axes parallel to *c* axis passing through tunnels. (right) Polyhedral and packing drawings of rutile (001) projection emphasizing 4_2 screw axes parallel to *c* in tunnels between octahedra.

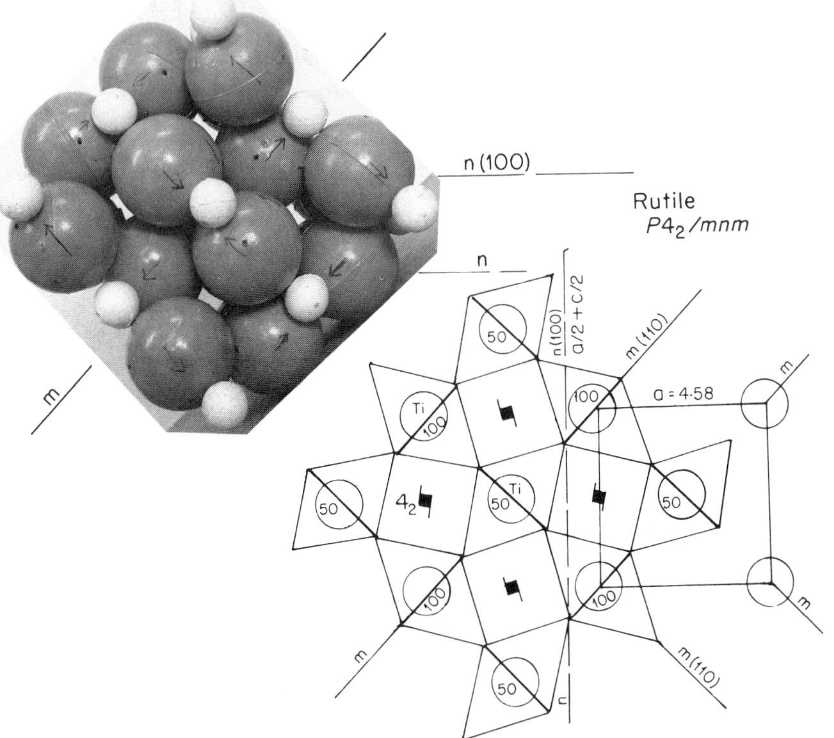

Anatase (see Figs. 18.11, 18.12)

$4/Ti^{VI}O_2^{III}$ Tetragonal $I4_1/amd$
$a = 3.78$ Å $c = 9.51$ Å $V_{uc} = 135.88$ Å^3
$M/\rho = V_m = 79.90/3.90 = 20.49$ cm^3
Electronegativity difference: Ti—O = 1.9 (59% ionic)
Hardness: 5.5–6.0 Luster: adamantine
CN: Ti^{VI} – distorted octahedron
O^{III} – triangle
A—X distance: Ti—O = 1.91,1.95 Å O—O (shared edge) = 2.43 Å O—O (unshared) = 2.80 Å
Specific refractivity: $K = 0.394$
Molecular polarizability: $\alpha_G = 12.48$ Å^3
Optical properties: $\omega = 2.561$ $\epsilon = 2.488$ $\omega - \epsilon = 0.073$ uniaxial (−)
Color (transmitted light): brown or greenish-blue
Pleochroism: weak, variable
Optical indicatrix model: oblate spheroid of revolution with maximum ellipsoidal $r = \omega \perp c$, and minimum $r = \epsilon \| c$

Discussion. Anatase consists of octahedral chains that zigzag along the a_1 and a_2 axes (Figs. 18.11, 18.12). Each TiO_6 octahedron shares four of its twelve edges with other octahedra along *a* axes while octahedral sites above and below these chains, along *c*, are empty. Ti atoms located at 100, 75, 50, and 25, spiral in clockwise and counterclockwise motion up and down *c* as 4_1 and 4_3 screw

Table 18.5. *Crystal chemical data for VI–III structures of the rutile type, $P4_2mnm$*

Phase	Comp.	a (Å)	c (Å)	ρ	ω	ϵ
Rutile	TiO_2	4.58	2.95	4.29	2.605	2.899
Cassiterite	SnO_2	4.74	3.19	6.99	2.001	2.098
Pyrolusite	MnO_2	4.39	2.87	5.22	Opaque	
Plattnerite	PbO_2	4.95	3.38	9.60	2.30	—
Stishovite	SiO_2	4.18	2.66	4.35	1.800	1.845
Sellaite	MgF_2	4.62	3.05	3.15	1.378	1.390
	GeO_2	4.39	2.86	6.30	1.99	2.05
	FeF_2	4.69	3.31	4.28	—	—
	MnF_2	4.87	3.31	3.93	—	—
	NiF_2	4.65	3.08	4.82	—	—
	ZnF_2	4.70	3.13	4.96	—	—
	PdF_2	4.93	3.37	5.85	—	—
	NbO_2	4.77	2.96	6.45	—	—
	CrO_2	4.41	2.91	5.28	—	—

axes (Fig. 18.11). Shortened shared edges of TiO_6 octahedra strung out parallel to a_1 and a_2 (Fig. 18.12) increase the electron density and index of refraction, ω, parallel to a_1 and a_2 over that, ϵ, parallel to c: velocity $\| a$ = slow; $\| c$ = fast; $\omega > \epsilon$ gives an optically negative mineral.

Brookite (Fig. 18.13)

$8/Ti^{VI}O_2^{III}$ Orthorhombic *Pbca*
$a = 9.14$ Å $b = 5.44$ Å $c = 5.15$ Å $V_{uc} = 256.06$ Å^3
$M/\rho = V_m = 79.90/4.14 = 19.30$ cm^3
Electronegativity difference: Ti—O = 1.9 (59% ionic)
Hardness: 5.5–6.0 Luster: adamantine
CN: Ti^{VI} – distorted octahedron
O^{III} – close to a right triangle
A—X distance: Ti—O = 1.87, 2.04 Å O—O (shared edge) = 2.49 Å
O—O (unshared) = 2.80 Å
Specific refractivity: $K = 0.390$
Molecular polarizability: $\alpha_G = 12.39$ Å^3
Optical properties: $\gamma = 2.700$ $\beta = 2.584$ $\alpha = 2.583$ $\gamma - \alpha = 0.117$
biaxial (+)

Figure 18.11. Packing model (A) and polyhedral and packing drawings (B) of anatase, TiO_2, (100) and (001) projections. Rods marked R and L are 4_1 and 4_3 screw axes around which Ti atoms can be seen translating one-fourth $c/90°$ or 100, 75, 50, 25, clockwise and counterclockwise.

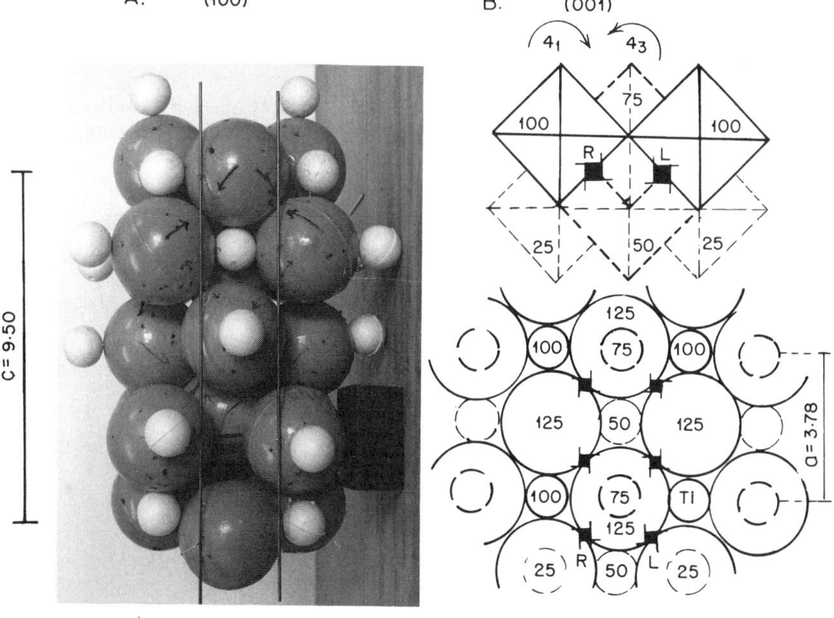

Figure 18.12. Polyhedral and packing drawing of very distorted octahedra of anatase, TiO_2, (100) projection. Heavy lines, SE, are shared octahedral edges, four per octahedron, running parallel to *a* axes, and slow vibration directions.

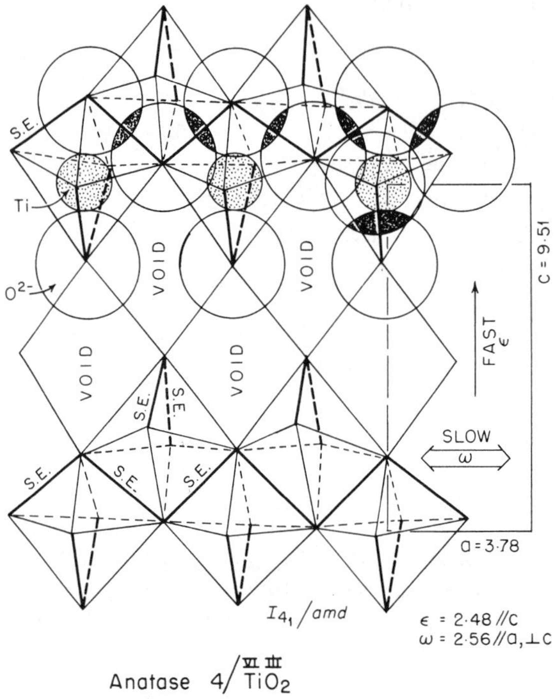

Figure 18.13. Unit cell of brookite, (100) projection (left), and polyhedral (octahedral) linkage of octahedra each sharing three edges (heavy lines).

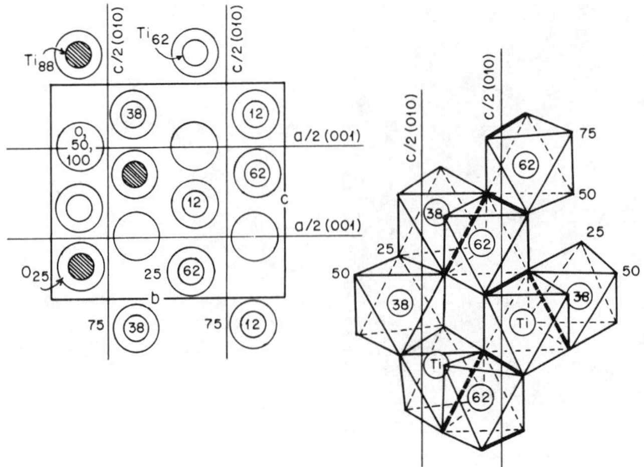

Color (transmitted light): yellow to dark brown
Pleochroism: weak to nil
Optical indicatrix model: prolate triaxial ellipsoid, almost a prolate spheroid of revolution for Na light, becoming increasingly triaxial with decrease in λ (e.g., blue light), maximum ellipsoidal $r = \gamma \| b$. This large change in ellipsoidal shape and dimension results from high dispersion (n for blue light greater than n for red light), endowing brookite with anomalous optical properties.

Brookite is a comparatively rare mineral.

VI–VI–IV–III–IV structures

Pseudobrookite (see Fig. 18.14)

$4/Fe_2^{VI}Ti^{VI}①^{IV}②_2^{III}③_2^{IV}$ Orthorhombic *Bbmm*
$a = 9.77$ Å $b = 9.95$ Å $c = 3.72$ Å $V_{uc} = 361.63$ Å^3
$M/\rho = V_m = 239.60/4.40 = 54.45$ cm^3
Electronegativity difference: Fe^{3+}—O = 1.7 Ti—O = 1.9
Integrated difference for pseudobrookite = 1.85 (57% ionic)
Hardness: 6.0 Luster: submetallic
CN: Fe^{3+VI} and Ti^{VI} – both very distorted octahedra
O-(1)^{IV} and O-(2)^{IV} – both tetrahedral
O-(3)^{III} – triangular
A—X distance: Ti—O = 1.93 Å Fe—O = 1.90 Å(4), 2.25 Å(2)
Specific refractivity: $K = 0.323$
Molecular polarizability: $\alpha_G = 30.65$ Å^3
Optical properties: $\gamma = 2.42$ $\beta = 2.39$ $\alpha = 2.38$ all for Li light, biaxial (+)
Color: dark brown-black, nearly opaque
Pleochroism: weak in red-brown with **Z** > **Y** > **X**
Optical indicatrix model: prolate triaxial ellipsoid with maximum ellipsoidal $r = \gamma \| b$

Discussion. Pseudobrookite consists of separate strings of TiO_6 and FeO_6 octahedra that run parallel to the *c* axis, and which are cross-linked laterally by additional FeO_6 octahedra. Vertices of TiO_6 octahedral strings are designated oxygen ① or O-(1), vertices of FeO_6 octahedral strings are designated oxygen ③ or O-(3), and lateral-linking oxygens are designated oxygen ② or O-(2). There are two each of oxygens ② and ③ and one of oxygen ① per formula of

$Fe_2^{3+}TiO_5$. Oxygens ① and ③ are overbonded and oxygen ② is underbonded, giving an EBS distribution as follows:

$$
\begin{array}{lrl}
\text{① Ti Ti Fe Fe} = \tfrac{14}{6} \times 1 = & \tfrac{14}{6} \\
\text{② Ti Fe Fe} = \tfrac{10}{6} \times 2 = & \tfrac{20}{6} \\
\text{③ Fe Ti Fe Fe} = \tfrac{13}{6} \times 2 = & \underline{\tfrac{26}{6}} \\
 & +\tfrac{60}{6} = +10 \text{ per } 5\ O^{2-}
\end{array}
$$

This solution is based on the assumption that all Ti occupies octahedra designated M-1, and Fe, M-2, of Figure 18.14.

An unoxidized version of pseudobrookite returned from the historic Apollo lunar mission contained all the iron as Fe^{2+} and contained an equal amount of Mg coupled to twice the amount of Ti of pseudobrookite. The formula is $Fe_{0.5}^{2+}Mg_{0.5}Ti_2O_5$, and the mineral was named armalcolite after the astronauts Armstrong, Aldrin, and Collins, who returned it from space. In this structure, the sites designated M-2, (two per formula, eight per unit cell) are occupied by Ti, and the sites designated M-2 are randomly occupied by Mg and Fe^{2+}.

XII–VI–VI structures

Perovskite (ideal cubic structure, Fig. 18.15)

$1/Ca^{XII}Ti^{VI}O_3^{VI}$ Isometric *Pm3m*

$a = 3.83$ Å: below 900 °C the true unit cell is distorted to orthorhombic symmetry. V_{uc}(ideal) = 56.02 Å^3

Figure 18.14. Unit cell of pseudobrookite, analogous to armalcolite, (001) projection, showing very distorted Ti and Fe octahedra, designated M-1 and M-2, respectively.

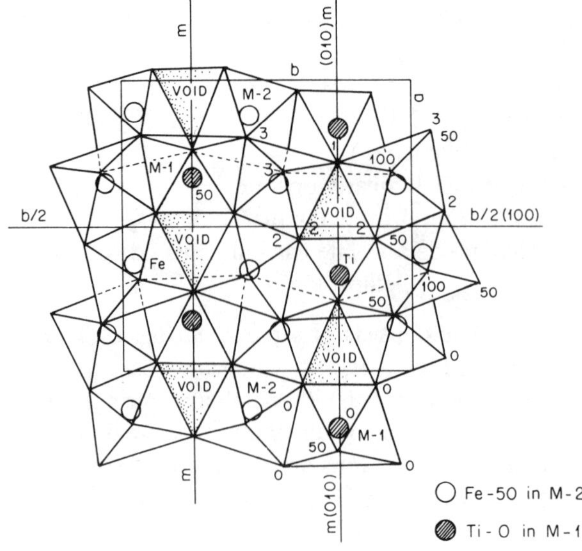

$M/\rho = V_m = 135.98/4.03 = 33.74$ cm^3
Electronegativity difference: Ca—O = 2.5 Ti—O = 1.9
Integrated perovskite = 2.3 (74% ionic)
Hardness: 5.5 Luster: adamantine
CN: Ca^{XII} – cube–octahedron
Ti^{VI} – regular octahedron,
O^{VI} – regular octahedron of Ca(4) + Ti(2)
A—X distance: Ti—O = 1.90–92 Å Ca—O, O—O = 2.70 Å
Specific refractivity: $K = 0.3474$
Molecular polarizability: $\alpha_G = 18.72$ Å^3
Optical properties: $n = 2.40$ Isotropic to weakly birefringent
Color (transmitted light): pale brown to dark brown
Optical indicatrix model: sphere of $r = n$

Discussion. One of the most close-packed structures known is that of ideal cubic perovskite, $Ca^{XII}Ti^{VI}O_3^{VI}$, in which oxygen and calcium ions form an almost perfect cubic close-packed array. The twelve oxygen anions that coordinate each calcium ion form a cube–octahedron bounded by six square and eight triangular faces. *All fourteen of these faces are shared.* Six are common to square faces of six bounding CaO_{12} cube–octahedra, and eight are common to triangular faces of eight bounding TiO_6 octahedra. Each TiO_6 octahedron is in turn joined to six additional TiO_6 octahedra by corner sharing. Perovskite may thus be regarded as an octahedral framework structure or tektotitanate of composi-

Figure 18.15. Packing and polyhedral drawing of several unit cells of ideal perovskite, $CaTiO_3$, analogous to lueshite, $NaNbO_3$.

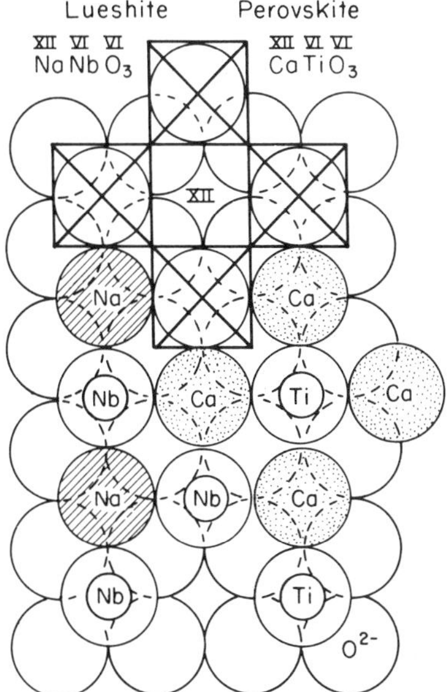

tion TiO_3 containing interstitial Ca. The structure of perovskite may be visualized by (Fig. 18.15):

1. placing eight Ti atoms on corners of a cube, one Ca^+ in the cube center, and O^{2-} ions on all edges
2. placing O^{2-} ions on eight corners and four faces of a face centered cubic array, replacing the top and bottom face center with a Ca^{2+} ion, and placing four Ti^{4+} ions on the four edges that lie vertically between cube corners. In effect, one of every four oxygens is replaced by a Ca^{2+} ion.

Examination of Figure 18.15 shows that O^{2-} ions are surrounded octahedrally by four Ca^{2+} ions in the plane of this illustration and two apical Ti ions above and below this plane. Placement of six cations around an anion leads to a very high packing index, almost perfectly economical in filling of space. It has been postulated that $Mg^{XII}Si^{VI}O_3^{VI}$ with the perovskite structure exists in the lower mantle where the enormous pressures operative would place Mg^{2+} in XII coordination. It has recently been synthesized under lower mantle conditions, Table 18.6. Because the Ca^{2+} ion is somewhat smaller than O^{2-}, the close-packed O—Ca planes of natural perovskite become warped, lowering the symmetry from cubic to orthorhombic or even monoclinic. At temperatures greater than 900°C, however, all perovskite-type structures become truly cubic. Some phases, such as synthetic $SrTiO_3$, are cubic at 26°C with the ideal perovskite structure, and $a = 3.90$ Å (Table 18.6).

In natural occurrences, particularly in magmatic rocks rich in alkalies and CO_2, the Ca^{2+} and Ti^{4+} of perovskite may be extensively proxied for by Na^+ and Nb^{5+}. The pure end member, $NaNbO_3$, isostructural with perovskite, occurs as black cubes in weathered carbonatite (CO_2-rich magmatic rock) from eastern Zaire; it has been named lueshite.

Data for several of the many substances that crystallize with the perovskite structure are given in Table 18.6.

IV–IV structures

Zincite (see Figs. 18.16)

$2/Zn^{IV}O^{IV}$ (polar) Hexagonal $P6_3mc$
$a = 3.249$ Å $c = 5.207$ Å $V_{uc} = 47.600$ Å^3
$M/\rho = V_m = 81.38/5.68 = 14.33$ cm^3
Electronegativity difference: Zn—O = 2.0 (63% ionic)
Hardness: 4.0 Luster: resinous to adamantine
Color: red
CN: Zn^{IV}–tetrahedron O^{IV}–tetrahedron
all tetrahedra are polar
A—X distance: Zn—O = 1.97 Å
Specific refractivity: $K = .179$

Molecular polarizability: $\alpha_G = 5.78$ Å^3
Optical properties: $\omega = 2.013$ $\epsilon = 2.029$ $\epsilon - \omega = 0.0016$
uniaxial (+)
Color (transmitted light): red
Pleochroism: none
Optical indicatrix model: prolate spheroid of revolution with maximum $r = \epsilon \| c$

Synthetic (see Fig. 18.17)

$2/Pd^{IV}O^{IV}$ Tetragonal $P4_2/mmc$
$a = 3.043$ Å $c = 5.337$ $V_{uc} = 49.420$ Å
$M/\rho = V_m = 122.4/8.22 = 14.89$ cm^3
Electronegativity difference: Pd—O = 1.3 (34% ionic)
Hardness: n.a. Luster: n.a. Color: n.a.
CN: Pd^{IV} – rectangular planar O^{IV} – tetrahedral
A—X distance: Pd—O = 2.00 Å (Pt—S = 2.32 Å)
Specific refractivity: K = n.a.
Molecular polarizability: α_G = n.a.
Optical properties: n.a.
(The abbreviation n.a. means not available.)

Discussion. There are three types or classes of IV–IV structures, some used by isostructural oxides and sulfides, and others by oxides alone.

1. *IV–IV (CCP).* The structure of sphalerite, ZnS (see Figure 3.4) is based on cubic close packing with both Zn and S enclosed in tetrahedra of one another.

Table 18.6. *Crystal chemical data for XII–VI–VI structures of the ideal perovskite type, Pm3m*

Phase	Comp.	a (Å)	b (Å)	c (Å)	ρ	n
Perovskite > 900°C	$CaTiO_3$	3.83			—	—
	$4/CaTiO_3$	(5.38	7.64	5.44)[a]	4.03	2.40
Lueshite	$NaNbO_3$	—	—	—	4.44	2.30
Neighborite > 900°C	$NaMgF_3$	3.96				
	$4/NaMgF_3$	(5.36	7.67	5.50)	3.06	1.364
	$BaTiO_3$	4.01			6.00	—
	$BaUO_3$	4.40			8.25	—
	$SrTiO_3$	3.90			5.14	—
	$BaZrO_3$	4.19			6.24	—
	$4/MgSiO_3$[b]	4.78	4.93	6.91	4.11	

[a] Parentheses indicate orthorhombic unit cell dimensions.
[b] Synthesized under lower mantle conditions with T = 1830°C and P = 27 kbars (Horiuchi, Ito, and Weidner 1987)

Sphalerite has the diamond structure, $Fd3m$, with half of the C atoms replaced by Zn, and the other half by S. The 4_1 and 4_3 screw axes and diamond glides d are eliminated by the ordering of four Zn and four S atoms into the eight C positions of diamond. Thus, sphalerite, otherwise isostructural with diamond, has space group $F\bar{4}3m$. ZnO does not crystallize in this structure.

2. *IV–IV (HCP polar).* The structure of zincite, ZnO (see Figure 18.16), is that of wurtzite, ZnS, polymorphous with sphalerite. In this structure, oxygen and zinc atoms are in hexagonal close packing in space group $P6_3mc$. Occupancy of four of the eight tetrahedral sites of hexagonal close packed arrays (see Chapter 4) again controls the structure. If Zn is placed in three coplanar sites around an oxygen, the fourth Zn must be placed beneath and below this oxygen, not above and on top. Placing the fourth Zn below the oxygen locates all four Zn atoms geometrically equidistant from one another at corners of a regular tetrahedron (Fig. 18.16), and in this arrangement all tetrahedra ZnO_4 and O_4Zn point in the same direction (toward the viewer in Fig. 18.16), and zincite is a polar structure. The same structure is used by both bromellite, BeO, and silicon carbide, SiC (Table 18.7), although this last phase is polymorphous. Placement of the fourth Zn atom of HCP on top of the central oxygen brings positive charges too close to one another. That ZnO uses only the hexagonal structure, $P6_3mc$, whereas ZnS uses both the isometric $F\bar{4}3m$ and hexagonal structures, indicates that zincite is more ionic (63%) and ZnS more covalent

Figure 18.16. The polar zincite structure, ZnO. (A) polyhedral (0001) projection; (B) packing drawing perpendicular to *c*; (C) packing drawing parallel to *c*.

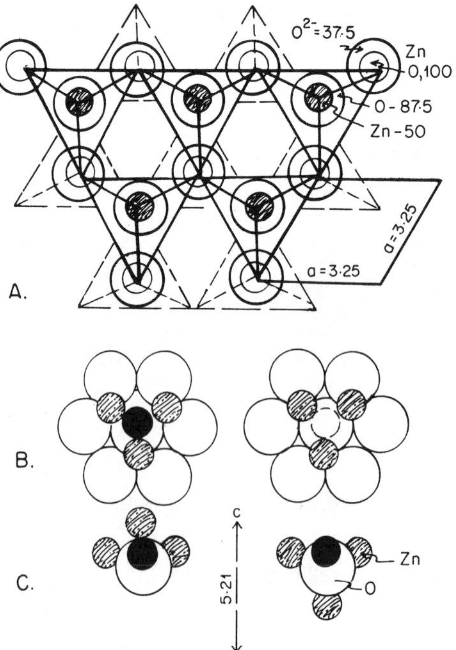

(78%), which would be predicted from electronegativity differences (see Chapters 2–4).

3. *IV–IV square planar.* The structure of synthetic palladous oxide, PdO (Pauling 1960) is identical with that of PtO and of cooperite, PtS (Fig. 18.17). Each Pd (or Pt) is surrounded by four O (or S) atoms at corners of a planar coordination site that is actually rectangular rather than square planar. Each O (or S) is in turn surrounded tetrahedrally by four Pd (or Pt) atoms. The square planar coordination cannot occur in ionic structures, and owes its formation to the use of dsp^2 hybrid bond orbitals on Pd (or Pt, see Chapter 3).

This structure is much easier to visualize from the doubled tetragonal $P4_2/mmc$ cell, $Z = 4$, which is base or c centered (although in crystallography tetragonal C is equivalent to P; see Chapter 6). The true unit cell is primitive, with $Z = 2$, and both cells are illustrated in Figure 18.17. The C-cell illustrates both types of IV-coordination site. The P-cell shows only the rectangular planar site, because the tetrahedral site enclosed by Pt_4 is empty. Note that the P-cell is remarkable, in that it is a hollow bounded by rectangles. Thus, both square or rectangular sites, PdO_4 (PtS_4), run in continuous planes at right angles to one another (100) and (010), and alternate with empty square sites along the c axis, whereas tetrahedral OPd_4 (SPt_4) sites are likewise alternately occupied and empty but along the a axes. 4_2 screw axes pass through the hollow cells or tunnels parallel to c of the P-cell. Note that both cells C and P have the same c dimension, and that (100) of the C-cell is equivalent to (110) of the P-cell. Both cells display the same symmetry, and the dimensions are such that the V_{uc} (Å^3) of the C-cell is just twice that of the P-cell.

Tenorite, CuO, and many other copper minerals display this same square planar hybrid coordination, verifying the use of dsp^2 hybrid bond orbitals by divalent Cu. This coordination is used by divalent Cu in libethenite,

Table 18.7. *Crystal chemical data for IV–IV structures of the zincite–wurtzite type, P6mc*

Phase	Comp.	a (Å)	c (Å)	ρ	ω	ϵ
Zincite	ZnO	3.25	5.21	5.68	2.013	2.029
Wurtzite	ZnS	3.81	6.23	3.98	2.36	2.38
Bromellite	BeO	2.70	4.38	3.02	1.719	1.732
	SiC	3.08	5.05	3.22	2.65	2.69
	AgI	4.58	7.49	5.69	2.218	2.229
	AlN	3.11	4.98	3.26		
	CdS	4.13	6.75	4.82	2.506	2.529
	CuH	2.89	4.61	6.41	—	—
	ZnSe	3.98	6.53	5.42	2.89	—
	ZnTe	4.27	6.99	5.81	—	—
	$(NH)_4F$	4.39	7.02	1.05	—	—

$Cu_2(OH)[PO_4]$, isostructural with andalusite, $Al_2O[SiO_4]$, azurite, $Cu_3(OH)_2(CO_3)_2$, and in many other minerals and synthetic compounds.

VIII–VI–IV–IV structures: the atopites

Pyrochlore (see Figs. 18.18, 18.19)

$8/Na,Ca^{VIII}Nb_2^{VI}O_6^{IV}F^{IV}$ Isometric $Fd3m$
(microlite with Ta > Nb)
$a = 10.397$ Å (10.402 Å–microlite) $V_{uc} = 1125.5$ Å^3
$M/\rho = V_m = 363.88/4.29 = 84.82$ cm^3
Electronegativity difference: Na—O = 2.6 Ca—O = 2.5
Nb—O = 1.9 Na—F = 3.1
Ca—F = 3.0
Integrated pyrochlore = 2.35 (75% ionic)
Hardness: 5.0–5.5 Luster: adamantine to dull, earthy

Figure 18.17. Structure of PdO, PtS (cooperite), or PtO showing planar dsp^2 hybrid bond orbitals around the platinum group metal atom. The smaller, hollow *P*-cell and the larger *C*-cell are compared.

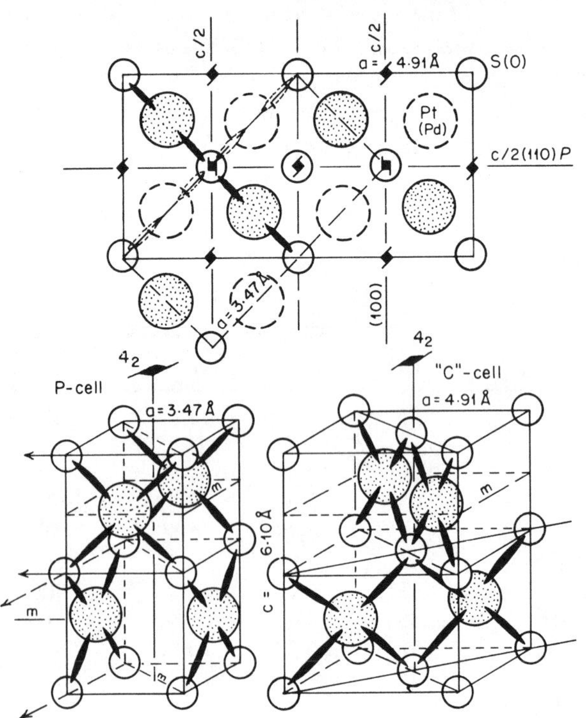

Color: dark brown
CN: (Na, Ca)^{VIII} – distorted cube
Nb^{VI} – distorted octahedron
O^{IV}, Na, Ca, Nb_2 – tetrahedral
F^{IV}, Na_2Ca_2 – tetrahedral
A—X distance: (Na,Ca)—O = 2.65 Å (Na,Ca)—F = 2.26 Å
Nb—O = 1.95 Å O—O = 2.76, 2.64 Å
Specific refractivity: $K = 0.224$
Molecular polarizability: $\alpha_G = 32.37$ Å^3
Optical properties: $n = 1.96$ (variable with composition 1.87–2.18) Isotropic
Color (transmitted light): light to deep brown
Optical indicatrix model = sphere of radius $r = n$

Discussion. An interesting group of minerals of the general formula $A_2^{2+}B_2^{5+}O_7$ includes a variety of niobates, tantalates, antimonates, titanates, and zirconates, to which various different group names have been applied. These were originally classed as atopites for a rare antimonate member, now a discredited species. The best-known member of this group of minerals, all of which crystallize in space group $Fd3m$, is pyrochlore, $NaCaNb_2O_6F$. It has now become the principal ore mineral of niobium, which is used in high-grade steel alloys. Pyrochlore, like perovskite, occurs principally in alkali-rich, CO_2-rich rocks called carbonatites, found in Zaire, Brazil, Norway, Canada, and Australia. The crystal structure, which uses space group $Fd3m$ (Fig. 18.18), is fascinating, because it represents a true framework of NbO_6 octahedra and edge sharing cubes in which all six oxygens of each octahedron are shared with corners of six other octahedra (Fig. 18.19). Voids in this framework of composition NbO_3 (or Nb_2O_6) are just large enough to be occupied by fluorite-like distorted cubic blocks of composition NaCaF. Each O^{2-} ion is surrounded by two Nb ions, one Ca and one Na (bond strength = +2.402); each F ion by two Na and two Ca (bond strength = +0.750) (Fig. 18.18). The six O^{2-} and one F^- ion of the formula thus have a cumulative EBS of + 13.00 (see Pauling Rule 2C of Chapter 5).

The tantalum analogue of pyrochlore, $NaCaTa_2O_6F$, is called microlite. It does not occur in carbonatites, but occurs only in pegmatites. The reason for this separation of the geochemically coherent elements Ta and Nb into two different types of geological association is an enigma.

Most natural samples of pyrochlore do not show the ideal formula, and up to 75% of the two large-ion CN^{VIII} sites may be empty or occupied by moles of H_2O. In addition, accompanying anion defects may occur, leaving large voids in the structure. The following improved formula may be used to take care of these various substitutions:

$$A_{2-x}^{2+}Q_xB_2^{5+}O_{7-x}R_x$$

Figure 18.18. Packing drawing (upper right) of the Ca—Na and F-cube sites of pyrochlore, 8/Na,Ca(Nb,Ta)$_2$O$_6$F, and schematic crisscross orientation (lower left) of (Nb,Ta)—O octahedral strips. Compare with structures of diamond (see Chapter 2) and spinel (see Figs. 18.22, 18.23).

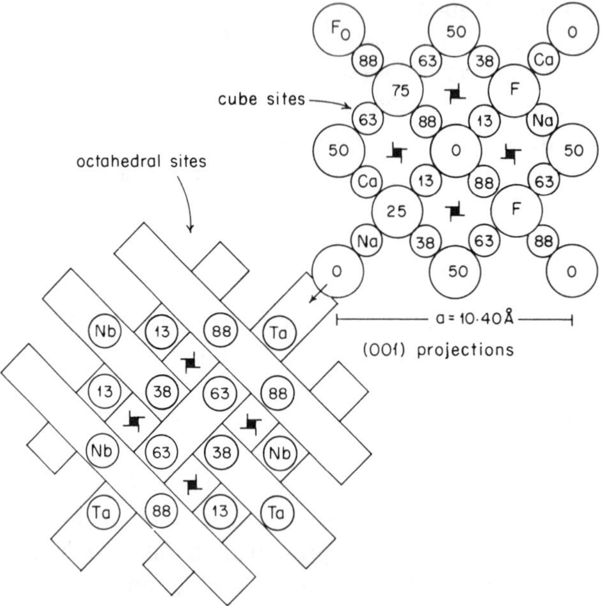

Figure 18.19. Pyrochlore structure viewed along a threefold axis: Nb—O octahedral framework (left) and Na—Ca—O—F cube layer (right). Octahedron number 1 shares all six corners with six other octahedra, numbers 2–7 inclusive. Cube layer fits over octahedral layer.

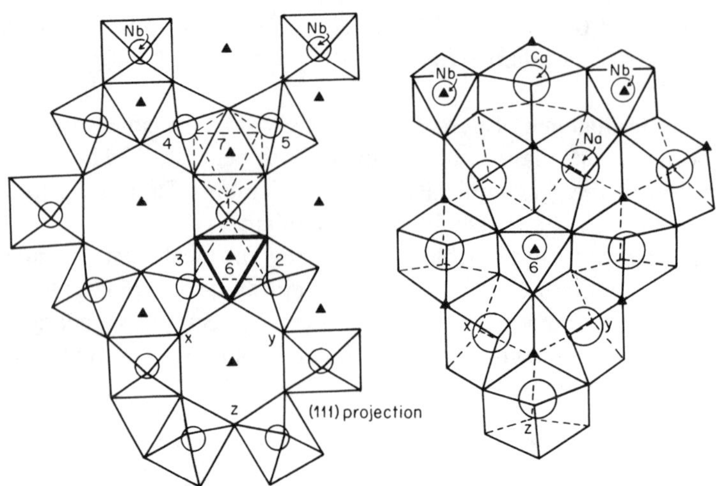

where

A = divalent ions Ca, Ba, and Sr
Q = Na^+, H_2O, or voids
B = pentavalent Nb and Ta, ±Ti, Fe^{3+}
O = O^{2-}
R = F, (OH), or voids

Where $x = 1$, the ideal formula of pyrochlore is obtained: $CaNaNb_2O_6F$. Where more extensive defects occur, as in the variety "pandaite" from Africa, x may reach 1.5, and the composition may be represented as follows:

$$Ba_{0.5}(H_2O)_{1.5}Nb_2O_{5.5}\square_{1.5} \quad \text{or} \quad Ba_{0.5}Nb_2O_4(OH)_3$$

Defects in the anion framework would leave gaping holes in an otherwise close-packed structure. Calculation of anion defects may be avoided by use of the second formula just given, discussed in Chapter 9.

Evidently, the large CN^{VIII} cation site can readily be base exchanged as in zeolites, and it was noted that separation of pyrochlore in the heavy liquid, Clerici's solution (thallous mallonate-formate) resulted in the exchange of Ba^{2+} ions in the mineral with Tl^+ ions from the Clerici liquid. Large commercial concentrations of Ba-rich pyrochlore in laterites of eastern Zaïre owe their concentration to weathering processes that caused the exchange and replacement of Na and Ca by the much heavier Ba ions. Many pyrochlores are radioactive because of extensive substitution of Th^{4+} ions in the CN^{VIII} sites.

VI–VI–III–III–III structures

Columbite–tantalite (see Figs. 18.20, 18.21)

4/(Fe,Mn)(Nb,Ta)$_2$①$_2$②$_2$③$_2$
Orthorhombic *Pcan*
$a = 5.73$ Å $b = 14.24$ Å $c = 5.08$ Å (ferrocolumbite containing a moderate % of Ta_2O_5)
$a = 5.76$ Å $b = 14.42$ Å $c = 5.09$ Å (manganotantalite containing some Nb_2O_5)
Mass: 337.67 (columbite) 513.83 (tantalite)
M/ρ $V_m = 64.94$ cm^3 (columbite) 64.23 cm^3 (tantalite)
$\rho = 5.20$ ferrocolumbite $\rho = 8.00$ manganotantalite
Electronegativity difference: Nb—O = 1.9 Ta—O = 2.0
Fe—O = 1.9 Mn—O = 2.0
integrated = 1.9–2.0 (59–63% ionic)
Hardness: 6 Luster: submetallic
CN: Fe, Mn, Nb, Ta: all VI–distorted octahedra
O: all III–very distorted obtuse triangles
A—X distances: (Nb,Ta)—O = 2.01 Å Mn—O = 2.18 Å

Specific refractivity: $K = 0.270$ (columbite) 0.167 (tantalite)
Molecular polarizability: $\alpha_G = 36.16$ Å^3 (columbite)
$\alpha_G = 33.94$ Å^3 (tantalite)
Optical properties: $\gamma = 2.44$ $\beta = 2.41$ $\alpha = 2.36$ (columbite) biaxial (+)
$\gamma = 2.43$ $\beta = 2.32$ $\alpha = 2.26$ (tantalite) biaxial (−)
$\gamma - \alpha = 0.08$ (columbite) 0.17(tantalite)
$\mathbf{Z} = c$ $\mathbf{Y} = b$ $\mathbf{X} = a$
Color (transmitted light): red-brown Pleochroism: strong
X = yellow **Y** and **Z** dark red brown **Z** > **Y** > **X**
Optical indicatrix model: triaxial ellipsoids, either prolate (columbite) to oblate (tantalite), maximum radius of indicatrix $r = \gamma \| c$

Discussion. Columbite, $(Fe,Mn)Nb_2O_6$, and its Ta analogue, tantalite, $(Fe,Mn)Ta_2O_6$, consist of approximately hexagonally close-packed oxygen arrays in which distorted octahedral sites are occupied by (Fe,Mn) and (Nb,Ta) in the manner illustrated in Figures 18.20 and 18.21. Chains of oxygen octahedra containing Fe or Mn and Nb or Ta zigzag parallel to *c*, sharing edges in the manner of brookite chains of Figure 14.13. Each such octahedral chain is completely ordered, so there are two chains containing only Nb or Ta and one chain containing only Fe or Mn. Because of this ordering, there are three

Figure 18.20. Packing drawing of columbite–tantalite $4/(Fe,Mn)(Nb,Ta)_2$-①$_2$②$_2$③$_2$. Numbers 25, 50, 75, 100 are atom elevations: 1, 2, 3 are oxygen designations. Lines connecting O^{2-} atoms designate shared octahedral edges.

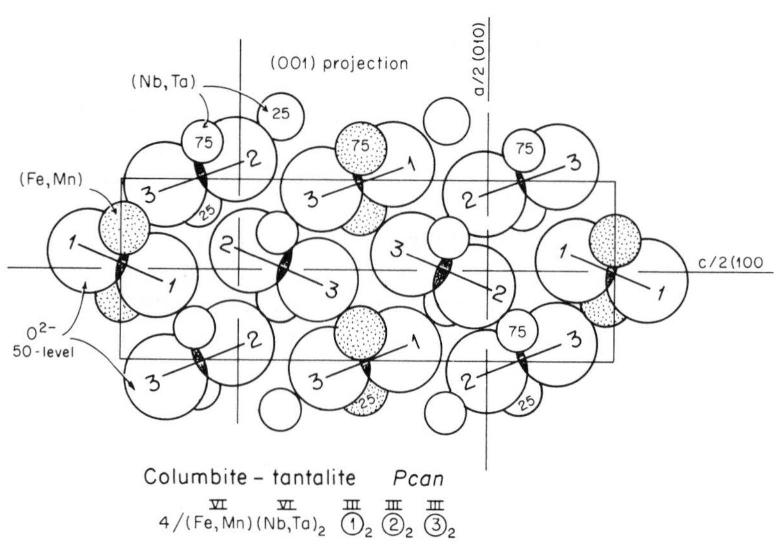

oxygen species present: O-(1), O-(2), and O-(3). The array of cation neighbors around each of them and the manner in which each of them satisfies EBS is illustrated in Figure 18.21. O-(1) is underbonded, O-(2) is overbonded, and O-(3) is neutral. Axial glides $c/2$ (100), $a/2$ (010) and *n* glides (001) characterize the primitive lattice as using orthorhombic space group *Pcan*. In Figure 18.20, *c* glides and *a* glides are labeled: *n* glides parallel to the plane of the projection are less obvious, but it will be noted by the careful student that all Fe and Nb atom pairs are related by the *n* glide (001) with translation $a/2 + c/2$.

Columbite, (Fe, Mn)Nb_2O_6, and its Ta analogue, tantalite, (Fe, Mn)Ta_2O_6, are known almost exclusively from granite pegmatites in which they may form huge, spectacular, but often rare, black crystals. Before the discovery in the 1960s of the large, rich deposits of pyrochlore of carbonatites, columbite was mined from pegmatites containing large crystals sporadic in occurrence. Tantalite of pegmatites remains a major source of tantalum metal, more in demand than niobium because of its high melting point and anticorrosive, nonreactive nature. Large reserves of tantalum occur in cassiterite, SnO_2, deposits of Malaysia, in which $Ta^{5+}Fe^{3+} \leftrightharpoons 2Sn^{4+}$.

VI–IV–IV structures – spinels

Spinel (see Figs. 18.22, 18.23)

$8/Al_2^{VI}[Mg^{IV}O_4^{IV}]$ Isometric $Fd3m$
$a = 8.103$ Å^3 $V_{uc} = 532.03$ Å^3
$M/\rho = V_m = 142.28/3.55 = 40.08$ cm^3
Electronegativity difference: Al—O = 2.0 Mg—O = 2.3
Integrated spinel: 2.08 (66% ionic)
Hardness: 8.0 Luster: vitreous

Figure 18.21. Ion array from Figure 18.20, showing EBS distributions of Fe and Nb ions around oxygens ①, ②, and ③.

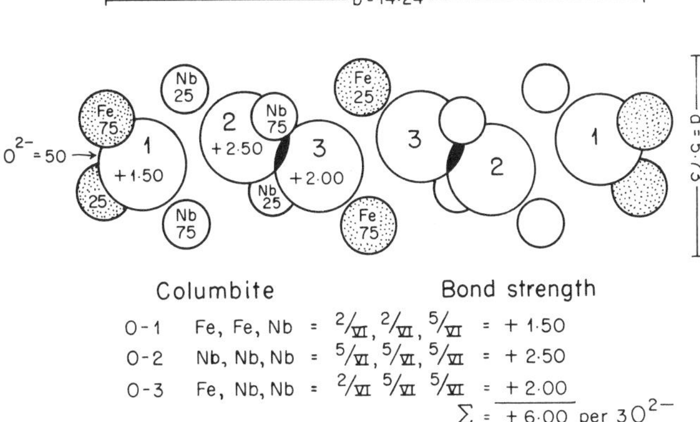

Color: dark green, black
CN: Al^{VI} – slightly elongated octahedron or orthorhombic bipyramid
Mg^{IV} – regular tetrahedron
O^{IV} – regular tetrahedron
A—X distance: Mg—O = 1.96 Å Al—O = 1.91 Å
Specific refractivity: $K = 0.202$
Molecular polarizability: $\alpha_G = 11.32$ Å^3
Optical properties: $n = 1.719$ Isotropic
Color (transmitted light): colorless, pale green
Optical indicatrix model: sphere of $r = n$

Discussion. Spinel-type structures cover a multitude of chemical compositions (Table 18.8) in cubic close-packed oxygen arrays, in which three of every six octahedral sites and one of every eight tetrahedral sites are occupied. This is the same site occupancy pattern as found in olivine (see Chapter 8), in which oxygen ions are in very approximate HCP. All spinel structures contain sixteen octahedral cations, eight tetrahedral cations, and thirty-two oxygens in one unit cell. The edge of this cell varies from 8.09–~8.60 Å. Spinels are commonly grouped into two classes: normal and inverse.

Table 18.8. *Crystal chemical data for selected VI–IV–IV structures of spinels*

Phase	Composition	μ	Type[a]	a (Å)	ρ	n
Spinel	$Al_2[MgO_4]$	.387	N	8.08	3.55	1.719
Synthetic	$Al_2[CoO_4]$	.390	N	8.11	4.44	—
Hercynite	$Al_2[Fe^{2+}O_4]$	.390	N	8.12	4.40	1.835
Gahnite	$Al_2[ZnO_4]$	.390	N	8.09	4.62	1.805
Galaxite	$Al_2[MnO_4]$	.390	N	8.26	4.04	1.92
Magnesiochromite	$Cr_2[MgO_4]$	.385	N	8.33	4.43	2.00
Chromite	$Cr_2[Fe^{2+}O_4]$	—	N	8.38	5.09	2.16
Synthetic	$Cr_2[CuS_4]$	.381	N	9.63	4.40	—
Synthetic	$Cr_2[CuTe_4]$	.379	N	11.05	6.67	—
Trevorite	$Fe^{3+}Ni[Fe^{3+}O_4]$	.381	I	8.36	5.26	2.30
Magnetite	$Fe^{3+}Fe^{2+}[Fe^{3+}O_4]$	.379	I	8.40	3.55	2.42
Synthetic	$MgTi[MgO_4]$	.390	I	8.44	3.55	—
Ulvöspinel	$Fe^{2+}Ti[Fe^{2+}O_4]$	.390	I	8.50	4.78	—
Synthetic	$ZnTi[ZnO_4]$	.380	I	8.48	5.32	—
Synthetic	$MgGa[GaO_4]$	.392	I	8.28	5.33	—
Synthetic	$NiLi[LiF_4]$	.381	I	8.31	3.44	—
Synthetic	$NiIn[InS_4]$	.384	I	10.485	4.80	—
Maghemite	$Fe^{3+}_{1.78}[Fe^{3+}_{0.88}O_4]$	—	D	8.34	4.88	2.65
Ringwoodite	$Mg_{1.48}Fe_{.52}[SiO_4]$	—	N	8.11	3.58	1.768
Synthetic	γ-$Mg_2[SiO_4]$	.368	N	8.06	3.56	—

[a] N, normal; I, inverse; D, disordered or defect.

Normal	$A_2^{3+VI}[B^{IV}O_4^{IV}]$
e.g., spinel	$Al_2^{VI}[Mg^{IV}O_4^{IV}]$
Inverse	$A^{VI}B^{VI}[A^{IV}O_4^{IV}]$
e.g., magnetite	$Fe^{3+}Fe^{2+}[Fe^{3+}O_4]$

These have been defined in different ways. Here, we define a normal spinel as one containing both (two) trivalent ions in octahedral sites and one divalent ion in tetrahedral sites. An inverse spinel has ions of the same valence, usually trivalent: one each in octahedral and tetrahedral sites, with the other octahedral site occupied by an ion with lower or higher valence.

Ordering of ions in spinel structures depends on

1. Closeness of cubic packing, expressed by a parameter, discriminant, or elevation designated μ
2. Crystal field stabilization energy (see Chapter 7)
3. Ionic radii
4. Lattice energy
5. sp^3 hybridization use.

It has been noted earlier (see Chapter 7) that a small highly charged ion in an octahedral site will stabilize a structure and increase lattice energy, and that a transition metal ion with CFSE will increase its stabilization energy exponentially to the fifth power as the metal–oxygen distance decreases (see Chapter 7). Thus, the smaller the atom or ion that can occupy an octahedral site, the greater the stability. For this reason, Al^{3+}, $r^{VI} = 0.53$ Å, the smallest of the octahedral cations in spinel, is strongly favored in the CN VI sites. Mg^{2+}, Mn^{2+}, and Zn^{2+} all are larger in radius, and none of these has any CFSE, either through having no d electrons (Mg^{2+}), or all d orbitals have single or double occupancy ($Mn^{2+}[3d^5]$ and $Zn^{2+}[3d^{10}]$). Thus divalent Mg, Mn, and Zn are relegated to tetrahedral sites, CN IV, in spinels. $Fe^{2+}[3d^6]$ does have CFSE $\Delta_0 = \frac{2}{5}$. This, however, is not sufficient to approach the lattice energy of the small-radius Al^{3+}, and Fe^{2+} must reside in tetrahedral sites, CN IV, giving a normal spinel. For this reason, all aluminate spinels are normal. Zn is strongly ordered into tetrahedral sites because it readily forms sp^3 hybrid bond orbitals.

It was shown in Chapter 7 that $Cr^{3+}[3d^3]$ has a maximum CFSE of $\Delta_0 = \frac{6}{5}$ and will displace all other high spin ions from octahedral occupancy. Thus all chrome spinels will be normal.

Magnetite is interesting because $Fe^{3+}[3d^5]$, CFSE $\Delta_0 = 0$, is associated with $Fe^{2+}[3d^6]$, CFSE $\Delta_0 = \frac{2}{5}$. Thus Fe^{2+} gains priority of admission into the octahedral site over Fe^{3+}. Because the formula contains two Fe^{3+} ions to one Fe^{2+} ion, there is only one divalent Fe ion for the octahedral sites, and trivalent Fe will occupy both sites, one ion per formula in CN VI and one in CN IV.

Returning now to the packing parameter or discriminant μ, we note that it has been found that, for values of $\mu > 0.375$, the tetrahedral site is larger than the octahedral site in spinels, and for $\mu < 0.375$ the tetrahedral site is smaller than the octahedral site. The μ parameter for ideal CCP is 0.375.

In addition, we note that for a spinel such as Cr_2MgO_4 (Fig. 18.22), placement of the trivalent ions in octahedral sites across shared octahedral edges will induce shared-edge shortening. It may be seen (Fig. 18.22) that this effect will contract octahedra, and open up and enlarge tetrahedra. Spinel defect structures such as maghemite, γ-Fe_2O_3, have a composition that may be written

$$(Fe^{3+VI}_{1.78}\square_{0.22})(Fe^{3+IV}_{0.88}\square_{0.12})O_4$$

implying that each type of site is only partially occupied (see Chapter 9).

The crystal structure of spinel is illustrated in Figures 18.22 and 18.23. It will be seen that [MgO_4] tetrahedra at elevation 100, 75, 50, and 25 rotate around 4_1 and 4_3 screw axes, as do AlO_6 octahedral site ions that rotate from elevations 87, 63, 37, and 13 around the same screw axes. Diamond glides, d, $a/4 + c/4$ (100) are perpendicular to screw axes and are oriented in planes that run between 4_1

Figure 18.22. Spinel structure type. Packing model of spinel $8/Al_2[MgO_4]$. Note contraction of octahedra and resulting expansion of tetrahedra.

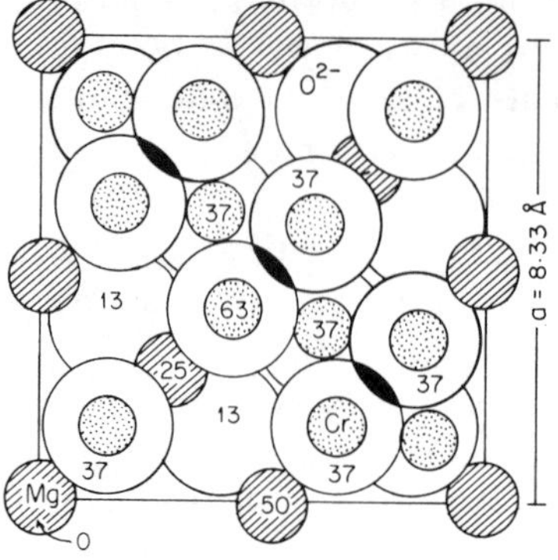

Magnesiochromite

and 4_3 axes (Fig. 18.23). The face centering of the lattice is well shown in the photograph (Fig. 18.22) in which the white spheres lie on corners and face centers of the (001) lattice projection. Mirror planes may be drawn diagonally across Figs. 18.22 and 18.23 through the corners and face centers of the cell. Adding all these elements gives once again the well-known space group $Fd3m$, favored by diamond and pyrochlore as well as by the spinels.

IV–I structure

Synthetic $2/Si^{IV}F^{I}_4$ *(Fig. 18.24)*

Isometric $I\bar{4}3m$ $a = 5.41$ $V_{uc} = 158.34$ Å^3
$M/\rho = V_m = 104.09/2.18 = 47.75$ cm^3
Electronegativity difference: Si—F = 2.2 (70% ionic)
Hardness: n.a. Luster: n.a. Color: white
CN: Si^{IV}–regular tetrahedron F^{I}–point
A—X distance: Si—F = 1.56 Å
Specific refractivity: $K \cong 0.138$
Molecular polarizability: $\alpha_G = 5.69$ Å^3
Optical properties: $n \cong 1.300$, est. from n for Na_2SiF_6 ($\omega = 1.312$, $\epsilon = 1.309$)
Optical indicatrix model: sphere of $r = n$

Figure 18.23. Diagrammatic representation of symmetry elements of the spinel structure $Fd3m$. Mg ions occupy the positions of C atoms of diamond. Note rotation of Mg—O tetrahedra and Al ions around 4_1 and 4_3 screw axes.

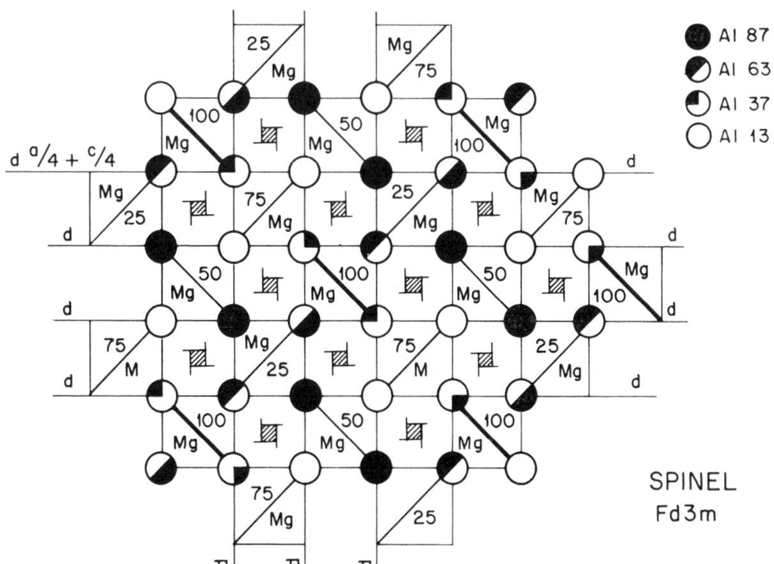

VI–II–I structure

Synthetic tin tetrafluoride (see Fig. 18.25)

Tetragonal: $I4/mmm$
$2/Sn^{VI}F\text{-}1_2^{II}\,F\text{-}2_2^{I} = 2/SnF_4$
$a = 4.05$ Å $c = 7.93$ Å $V_{uc} = 129.9$ Å^3
$M/\rho = V_m = 194.69/4.975 = 39.13$ cm^3
Electronegativity difference: Sn—F = 2.2 (70% ionic)
Hardness: n.a. Luster: n.a. Color: n.a.
CN: Sn^{VI} – regular octahedron $F(1)^{II}$ – linear $F(2)^{I}$ – point
A—X distance: Sn—F = n.a. F—F = n.a.
Specific refractivity: $K \cong 0.100$
Molecular polarizability: $\alpha_G \cong 7.72$ Å^3
Optical properties: $n \cong 1.500$ (mean n) uniaxial (−) estimated
Optical indicatrix model: oblate spheroid of revolution with maximum $r = \omega \perp c$

Discussion. The structure of synthetic SiF_4 is shown here because it may be one of a kind. Each Si is enclosed in a regular tetrahedron of F ions, but only one Si is in contact with each F ion (Fig. 18.25). Thus the tetrahedron can be bonded only by weak van der Waals bonds.

The structure of SnF_4 or, more properly, $SnF(1)_2F(2)_2$ is also unusual. Careful study will show that the F anions are packed in virtually the same manner as the O^{2-} ions in anatase, $Ti^{VI}O_2^{III}$. In tin tetrafluoride, however, only half as many octahedral sites per unit cell are occupied, and we may compare unit cell

Figure 18.24. Packing model of synthetic $Si^{IV}F_4^{I}$; tetrahedra are linked solely by van der Waals bonds.

contents as follows:

Sn_2F_8
Ti_4O_8

It will be seen that F(1) contacts two Sn ions and is overbonded (EBS) = +1.334), where F(2) contacts only one Sn ion and is underbonded (EBS) = +0.666). Equal numbers of each anion balance the negative charge on the F ions. The *n* glide, $a/2 + c/2$, parallel to the rod of the photograph (Fig. 18.20), is particularly instructive.

Figure 18.25. Packing model of synthetic $Sn^{VI}F_4$ or $Sn^{VI}F(1)_2F(2)_2$. N–S oriented rod delineates trace of *n* glide of SnF_6 octahedra (apical F ions removed for clarity).

Figure 18.26. Packing model (using covalent radii) of cuprite, $2/Cu_2O$, showing linear *s–p* hybrid covalent bonds around Cu atoms.

II–IV structure

Cuprite

$2/Cu_2^{II}O^{IV}$ Isometric *Pn3m*
$a = 4.27$ Å $V_{uc} = 77.85$ Å^3
$M/\rho = V_m = 143.092/6.70 = 21.36$ cm^3
Electronegativity difference: Cu—O = 1.6 (47% ionic)
Hardness: 3.5 Luster: adamantine (crystals)
Color: dark red dull, earthy (massive)
CN: Cu^{II}-linear O^{IV}-tetrahedral
A—X distance: Cu—O = 1.84 Å O—O = 3.69 Å
Cu—Cu = 3.01 Å
Specific refractivity: $K = 0.2975$
Molecular polarizability: $\alpha_G = 15.85$ Å^3
Optical properties: $n = 2.705$ Isotropic
Color (transmitted light): red
Optical indicatrix model: sphere of $r = n$

Discussion. Cuprite provides a remarkable confirmation of covalent bonding in the form of a linear *s–p* hybrid bond orbital (see Chapter 3). The univalent Cu atom is coordinated by only two O atoms, and the Cu atom is larger than the O atom in the photograph (Fig. 18.26). Because the bond is clearly covalent, appropriate covalent radii of 1.18 Å for Cu and 0.66 Å for O should better explain the packing of atoms than fallacious ionic radii. Every oxygen lies at the center of a tetrahedron of copper atoms.

The extremely high index of refraction, specific refractivity *K*, and polarizability α_G/V_m (Å^3) = 15.85/38.92 = 0.407 or 41% polarized may be compared with values for diamond: 1.925/5.68 = 0.339 (34%). Thus the polarizability per molar volume for cuprite exceeds that of diamond (41% to 33%), and the index of refraction of each bears this out: cuprite, $n = 2.705$, diamond, $n = 2.418$.

Bibliography

Akimoto, S., Matsui, Y., and Syono, Y. (1976). High pressure crystal chemistry of orthosilicates and the formation of the mantle transition zone. In *The physics and chemistry of minerals and rocks,* ed. R. J. G. Strens. Wiley, New York, pp. 327–63.

Atiji, M., and Lipscomb, W. N. (1954). The structure of SiF_4. *Acta Crystallogr.,* 7: 597.

Bannister, F. A., and Hey, M. (1932). Determination of minerals in platinum concentrates from the Transvaal by X-ray methods. *Min. Mag.,* 23: 188–206.

Barth, T. F. W., and Posnjak, E. (1932). Spinel structures: with and without variate atom equipoints. *Z. Krist.,* 82: 325.

(1934). Notes on some structures of the ilmenite type. *Z. Kryst.,* 88: 265–71.

Binns, R. A., Davis, R. J., and Reed, S. J. B. (1969). Ringwoodite, natural $(Mg,Fe)_2SiO_4$ spinel in the Tenham meteorite. *Nature,* 221: 943–4.

Bragg, W. H., and Bragg, W. L. (1916). *X-rays and crystal structure.* G. Bell, London.

Bragg, W. L. (1914). The analysis of crystals by the X-ray spectrometer. *Proc. Roy. Soc. A,* 89: 468.
(1920). The crystalline structure of zinc oxide. *Philos. Mag.,* 39: 647.
Cummings, J. P., and Simonsen, H. (1970). The crystal structure of $CaNb_2O_6$. *Am. Mineral.,* 55: 90–97.
Dachs, H. (1956). Die Kristallstruktur des Bixbyits. *Z. Krist.,* 107: 370–95.
Geller, S. (1971). Structures of Mn_2O_3, $(Mn_{0.983}Fe_{0.017})_2O_3$ and $(Mn_{0.37}Fe_{0.63})_2O_3$ and relation to magnetic ordering. *Acta Crystallogr. B,* 27: 821–8.
Gibbs, R. E. (1920). The structure of α-quartz. *Proc. Roy. Soc. A,* 110: 443.
Grice, J. D., Ferguson, R. B., and Hawthorne, F. C. (1976). The crystal structures of tantalite, ixiolite and woodginite from Bernic Lake, Manitoba. I. Tantalite and ixiolite. *Can. Mineral.,* 14:540–9.
Grønveld, F., and Røst, E. (1956). On the sulfides, selenides, and tellurides of palladium. *Acta Chem. Scand.,* 10: 1620.
Hill, R. J., Craig, J. R., and Gibbs, G. V. (1979). Systematics of the spinel structure type. *Phys. Chem. Minerals,* 4: 317–39.
Hoppe, R., and Dähne, W. (1962). The crystal structure of SnF_4 and PbF_4. *Naturwissenschaften,* 49: 254.
Horiuchi, H., Ito, E., and Weidner, D. J. (1987). Perovskite-type $MgSiO_3$: Single crystal X-ray diffraction study. *Am. Mineral.,* 72: 357–60.
Hull, A. W. (1919). The positions of atoms in metals. *Proc. Am. Inst. Elect. Engr.,* 38: 1171.
Hutton, J., Nelmes, R. J., and Scheel, H. J. (1981). Neutron diffraction study of $SrTiO_3$ with ideal perovskite structure. *Acta Crystallogr. A,* 37: 916–20.
Kamb, B. (1968). Structural bases of the olivine-spinel stability relation. *Am. Mineral.,* 53: 1439–55.
Krause, I. (1934). Uber keramische Farbkörper. *Ber. Deutsche Keram. Ges.,* 15: 101–110.
Lindsley, D. H. (1976). The crystal chemistry and structure of oxide minerals as exemplified by the Fe–Ti oxides. *Min. Soc. Am. Rev. Mineral.,* 3: L1–L55.
McCullough, J. D., and Trueblood, K. N. (1959). The crystal structure of baddeleyite (monoclinic ZrO_2). *Acta Crystallogr.* 12: *Pt.7,* 507–11.
Mazzi, F., and Munno, R. (1983). Calciobetafite (new mineral of the pyrochlore group) and related minerals from Campi Flegri, Italy; crystal structures of polymignite and zirkelite: comparison with pyrochlore and zirkelite. *Am. Mineral.,* 68: 262–76.
Megaw, H.D. (1946). Crystal structure of double oxides of the perovskite type. *Proc. Phys. Soc.,* 58: 133.
Moore, W. J., Jr., and Pauling, L. (1941). The crystal structure of the tetragonal monoxides of Pb, Sn, Pd, and Pt. *J. Am. Chem. Soc.,* 63: 1392.
Náray-Szábo, St. v. (1943). The structural type of perovskite. *Naturwissenschaften,* 31: 202.
Natta, G. (1930). Structure of silicon tetrafluoride. *Gazz. Chim. Ital.,* 60: 911.
Newnham, R. E., and deHahn, Y. M. (1962). Refinement of the $-Al_2O_3$, Ti_2O_3, V_2O_3 and Cr_2O_3 structures. *Z. Krist.,* 117: 235.
Nishikawa, S. (1915). The structure of some crystals of the spinel group. *Proc. Tokyo Math. Phys. Soc.,* 8: 199.
Pauling, L. (1930). The crystal structure of pseudobrookite. *Z. Krist.,* 73: 97–112.
(1960). *The nature of the chemical bond,* 3d ed., Cornell Univ. Press, Ithaca, New York, chap. 5, pp. 168–172.

Pauling, L., and Hendricks, S. B. (1925). The crystal structures of hematite and corundum. *J. Am. Chem. Soc.,* 47: 781.

Pauling, L., and Shappell, M. D. (1930). The crystal structure of bixbyite and the *C*-modification of the sequioxides. *Z. Krist.,* 75: 128.

Pauling, L., and Sturdivant, J. H. (1928). The crystal structure of brookite. *Z. Krist.,* 68: 239–56.

Ringwood, A. E. (1975). *Composition and petrology of the earth's mantle.* McGraw-Hill, New York.

Ringwood, A. E., and Major, A. (1966). Synthesis of Mg_2SiO_4–Fe_2SiO_4 spinel solid solutions. *Earth. Planet. Sci. Lett.,* 1: 241–5.

Sturdivant, J. H. (1930), The crystal structure of columbite. *Z. Krist.,* 75: 88–108.

Vegard, L. (1916). Results of crystal analysis. *Philos. Mag.,* 32: 65.

Verwey, E. J. W., and Heilmann, E. L. (1947). Physical properties and cation arrangement of oxides with spinel structures. I. Cation arrangement in spinels. *J. Chem. Phys.,* 15: 174–180.

Walitzi, E. M. (1963). Strukturverfeinerung von Libethenit, $Cu_2(OH)PO_4$. *Tschermaks Min. Pet. Mitt.,* 8: 614–24.

Wechsler, B. A., Prewitt, C. T., and Papike, J. J. (1976). Chemistry and structure of lunar and synthetic armalcolite. *Earth Planet. Sci. Lett.,* 29: 91–103.

White, T. J. (1984). The microstructure and microchemistry of synthetic zirconolite, zirkelite, and related phases. *Am. Mineral.,* 69:1156–72.

Wyckoff, R. G. W. (1968). *Crystal structures, 3,* 2d ed. Wiley, New York, Vol. 3, pp. 73–82, Vol. 2, pp. 390–401.

19 Structures built of complex anions: sulfates, tungstates, phosphates, carbonates, and fluocarbonates

This chapter considers those structures built from a packing of complex anions:

1. $[XO_4]^{n-}$, such as sulfates, tungstates, phosphates, and vanadates.
2. $[XO_3]^{n-}$, such as simple carbonates and nitrates.
3. More complex $[CO_3]^{2-}F^-$ fluocarbonates.

Borates, some of which fall into group $[XO_3]^{n-}$, have already, because of their more complex polymerization types, been treated separately in Chapter 17 and need not be repeated here. Nitrates with $[NO_3]^-$ tend to build structures analogous to carbonates, using Na^+ in place of Ca^{2+}.

Just as nesosilicates may be treated as a packing of isolated $[SiO_4]_n^{4n-}$ tetrahedra held together by cations with positive charge $4n^+$, so too may sulfates be treated as a packing of $[SO_4]_n^{2n-}$, phosphates as $[PO_4]_n^{3n-}$, and carbonates as $[CO_3]_n^{2n-}$, held together by cations with positive charge equal to that of the negatively charged complex anion group. Thus, by analogy, in place of nesosilicates we have, in essence, nesosulfates, nesophosphates, nesocarbonates, and so on. What then of groups where complex anions are polymerized, as in the silicates? They do not exist, because Pauling's Rule 4 (see Chapter 4) states that cations with large charge and small radius tend not to share their polyhedral elements. In particular, those cations with formal charge greater than 4^+ in value will build polyhedral complex anions with oxygen. These cations will neither share corners nor polymerize with one another. Thus $[PO_4]$, $[SO_4]$, $[WO_4]$, $[CO_3]$, and so on, will all behave as does the $[SiO_4]$ group in nesosilicates.

Thus, in anhydrite, $Ca[SO_4]$, scheelite, $Ca[WO_4]$, and zircon, $Zr[SiO_4]$, we are dealing with the packing of isolated tetrahedra of charge 2^- or 4^- bonded by atoms with charge 2^+ or 4^+. These three chemically very different minerals do, in fact, build very similar crystal structures that also show strikingly similar properties of refractivity. Carbonates are a special breed because their anions are planar and do not pack in the same manner as tetrahedra.

To carry this discussion to the next logical step, it should be apparent that tetrahedra containing atoms ("cations") with valency ("formal charge") greater than 4^+ form essentially covalent groups (see Chapters 2 and 3). Although the simple arithmetic of balancing positive and negative charges may be done by adding $Ca^{2+}S^{6+}$ and O_4^{2-}, the S atom has an electronegativity of 2.5, which, when subtracted from that of O, 3.5, indicates that the S—O bond in the sulfate anion is 78% covalent. Similarly, the P—O bond is 61% covalent in $[PO_4]$, and the C—O bond in $[CO_3]$ is 78% covalent.

Anhydrite

$4/Ca^{VIII}[S^{IV}O_4^{III}]$ Orthorhombic *Amma*
$a = 6.99$ Å $b = 6.996$ Å $c = 6.24$ Å $V_{uc} = 301.74$ Å^3
$M/\rho = V_m = 136.14/2.996 = 45.44$ cm^3
Electronegativity difference: Ca—O = 2.5 S—O = 1.0
Integrated anhydrite = 2.0 (63% ionic)
Hardness: 3.5 Luster: vitreous
Color: white, grey
CN: Ca^{VIII} – triangular dodecahedron
S^{IV} – tetrahedron
O^{III} – isosceles triangle
A—X distances: Ca—O = 2.32–2.55 Å S—O 1.65 Å
Specific refractivity: $K = 0.195$
Molecular polarizability: $\alpha_G = 10.52$ Å^3
Optical properties: $\gamma = 1.609$ $\mathbf{Z} = c$ $\beta = 1.574$ $\mathbf{Y} = a$ $\alpha = 1.569$
$\mathbf{X} = b$ $\gamma - \alpha = 0.040$ biaxial (+) 2V = 42°
Color (transmitted light) colorless, nonpleochroic
Optical indicatrix model: prolate triaxial ellipsoid

Discussion. The structure of anhydrite, VIII–IV–III type, is similar to that of zircon (see Figs. 12.5, 12.6). Each oxygen is surrounded by two Ca^{2+}/VIII and one S^{+6}/IV for an EBS of +2.00. Alternatively, each $[SO_4]^{2-}$ is surrounded by eight Ca for an EBS of +2.00. The presence of staggered chains

$$O_2—S—O_2—Ca—O_2---$$

parallel to the *c* axis (Figs. 19.1, 19.2) yields highest electron density, polarizability, and index of refraction parallel to *c*, and accounts for the optically positive sign, as in zircon (see Figs. 11.7, 12.5). It is incorrect to think of anhydrite as a layer lattice of tetrahedra connected by Ca^{2+} ions. Layers of tetrahedra run parallel to (011) (Figs. 19.1, 19.2) but there is no (011) cleavage. Cleavages parallel to (010), (100), and (001) all cut across the tetrahedral (011) layering direction. The *c* axis of anhydrite, 6.24 Å, is slightly longer than the *c* axis of zircon, 6.02 Å, which would be predicted from $r_{Ca} = 1.12$ Å and $r_{Zr} =$ 0.84 Å, both values for CN^{VIII}. In both structures pairs of O ions form shortened shared edges of opposed tetrahedral keels, bounding either the Ca or the Zr atom along the *c* axis (Figs. 19.1, 19.2; see also Fig. 12.5).

Scheelite

$4/Ca[WO_4]$ Tetragonal $I4_1/a$
$a = 5.23$ Å $c = 11.35$ Å $V_{uc} = 310.45$ Å^3
$M/\rho = V_m = 287.93/6.16 = 46.74$ cm^3
Electronegativity difference: Ca—O = 2.5 W—O = 1.1
Integrated scheelite = 2.0 (63% ionic). Note the strong contrast between Ca—O bonds (79% ionic) and W—O bonds (74% covalent).

Figure 19.1. Packing model (upper) and polyhedral drawing of anhydrite, $4/Ca[SO_4]$. Large white spheres are Ca ions; small white spheres are S atoms; large rubber balls are O^{2-} ions. Rod at left parallels *m* plane (010) and rod at right a *c*/2 axial glide plane not used in space group notation. Black spots outline staggered row of O_2—Ca—O_2—S. The lower part of the figure shows that lineup of tetrahedral $[SO_4]$ along (011) groups is unrelated to cleavage.

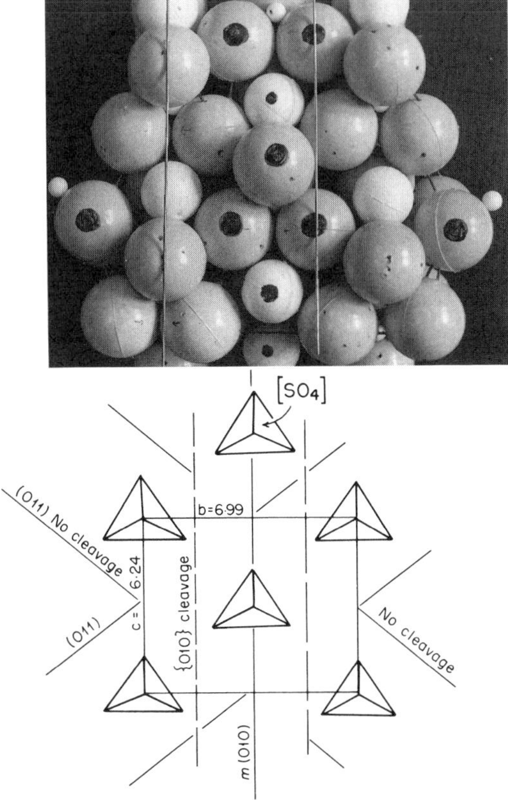

Hardness: 4.5 Luster: resinous adamantine
Color: white
Fluorescence: intense bluish-white under short-wave ($\lambda = 2800$ Å) ultraviolet source lamp, sufficiently brilliant to be used for prospecting for and evaluating grade of potential tungsten ore. Fluorescent color becomes yellowish with increasing substitution of Mo for W as scheelite grades to powellite ($Ca[MoO_4]$).
CN: Ca^{VIII} – very distorted triangular dodecahedron approximating a square antiprism (Fig. 19.3)
W^{IV} – tetrahedral with bounding oxygens flattened parallel to the *c* axis
A—X distances: Ca—O = 2.44–2.48 Å W—O = 1.788 Å
Specific refractivity: $K = 0.152$
Molecular polarizability: $\alpha_G = 17.34$ Å^3

Figure 19.2. Polyhedral representation of anhydrite structure showing staggered rows of O_2—Ca—O_2—S parallel to *c*, the direction of the slow vibration velocity vector **Z** (see Chapter 11 and Fig. 12.5 for zircon).

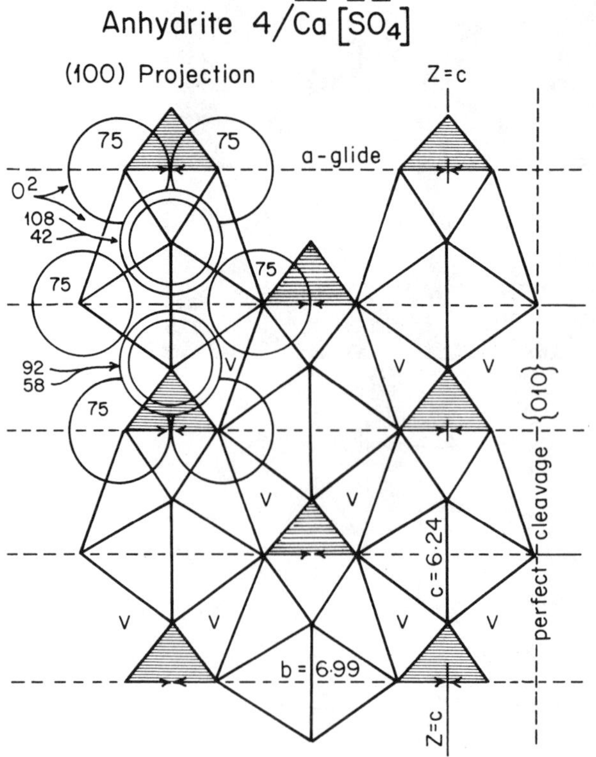

Optical properties: $\omega = 1.930$ $\epsilon = 1.947$ $\epsilon - \omega = 0.014$
uniaxial (+)
Color (transmitted light): colorless and nonpleochroic
Optical indicatrix model: prolate spheroid of revolution with r of circular section ω, the smaller index of refraction

Discussion. The structure of scheelite, like that of anhydrite, is also closely related to that of zircon (compare Figs. 19.3, 19.1, 19.2, 12.5). In scheelite, each oxygen is surrounded by two Ca and one W atom, satisfying EBS in the same manner as two Ca and one S in anhydrite, and two Zr and one Si in zircon, equal in all cases to +2.00. The optically positive sign of scheelite results from higher electron density, polarizability, and index of refraction induced by staggered rows of O_2—W—O_2 parallel to c. If the c dimension of scheelite, doubled with respect to that of zircon and anhydrite, be halved, the value becomes 5.67 Å, comparable to zircon, $c = 6.02$ Å, and anhydrite, $c = 6.24$ Å. Once again, it is evident that the *magnitude* of polarizability is a function of the *chemical bond and CN of bonded atom pairs,* but it is the *crystal structure* that

Figure 19.3. Polyhedral drawing of the scheelite structure $4/Ca[WO_4]$, $I4_1/a$; (A) left (100) projection; (B) right (001) projection. Compare with Figures 12.5 for zircon and 19.1, 19.2 for anhydrite.

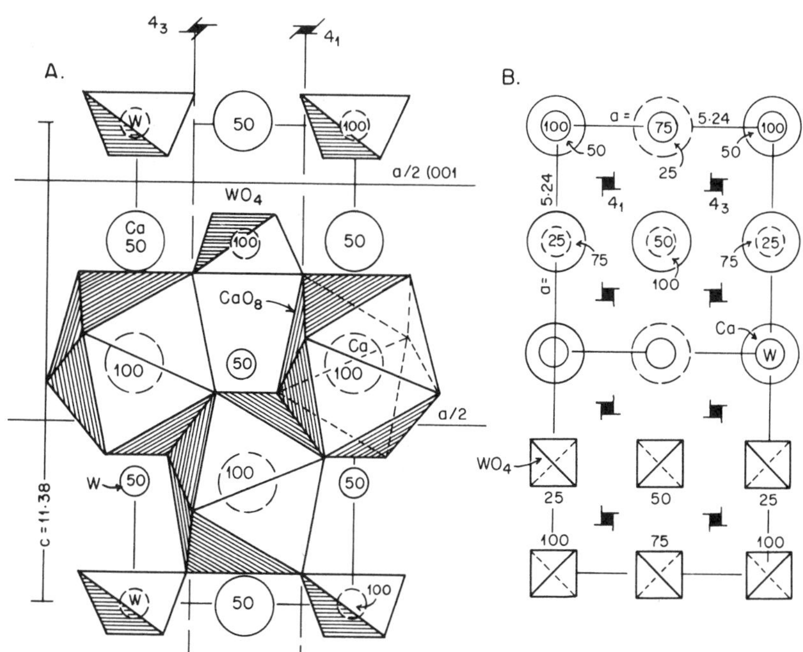

enhances and maximizes electron density, polarizability, and index of refraction, more in one crystallographic direction, and less in another.

Apatite

$2/Ca_3^{VII}Ca_2^{IX}[P^{IV}O_4^{IV}]_3F^{III}$ Hexagonal $P6_3/m$
$a = 9.37$ Å $c = 6.88$ Å $V_{uc} = 526.58$ Å^3
$M/\rho = V_m = 504.31/3.18 = 158.59$ cm^3
Electronegativity difference: apatite
Integrated 2.4 (76% ionic)
Hardness: 5 Luster: vitreous
Color: dark brown, pale yellow-green, blue, rarely white
CN: Ca(1)^{VII} – Each of these ions is one of a triad surrounding and coplanar with an F anion at corners of the unit cell (Figs. 19.4, 19.5). The CN polyhedron is very irregular and approximates

Figure 19.4. Packing model of fluorapatite structure in (0001) projection perpendicular to c (upper) and parallel to c (lower). Smallest spheres are tetrahedral ions Si, P, or B; corks are Ca ions partly proxied by rare-earth ions: La (golf ball) and Y (striped sphere). Large white spheres are F$^-$, and large gray spheres are O^{2-}.

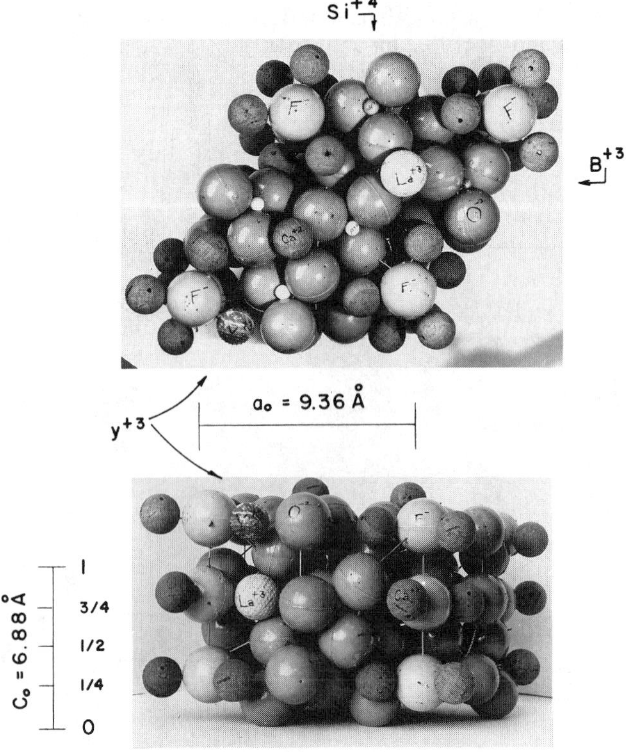

to VII, consisting of six O and one F. Ca(2) ions labeled Ca and La^{3+} in Figure 19.4 and lying inside of the unit cell delineated in Figure 19.5 have CN IX in a regular trigonal antiprism with $\bar{6}$ symmetry; Ca(2) ions are coplanar with three widely spaced O^{2-} anions that form a mirror plane bisecting a trigonal prism. These arrays place two Ca(1), one Ca(2), and one P around each O^{2-} (EBS = +2.044 × 12 = +24.528) and three Ca(1) around each F^- (EBS = +0.858 × 1 = +0.858), giving an EBS summation of +25.386 for twelve O^{2-} and one F^- of the apatite formula. The slight excess of charge is evidently balanced by occasional cation omission.

A—X distances: Ca—O = 2.39–3.03 Å Ca—F = 2.51–2.80 Å
P—O = 1.53–1.61 Å O—O = 2.49 – 2.71 Å

Specific refractivity: $K = 0.199$

Molecular polarizability: $\alpha_G = 39.72$ Å^3

Optical properties: $\omega = 1.633$ $\epsilon = 1.630$ $\omega - \epsilon = 0.003$
uniaxial (−)

Color (transmitted light): under the microscope, colorless and nonpleochroic; recognized by small well-formed hexagonal prisms and sections, almost isotropic, and with intermediate index of refraction. It is a trace mineral constituent of virtually all types of rock.

Figure 19.5. Polyhedral model of Figure 19.4, fluorapatite, $2/Ca_3Ca_2[PO_4]_3F$, illustrating CN IX and VII coordinations of Ca ions.

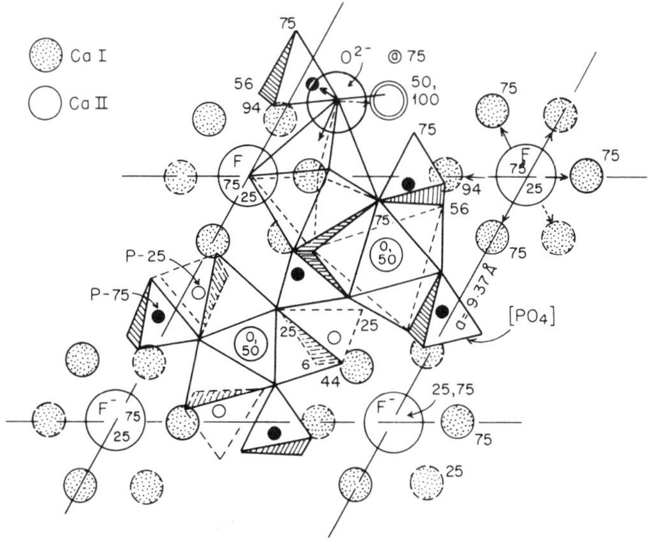

Optical indicatrix model: slightly oblate spheroid of revolution with r of circular section equal to ω approaching a sphere, because ω is nearly equal to ϵ

Discussion. Apatite is made up of $[PO_4]$ tetrahedra that are isolated or not linked to one another, which would be predicted from Pauling's Rule 4 (see Chapter 4). The $[PO_4]$ tetrahedra are joined together by ions of Ca(2) at levels 0, 50, 100 (Figs. 19.4, 19.5). These occupy trigonal antiprisms with $\bar{6}$ symmetry that share faces at levels 25 and 75 along c. These units are, in turn, interrupted and joined by equilateral triads of Ca(1) around F^- anions that lie on 6_3 screw axes along c. The close packing and nonpolymerized nature of the $[PO_4]$ groups inhibits the concentration of selective polarizability along axial or preferred directions, and, optically, apatite approaches a condition of isotropy. An interesting chemical feature of apatites occurs when Cl^- substitutes for F^- in the tunnels along the 6_3 axes. Whereas the F^- ions are at levels 25 and 75 along c, the Cl^- ions will occupy levels 0 and 50 with the Ca(1) triads remaining fixed in their original positions. This places an octahedron, rather than a triad, of Ca(1) ions around each Cl^-. This type of anion shift in the tunnels obviously will limit solid solution between F and Cl apatites, and it is more common to find either one alone, rather than a mixture of both. This anion shift is readily explained by the much larger radius of Cl^- compared with F^-.

A wide variety of compounds, both natural and synthetic, crystallize with the apatite structure. Natural apatites are characterized by the substitution of the larger Cl^- for the smaller F^- anion in the tunnels parallel to the c axis. $(OH)^-$ and, rarely, $[CO_3]^{2-}$ may also fit in here. The names fluorapatite, chlorapatite, hydroxylapatite, and carbonate–apatite are frequently used.

Minerals isostructural with apatite are as follows: britholite, $Ca_2Ce_3[SiO_4]_3F$ to $Ca_3Ce_2[Si_{0.7}P_{0.3}O_4]_3F$; spencite, $Y_3ThCa[Si_{0.7}B_{0.3}O_4]_3O$; vanadinite, $Pb_5[VO_4]Cl$; fermorite, $(Ca,Sr)_5[(P,As)O_4]_3(OH,F)$; johnbaumite; $Ca_5[AsO_4]_3(OH)$; mimetite, $Pb_5[AsO_4]_3Cl$; alfordite, $Ba_5[PO_4]_3Cl$; and belovite, $(Sr,Ce,Na,Ca)_5[PO_4]_3(OH)$. The commercial application of fluoridated toothpastes, now widespread, results in substitution of F ions for the (OH) ions of apatite in teeth, with a resulting increase in resistance to solution by bacterial acids in the mouth. Apatite is mined from large deposits in sedimentary rocks and from carbonate-rich magmatic rocks (carbonatites) for use in phosphate fertilizers.

Calcite

Trigonal–rhombohedral $R\bar{3}c$

$2/Ca[CO_3]$ $a_{rh} = 6.37$ Å $\alpha = 101.9°$ $V_{uc} = 122.12$ Å^3

$6/CaCO_3$ $a_{hex} = 4.99$ Å $c_{hex} = 17.06$ Å $V_{uc} = 367.87$ Å^3

$M/\rho = V_m = 100.08/2.71 = 36.93$ cm^3

Electronegativity difference: Ca—O = 2.5 C—O = 1.0

Integrated calcite 2.0 (63% ionic bonding)

Hardness: 3 Luster: vitreous
Color: colorless to white rarely blue or green
Cleavage: perfect $(10\bar{1}1)$ with twinning $(01\bar{1}2)$ common
CN: Ca^{VI} – octahedron C^{III} – triangle
O^{III} – triangle of C + 2 Ca
A—X distances: Ca—O = 2.36 Å C—O = 1.283 Å
Specific refractivity: $K = 0.222$
Molecular polarizability: $\alpha_G = 8.79$ Å^3
Optical properties: $\omega = 1.658$ $\epsilon = 1.486$ $\omega - \epsilon = 0.172$
uniaxial (−)
Color (transmitted light): colorless, nonpleochroic
Optical indicatrix model: very oblate spheroid of revolution with $r_{\omega\,\text{circ. sect.}} \gg r_{\epsilon\perp\text{circ. sect.}}$, a function of the extreme double refraction

Discussion. The structure of calcite (Fig. 19.6) is often described as that of a unit or cleavage rhombohedron derived from the halite structure by compression of the cube of NaCl along a threefold axis. In the calcite cleavage rhombohedron (a pseudocell with $4/CaCO_3$), ions of Ca lie on corners and face centers of the unit cell as do the Na (or Cl) ions in NaCl (Fig. 19.6). When viewed along c in an (0001) projection, each Ca ion is centered in an octahedron formed by the corner oxygens of six $[CO_3]$ groups (Fig. 19.6). It is important to see that each such O^{2-} anion is coordinated by one C (EBS = 4^+/III) and two Ca ($2 \times 2^+$/VI) = +2.00. Alternatively, and more correctly, each $[CO_3]^{2-}$ anion is coordinated by six Ca ions, making the covalency more apparent and the analogy with the NaCl structure more correct.

Aragonite

Orthorhombic *Pbnm*
$4/Ca[CO_3]$ $a = 7.97$ Å $b = 5.74$ Å $c = 4.96$ Å $V_{uc} = 226.85$ Å^3
$M/\rho = V_m = 100.08/2.93 = 34.16$ cm^3
Integrated = 2.0 (63% ionic)
Hardness: 3.5 Luster: vitreous
Color: colorless or white
CN: Ca^{IX} – complex polyhedron bounded by ten triangular and two rectangular faces (Fig. 19.6)
C^{III} – triangular, and slightly but significantly nonplanar (010) and nonequilateral
O^{IV} – flattened tetrahedron of one C and three Ca
A—X distances: Ca—O = 2.45–2.66 Å C—O = 1.280–1.287 Å
Specific refractivity: $K = 0.2156$
Molecular polarizability: $\alpha_G = 8.55$ Å^3
Color (transmitted light): colorless and nonpleochroic

Optical properties: $\gamma = 1.685$ $\mathbf{Z} = a$ $\beta = 1.680$ $\mathbf{Y} = c$, $\alpha = 1.530$
$\mathbf{X} = b$ $\gamma - \alpha = 0.155$ $2V = 18°$ biaxial (−)
Optical indicatrix model: oblate triaxial ellipsoid not far removed from an oblate spheroid of revolution

Discussion. Aragonite is the high pressure polymorph of $CaCO_3$, which would be expected from the higher CN of Ca, IX, compared with VI in calcite, and its higher density, 2.93 compared with 2.71 of calcite. It occurs in rocks that have been metamorphosed at very high pressures and relatively low temperatures, such as are found along subducted lithospheric plates. Although many organisms build their shells of aragonite, this mineral eventually inverts to calcite, which is more stable at atmospheric pressure.

Figure 19.6. Polyhedral model of calcite (upper left) showing octahedra of Ca and $[CO_3]$ triangles projected perpendicular to *c*. Diagrams (right) show cleavage rhomb of calcite as a deformed NaCl cube. Diagrams (lower) show structure of aragonite in (010) projection for *Pbnm*.

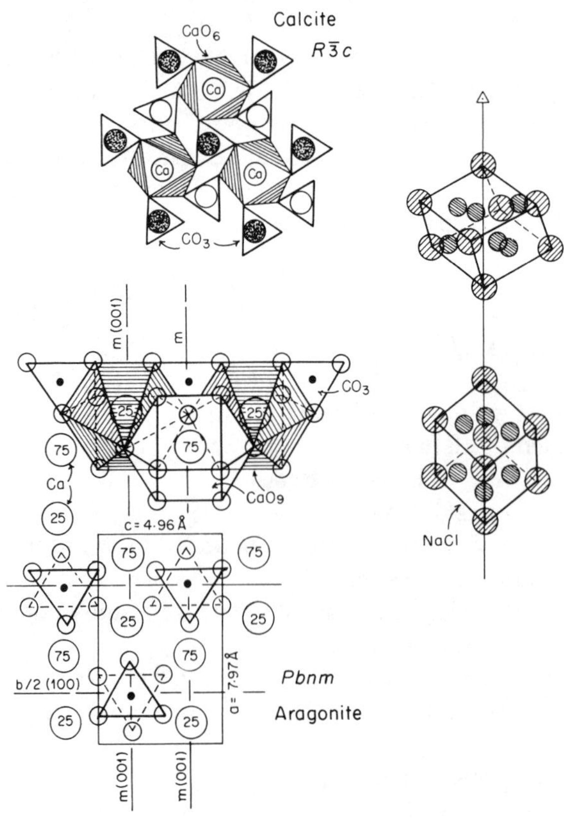

Bastnaesite (also bastnäsite)

$P\bar{6}2c$
6/(Ce,La,Nd . . .)F[CO_3]
$a = 7.09$ Å $c = 9.72$ Å $V_{uc} = 423.13$ $Å_3$
$M/\rho = V_m = 219.43/5.166 = 42.475$ cm^3
Electronegativity difference: Ce—O = 2.4 Å Ce—F = 2.9 Å
C—O = 1.0
Integrated bastnaesite = 2.2 (70% ionic)
Hardness: 4 Luster: dull, resinous
Color: light to dark brown
Habit: stout, hexagonal crystals
CN: Ce^{IX} – inside trigonal polyhedron of six O^{2-}, three F^-
C^{III} – triangle O^{III} – triangle of two Ce, one C
F^{III} – triangle of three Ce
A—X distances: Ce—O = 2.49–2.74 Å Ce—F = 2.39 Å
C—O = 1.27 Å O—O = 2.70 Å F—F = 2.74 Å
Specific refractivity: $K = 0.145$
Molecular polarizability: $\alpha_G = 12.63$ Å^3
Color (transmitted light): very pale brown, occasionally weakly pleochroic with $\epsilon > \omega$
Optical properties: $\omega = 1.717$ $\epsilon = 1.818$ $\epsilon - \omega = 0.101$ uniaxial (+)
Optical indicatrix model: prolate spheroid of revolution of $r_{circ. sect.} = \omega$, the low index of refraction and the smaller r of the indicatrix model

Discussion. Bastnaesite, once considered to be rare, is now the principal ore mineral of the lanthanide or light rare earth elements, because of its discovery in and exploitation of a huge deposit at Mountain Pass, California. The structure of this mineral may best be described as consisting of a hexagonal cage containing two kinds of elevenfold sites: (1) void centered inside cavities of nine O^{2-} ions, formed by three CO_3 groups capped by an F^- ion above and below along c, and (2) the Ce^{IX} sites also centered within eight O^{2-} and three F^- anions, but coordinated only by nine of these anions. The tenth and eleventh anions are considered to be avoided by Ce because they are repelled across a C—O_2—Ce shared edge. It will be seen (Figs. 19.7, 19.8) that threefold axes pass through F^- anions forming (0001) sheets above and below C_3O_9 groups made up of triangles of O_3 that alternate as empty and filled around c, for example, [CO_3]–void O_3–[CO_3]–void O_3–[CO_3]–void O_3. This pattern of alternating occupancy of O_3 triangles gives rise to the twofold axis of rotation and the $c/2$ axial glide of Figures 19.7 and 19.8. Note that the m plane bisects the [CO_3] groups parallel to (0001) and perpendicular to c and recall that in crystallographic language, $3 \perp m = \bar{6}$, hence the space group $P\bar{6}2c$. Finally, the optically positive sign for a

Figure 19.7. Packing models of bastnaesite 6/(Ce,La,Nd, ···)F[CO_3] perpendicular to c (left) and parallel to c (right) [CO_3]. Triangles are parallel to c, giving optically positive sign, unusual for carbonates.

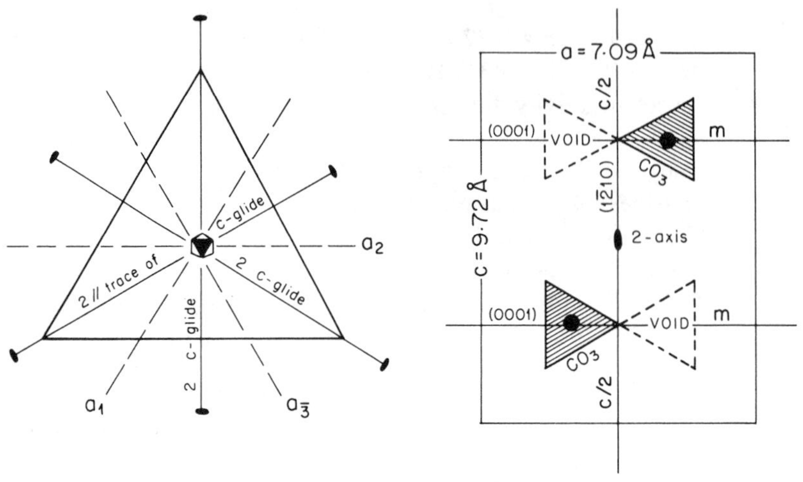

Figure 19.8. Structure of bastnaesite showing symmetry elements (left) and alternating C filled and empty O_3 triangles (right) delineating $c/2$ axial glides and twofold axis of rotation.

carbonate mineral is a verification of the refractivity and polarizability data presented earlier. Why an optically positive carbonate mineral when virtually all others are optically negative? In bastnaesite, the $[CO_3]$ groups are parallel to rather than perpendicular to the *c* crystal axis!

Azurite

Monoclinic $P2_1/c$
$2/CuCu_2(OH)_2[CO_3]_2$
$a = 5.00$ Å $b = 5.85$ Å $c = 10.35$ Å $\beta = 92.3°$
$V_{uc} = 302.51$ Å^3
$M/\rho = V_m = 344.66/3.78 = 91.18$ cm^3
Electronegativity difference: Cu—O = 1.6 C—O = 1.0
H—O = 1.4

Integrated azurite = 1.4 (39% ionic or 61% covalent bonding). The large covalency is to be expected from the square planar coordination of Cu atoms illustrated in Figure 19.9: we are dealing with dsp^2 hybrid bond orbitals (see Figs. 3.9, 18.17)

A—X distances and CN: There are four O^{2-} designations: O(1), O(2), and O(3) bound in CO_3 groups, and O(4), the (OH) ion (Fig. 19.9). Two site designations for Cu atoms are Cu(1) atoms that lie at O-levels and on face centers of the *a* axis projection of the unit cell, and Cu(2) atoms that lie at 25- and 75-levels and link Cu(1)–O_4 squares to $[CO_3]$ groups.

Cu(1)—O = 1.88 Å Cu(1)—(OH) = 1.98 Å

Cu(2)—O, (OH) = 1.92–2.04 Å C—O = 1.25–1.30 Å (nonequilateral) (OH)—O = 0.97 Å. Note that both Cu(1) and Cu(2) atoms lie within squares of two (OH) ions and two O^{2-} ions of different $[CO_3]$ groups; in Cu(1), the (OH) ions lie on opposite corners of a square, whereas in Cu(2), they lie on the same side of a square and form a shared edge between two such squares. A "fifth" oxygen, considered by some to coordinate Cu(2), lies at the long distance Cu—O = 2.38 Å and is here regarded as not coordinating Cu(2). Thus each O^{2-} of a $[CO_3]$ group contacts only one C and one Cu atom, and each (OH) contacts one Cu(1) and two Cu(2). Note the reflection and axial glide translation, *c*/2 (010) of all atoms: Cu(1), Cu(2), C, O(1), O(2), O(3), and O(4) and its H^+, illustrated in Figure 19.9.

Specific refractivity: $K = 0.205$
Molecular polarizability: $\alpha_G = 28.00$ Å^3
Optical properties: $\gamma = 1.838$ $\beta = 1.758$ $\alpha = 1.730$
$\mathbf{Z} \wedge c = -13°$ $\mathbf{Y} \wedge a = 15°$ $\mathbf{X} = b$
$\gamma - \alpha = 0.108$ biaxial (+) 2V = 68°

Color (transmitted light): blue
Pleochroism: marked with **Z** = deep blue-violet, **Y** = azure blue,
X = clear, lighter blue
Optical indicatrix model: prolate triaxial ellipsoid

Discussion. To understand the optical properties and polarizabilities operable in azurite, one must consider Figure 19.10 as well as Figure 19.9. Note that the planes of the $[CO_3]$ triangles are *not* parallel to the $b - c \sin \beta$ plane of

Figure 19.9. Packing model (upper) and polyhedral model (lower) of complex structure of azurite, $2/CuCu_2[CO_3]_2$, illustrating linkage of $[CO_3]$ triangles to dsp^2 hybrid bonded Cu atoms in square planar coordination, (100) projection.

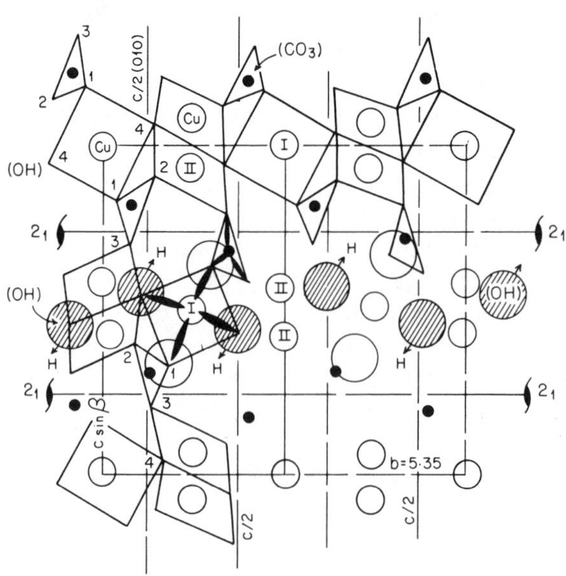

Figure 19.9. Note also that square planes, Cu(1)—$O_2(OH)_2$ are *not* parallel to the plane of Figure 19.9.

Because Figure 19.9 is projected along the *a* axis, and *a* has the shortest repeat distance, 5.00 Å, the orientation of these squares is deceptive. Using Figure 19.10, however, we may see in (010) projection, the squares, and their constituent atoms *on end* while the $[CO_3]$ groups now lie with their planes *parallel to (010),* the plane of the projection. As happens in bastnaesite (Figs. 19.7, 19.8), the planes of $[CO_3]$ lie parallel to *c* and in the *a-c* (010) plane. We should thus also expect the highest indices of refraction to lie in this *a-c* plane. This is indeed the case, as the **Z** (slow velocity) and **Y** (intermediate velocity) vibration directions lie in the same plane as the $[CO_3]$ triangles (Fig. 19.10).

It will also be seen that the angle of −13° made by the **Z** vibration direction with the *c* axis coincides with the staggered alignment of [O—H—O—Cu(1)—O—H—O] units consisting of Cu—O_2—$(OH)_2$ squares on end attached to corners of planar $[CO_3]$ groups. Thus, the complicated and unusual structure of azurite underscores all the principles of polarizability and refractivity espoused in this book. Planar elements are attached to staggered rows within which atoms are bonded by dsp^2 covalent bonds and (OH) bonds (see Chapters 3, 12).

Figure 19.10. Azurite (010) projection showing that staggered lines of O—(OH)—Cu—(OH)—O oriented at 13° to *c* are parallel to the slow velocity vibration vector **Z**.

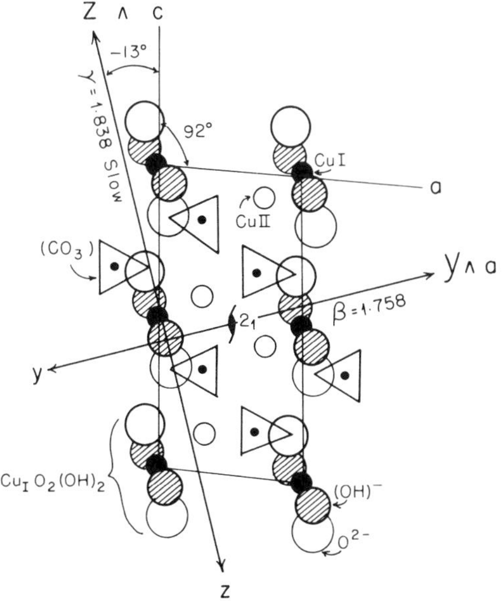

Bibliography

Borisov, S., and Klevcova, R. F. (1963). The crystal structure of RE-Sr-apatite. *Z. Strukt. Khim.,* 4: 629–63.

Cheng, G. C. H., and Zussman, J. (1963). The crystal structure of anhydrite ($CaSO_4$). *Acta Crystallogr.,* 16: *Pt. 8,* 767–69.

Chessin, H., Hamilton, W. C., and Post, B. (1965). Positional and thermal parameters of oxygen in calcite. *Acta Crystallogr.* 18: *Pt. 4,* 689–93.

Dal Negro, A., and Ungaretti, L. (1971). Refinement of the crystal structure of aragonite. *Am. Mineral.,* 56: 768–72.

de Villiers, J. P. R. (1971). Crystal structures of aragonite, strontianite, and witherite. *Am. Mineral.,* 56: 758–67.

Donnay, G., and Donnay, J. D. H. (1953). The crystallography of bastnaesite, parisite, roentgenite, and synchisite. *Am. Mineral.,* 38: 932.

Gatton, G., and Zeeman, J. (1958). Redetermination of the crystal structure of azurite $Cu_3(OH)_2(CO_3)_2$. *Acta Crystallogr.,* 11: *Pt. 12,* 866–71.

Hawthorne, F. C., and Ferguson, R. B. (1975). The crystal structure of anhydrite. *Can. Mineral.,* 13: 289–92.

Jaffe, H. W., and Molinski, V. J. (1962). Spencite, the yttrium analogue of tritomite from Sussex County, New Jersey. *Am. Mineral.,* 47: 9–25.

Kay, M. I., Frazer, B. C., and Almodovar, I. (1964). Neutron diffraction refinement of $CaWO_4$. *J. Chem. Phys.,* 40: 504–6.

Oftedahl, I. (1929). The crystal structure of bastnäsite. *Z. Krist.,* 72: 239.

Wyckoff, R. W. G. (1968). *Crystal structures,* Vol. 2, 2d ed. Wiley, New York.

Zalkin, A., and Templeton, D. H. (1964). X-ray diffraction refinement of the calcium tungstate structure. *J. Chem. Phys.,* 40: 501–4.

20 Sulfides, arsenides, and related compounds

Because most sulfides are in part metallic, and opaque, they contribute little to a knowledge of refractivity and polarizability. Sulfides and the related arsenides and antimonides are important species in the mineral kingdom and provide valuable information on chemical bonding and associated physical and chemical properties. These have previously been discussed in this book, particularly in Chapters 3 and 18. The crystal chemistry, structure, and properties of pyrite, FeS_2, cooperite, PtS, sphalerite, ZnS, and galena, PbS, were described in Chapter 3, and those of PtS and ZnS (ZnO structure type), in Chapter 18.

To all of these we add here the structures of niccolite, NiAs, and pyrargyrite, $Ag_3[SbS_3]$. Both provide additional and contrasting data to the earlier discussion of metallic and covalent bonding (see Chapter 3).

Niccolite (=nickeline)

Hexagonal $P6_3/mmc$
$2/Ni^{VI}As^{VI}$
$a = 3.61$ Å $c = 5.03$ Å $V_{uc} = 56.767$ Å^3
$M/\rho = V_m = 133.63/7.817 = 17.095$ cm^3
Electronegativity difference: $Ni_{1.8}$—$As_{2.0} = 0.2$ (1% ionic bonding)
A—X distances: Ni—As = 2.36–2.48 Å (calc.)
Ni—Ni = 2.31–2.60 Å (calc.)
Hardness: 5 Luster: metallic
Color: pinkish grey
Electrical conductivity: large parallel to *c*
CN: Ni^{VI} – octahedron As^{VI} – trigonal prism
Optical properties: opaque, with reflectivity on polished surfaces = 60%; soluble in aqua regia

Discussion. Although octahedral (VI) coordination is widespread in ionic structures built on packing geometry of charged ions, and also in covalent structures built by a predominance of d^2sp^3 octahedral hybrid bond orbitals, trigonal prismatic coordination (VI) is rarely met with in ionic structures. Similarly, octahedral face sharing is rare in ionic structures (though it does occur in corundum; see Fig. 18.2) but common in covalent structures. Note the octahedral face sharing of the triads of As atoms around Ni (Fig. 20.1).

Ni is sufficiently large to pack in trigonal prisms around As atoms and to form overlapped Ni—Ni metallic orbitals and consequent smeared-out metallic orbitals parallel to *c* (Fig. 20.1), resulting in the high electrical conductivity directed along the *c* axis.

Niccolite and other isostructural compounds are, for the most part, characterized by short *c*/*a* ratios, and Pauling (1960) suggests that many metallic minerals use this structure because of the ease of formation of metal–metal

Figure 20.1. Packing model (upper) and polyhedral model (lower) of the structure of niccolite, NiAs, built using metallic covalent radii of Pauling (1960). Features are two different types of CN VI, octahedral around Ni and triangular prismatic around As, and covalent Ni—As and metallic Ni—Ni bonds.

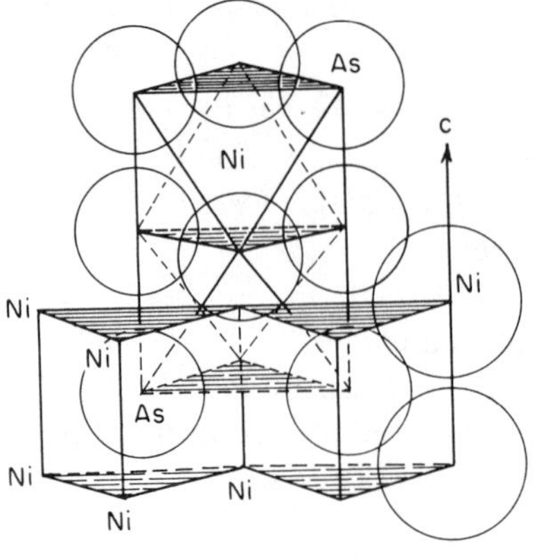

overlap of the larger, more metallic partner; for example,

Ni—Ni in NiAs $(c/a = 1.393)$
Au—Au in AuSn $(c/a = 1.278)$
Pt—Pt in PtB $(c/a = 1.208)$

By contrast, most pure metals use a hexagonally close-packed structure in which an ideal ratio $c/a = 1.633$. Metallic Mg, for example, has a ratio $c/a = 1.626$, because all atoms are identical and can form molecular metallic orbitals more readily with this structure.

From its electronegativity difference of 0.2 (1% ionic bonding), short c/a ratio, high reflectivity and conductivity, and solubility in aqua regia, niccolite should be predominantly bonded by Ni—Ni metallic bonds and subsidiary covalent Ni—As bonds.

Pyrargygrite

Trigonal $R3c$
$6/Ag_3^{II}[Sb^{III}S_3^{III}]_{hex}$
$a = 11.02$ Å $c = 8.73$ Å$_{(hex. cell)}$ $V_{uc} = 918.11$ Å^3
$M/\rho = V_m = 541.55/5.875 = 92.18$ cm^3
Electronegativity differences: $Ag_{1.9}—S_{2.5} = 0.6$ (9% ionic, 91% covalent)
$Sb_{1.9}—S_{2.5} = 0.6$ (9% ionic)
Hardness: 2.5 Luster: adamantine

Figure 20.2. Packing model of pyrargyrite, $6/Ag_3SbS_3$ projected nearly along c, (0001) projection. Note covalently bonded flattened pyramidal $[SbS_3]$ groups linked by Ag atoms and the typical openness of the covalent structure.

Color: deep ruby red
Electrical conductivity: low   Solubility: nitric acid
CN: Sb^{III} – in flattened triangular pyramid of $[SbS_3]$ (Fig. 20.2) with Sb at apex above a triangle of S atoms. All $[SbS_3]$ groups have their apical Sb oriented or pointing in the same direction, imparting a polarity to the structure (as in zincite and wurtzite, see Fig. 18.16, and tourmaline, see Fig. 14.1).
S^{III} – triangle of two Ag + one Sb
Ag^{II} – linear, coordinated by two S atoms
A—X distances: Ag—S = 2.38 Å (calc.)
Sb—S = 2.40 Å (calc.)
Specific refractivity: $K = 0.343$
Molecular polarizability: $\alpha_G = 73.67$ Å^3
Optical properties:
Pyrargyrite $\omega = 3.084$ $\epsilon = 2.881$ $\omega - \epsilon = 0.203$ uniaxial (−)
Hematite $\omega = 3.22$ $\epsilon = 2.94$ $\omega - \epsilon = 0.280$ uniaxial (−)
Color (transmitted light): red   Pleochroism: weak
Optical indicatrix model: oblate spheroid of revolution with

$$r_{\text{circ. sect.}} = \omega$$

Discussion. Pyrargyrite, from its properties, transparency, open structure, deep color, absence of metallic luster, high indices of refraction, solubility in HNO_3, and electronegativity difference 0.6 (9% ionic bonding) is evidently mostly (91%) covalently bonded. It provides a good contrast in chemical bonding with niccolite, pyrite, FeS_2, and cooperite, PtS, which were described earlier. With regard to its properties, pyrargyrite is closer to sphalerite than to niccolite or pyrite.

Many sulfide minerals adopt variations of the sphalerite structure; others, such as galena, use the halite structure or, less often, the pyrite structure.

It is beyond the scope of this book to discuss the details of sulfide structures further. To gain a greater insight into the crystal chemistry of sulfides, it becomes necessary to pursue studies of the more advanced molecular orbital and band theories. Briefly, covalent overlap of orbitals of metal and ligand in sulfides commonly permits closely spaced energy levels to coalesce into valence and conduction bands separated by energy gaps. Closely spaced energy gaps give rise to enhanced conductivity and metallic properties, whereas larger gaps promote formation of semiconductors.

The reader who would specialize in sulfide structures should begin with the many superb descriptions in Pauling's (1960) classic text and then refer to papers by specialists in the field: Goodenough (1972), Jellinek (1970), Prewitt and Rajmani (1974), Vaughan and Craig (1978), Vaughan et al. (1974), Vaughan and Tossell (1981), and Wuensch (1974).

Bibliography

deJong, W. F., and Willems, H. W. V. (1927). Compounds of the lattice type of pyrrhotite (FeS). *Physica,* 7: 74.

Faber, W. (1933). Niccolite. *Z. Krist.,* 84: 408.

Goodenough, J. B. (1972). Energy bands in TX_2 compounds with pyrite, marcasite and arsenopyrite structures. *J. Solid State Chem.,* 5: 144–52.

Gossner, B., and Mussgnug, F. (1928). The crystal structure of pyrargyrite. *Central. Mineral. Geol. 1928 A:* 65.

Hocart, R. (1937). The structure of proustite and pyrargyrite. *Compt. Rend. 20 S:* 68.

Jellinek, F. (1970). In *Inorganic sulfur chemistry,* ed. G. Nickless. Elsevier, Amsterdam, pp. 669–748.

Pauling, L. (1960). *The nature of the chemical bond,* 3d ed. Cornell Univ. Press, Ithaca, New York, Chap. 11, p. 423, Chap. 13, p. 562.

Prewitt, C. T., and Rajmani, V. (1974). Electron interactions and chemical bonding in sulfides. *Mineral. Soc. Am. Rev. Mineral.,* 1: PR-1–PR-41.

Vaughan, D. J., and Craig, J. R. (1978). *Mineral chemistry of metal sulphides.* Cambridge University Press, Cambridge.

Vaughan, D. J., and Tossell, J. A. (1981). Electronic structure of thiospinel minerals: results from MO calculations. *Am. Mineral.,* 66: 1250–3.

Vaughan, D. J., Tossell, J. A., and Johnson, K. H. (1974). The bonding of ferrous iron to sulphur and oxygen: a comparative study using SCF-X scattered wave molecular orbital calculations. *Geochim. et Cosmochim. Acta,* 38: 993–1005.

Wuensch, B. J. (1974). (1) Determination, relationships, and classification of sulfide mineral structures; (2) Sulfide crystal chemistry. *Mineral. Soc. Am. Rev. Mineralogy,* 1: W-1–W-44.

Mineral index

(Minerals for which essential crystal chemical data are given are paginated in **boldface**.)

acmite, **204**
adamite, 164
åkermanite, 115, **180–3**
alabandite, 104, **268**
albite, 79, 111, 137, **237–40**
alfordite, 312
almandine, 137, **158**
amphibole group, **215–21**
analcime, 80, 111, 127
anatase, 79, 115, 137, **280–2**
andalusite, 79, 105, 111, 115, 125, 137, **162–5**
  chiastolite var., 165
andesine, 239
andradite, 115, 138, **158**
anhydrite, 79, 141, **306–8**
anorthite, 79, 111, 115, 137, **237–41**
anthophyllite, 79–80, 111, 129, 148, **218–19**
antigorite, **227–8**
apatite, 79, **310–12**
aragonite, 20, 115, 137, **313–14**
armalcolite, **284**
augite, 79, 111, **204**
axinite, 126
azurite, **317–19**

baddeleyite, **274–6**
barite, 80, 98
bastnaesite, 65–6, 79, 140, **315–17**
belovite, 312
benitoite, 79
berlinite, 46
beryl, 64, 102, 111, 126, **187–9**
berzeliite, **158**
biotite, 228
bixbyite, **273–5**
borax, 137, **262**
britholite, 312
bromellite, 50, **289**
bronzite, **212–13**
brookite, 115, 137, **281–3**
brucite, **141–2**, 227
buddingtonite, **237**
bunsenite, **104**, **268**
bytownite, 79

cahnite, 252
calcite, 46, 79, 115, 121, 140–1, **312–14**
carbonate-apatite, 312
carnotite, 40
carobbiite, **268**
cassiterite, 137, **280**
celsian, **237**
cerargyrite, 104
chlorapatite, 312
chlorite, 127, **142**, **226**, **228–9**
chloritoid, 126, **170–3**
chromite, 91–2, **296–8**
chrysoberyl, 80
clay mineral group, **228–9**
clinochrysotile, 229
clinopyroxene, 126, **204–8**
coesite, 137
colemanite, **252**, **261**
columbite, **293–5**
cooperite, 35–40, 95, **289–90**
cordierite, 11, 126, **189–91**
corundum, 49, 79, 111–12, 137, **267–71**
cristobalite, **233–5**
cryolithionite, **158**
cubanite, 80
cummingtonite, **218–21**
cuprite, 40, 80, **301–2**

danburite, 80
datolite, 79, 126, **160–2**
diamond, 2, **30–1**, 35, 80, 95, **112–13**, 119, 137
diaspore, 80
diopside, 61, 68–9, 79, 95–6, **204–8**
dolomite, 79, 255

elbaite, 196
enstatite, 204, 216
epidote, 79, 126, **183–5**
eskolaite, **271**

fayalite, 105, 115
feldspar group, **235–49**
fermorite, 312
fluorapatite, **312**
fluorite, 61–2, 80, 137, **271–3**
forsterite, 79, **81**, **97**, 105, 115, **137**

gadolinite, 162
gahnite, **296**
galaxite, **296**
galena, 80, **104**, **268**
garnet group, 62–3, 80, 96, 125, **156–9**
gedrite, **218–19**
gehlenite, **183**
geikilite, **271**
gibbsite, 141–2
glaucophane, 129, **218–19**
goethite, 80
gold, **112–15**
goldmanite, **158**
graphite, 1, 2, **35–6**, 79, 95
grossular, 115, 137, **158**
grunerite, 129

halite, 50, **71–3**, 104, **112–13**, 137, **266–8**
halloysite, 229
hambergite, 252
hardystonite, 183
hauerite, **38–9**
hausmannite, 92
hedenbergite, **204–5**
hematite, 79, 111, **270–1**
hercynite, **296**
herderite, 162
hornblende, 47, 129, **217–18**
humite, 80, 125
hydrogrossular, **158**
hydroxyapatite, **312**
hypersthene, 79, 111, **204**, **212**

ice, 109
ilmenite, 79, **270–1**
inyoite, **252**, **261**

jadeite, 138, **204–5**
johannsenite, **204–5**
johnbaumite, 312

kamacite, **112–13**
kaliborite, **252**
kaolinite, 79, 111, **142**, **224**, **226**, **228**, **229**
karelianite, **271**
kernite, **252–3**, **263**
kimseyite, **158**
K-richterite, **218–19**
kyanite, 111, 125, 137, **165–8**

labradorite, 239, **247–9**
lawsonite, **177–80**
leucite, 111
libethenite, 164
ludwigite, 252, **256–9**
lueshite, **285–7**

maghemite, **102**, 137, **296–8**
magnesioarfvedsonite, 222
magnesiochromite, **296–8**
magnetite, 80, 92, **296–7**
manganosite, **268**
melilite, 79, 111, 126, **180–3**
meliphanite, 183
metahalloysite, 228
mica group, 127, 142, **224–9**
microcline, 79, **237–40**
microlite, **291–2**
mimetite, 312
molybdenite, 79
monalbite, **237**
monazite, 101
montmorillonite, **227–8**
muscovite, 79, 105, 111, 202, **226**, **228–9**

natrolite, 127
neighborite, **285–7**
nepheline, 127
niccolite (= nickeline), **321–3**
nordenskioldine, **252–5**

oligoclase, 239
olivine, 63, **81**, **95–8**, 111, 125
oldhamite, **268**
olivenite, 164
omphacite, **204–5**
orthoferrosilite, **204–5**
orthoclase, 101, **237–40**
orthopyroxene, **126**, **204–5**

pandaite, **292–3**
pargasite, **202**
paragonite, **228–9**
P-cummingtonite, **218**, **222**
periclase, 49, 98, 137, **266–8**
perovskite, **284–7**
phlogopite, 79, 148, **202**, **226**, **228–9**
pigeonite, 79, **204–5**, **210–13**
powellite, **308–9**
pseudobrookite, **283–4**
pyrargyrite, **323**

pyrite, **36–8**, 80, 95, 98
pyrochlore, 102, **290–3**
pyrope, **158**
pyrophanite, **271**
pyrophyllite, 141, **228–9**
pyroxene group, **201–15**
pyroxenoid, **127**, **203**
pyrrhotite, 102

quartz, 1, 109, 114–15, 121, 136–7, **232–6**
  α-quartz, **233–6**
  β-quartz, **233–4**

ramsdellite, 80
rare earth garnet, **158**
reedmergnerite, **237**
rhodochrosite, 137
rhodonite, 129
riebeckite, **218–19**
ringwoodite, **296**
roweite, 252, **261**
rutile, 61–2, 79, 115, 137, **276–80**

sahamalite, 132–3
sanidine, 138, **237–43**
sassolite, 140, **252–5**
scapolite, 127
scheelite, **307–9**
schorl, 196
schorlomite, **158**
schroeckingerite, **133–4**
sellaite, **135–7**
siderite, 137
sillimanite, 79, 105, 111, 115, 125, 137, **162–3**
sinhalite, 80, 252, **259–60**
soda-niter, 80, **140**
sodalite, 80
spencite, 312
spessartine, 137, **158**
sphalerite, **31–3**, 58, 115, 137
sphene (titanite), 101, 125, **151**
spinel, 80, **91–3**, 115, **295–9**
spodumene, **204–5**
staurolite, 125, **169–75**
stibnite, 80
stishovite, 1, **114–15**, **137**, **280**
sulfohalite, **66–7**
sulfur, 79, 137
sylvite, 104, **268**

talc, **65**, 110–11, 127, 142, 148, **226**, **228–9**
tantalite, **293–4**
thorianite, **271–3**
thorite, **153–6**
thortveitite, 63–4, **150**
titanite, (sphene), 101, **150–1**
topaz, 80, 125
tourmaline, 47, 79, 110–11, 126, **194–9**
tremolite, 102, 111, 201, **218–19**
trevorite, **296**
tridymite, **233–5**

ulvöspinel, **296**
uraninite, **272–3**
ureyite, **204–5**
uvarovite, **158**

vanadinite, 312
vermiculite, **228–9**
villiaumite, 104, 137, **267–8**

wadeite, **191–3**
wollastonite, 129, 203
wurtzite, **288–9**
wüstite, 173, **268**

xenotime, 62, **153–6**

yttrian spessartine, **105**, **158**
yttrium garnet, **105**, **158**

zeolites, 143
zincite, **286–8**
zircon, 62, 79, 101–3, 125, 141, **153–6**
zoisite, 129

Subject index

actinides, 10–11
admission, 101
a-glide, *a*/2
  anatase, 281
  bixbyite, 274
  brookite, 282
  columbite, 294
  datolite, 161
  garnet, 157
  gold, 112
  halite, 112
  ludwigite, 256
  pseudobrookite, 284
  pyrite, 37
  scheelite, 309
alkali elements, 10–11
alkaline earth elements, 10–11
Al—O—Si bow tie, 163
α_G formula, 136
alpha particle decay, 5
Al-rich orthopyroxene, 212–14
Al—Si ordering
  feldspars, 239–41, 246
  sillimanite, 151–2
aluminosilicates, 151–74
aluminum avoidance principle, 241
amphiboles
  A-, M-1, -2, -3, -4, and T-1, -2 sites, 217–19
  cleavage, 208–9
  exsolution, 220–2
  I vs. *C* cell, 220–2
  optical properties and density, 218–19
  pyroxene- and mica-like segments, 201–2, 211
  unit cell data, 216–19
analyzer, 118
ångstrom units, Å, 42, 109
anion, 22, 42
anion CN, 61–8
anisotropy, 120
anorthosite, 212
antiperthite, 244–5
apical oxygen, 65
atom, 2
atomic number, 2
atoms per unit cell, 110–11
aug lamellae, 211–12
Avogadro's number, *N*, 108
A—X distance, 42, 52

beta decay (β^-) process, 5
b glide, *b*/2
  aragonite, 314
  forsterite, 81
  ludwigite, 256
  pseudobrookite, 284
  sillimanite, 163
biaxial positive and negative, 120–1
birefringence, B, 140
body-centered lattice, *I*, 40
Bøggild lamellae, 246–9
Bohr magnetons, 38
bond dissociation energy, *D*, 22–4
bond energy variation
  atomic radius, 29
  interatomic distance, 29
  single, double, triple bond, 29
borates, 251–64
bow tie chains, 152
Bragg reflection, 76–8
Bragg's Law, 76–8
Bravais space lattice, 71–2
bridging oxygen, 63
brucite-like layer, 174

camouflage, 101
capture, 101
cation, 22, 42
cation arrays, 61–2
cations, uncentered, 68–9
CFSE, 89–94

c-glide, *c*/2
  azurite, 318
  brookite, 282
  chloritoid, 172, 174
  clinopyroxene, 206, 218
  columbite, 294
  cooperite, 290
  lawsonite, 178–80
  muscovite, 202
  phlogopite, 202
circular section, 120–2
clay minerals, 224–9
cleavage, pyroxene, amphibole, 208–9
complex anions, 44, 305–19
conduction bands, 324
CN, Ca^{2+}
  V–VI - lawsonite, 178–80
  VI - calcite, 313
  VII - anorthite, 238
  VII - apatite, 310
  VIII - åkermanite, 180–2
  VIII - amphibole, 217–18
  VIII - clinopyroxene, 204–8
  VIII - fluorite, 272
  VIII - garnet, 156–7
  VIII - zircon, 154–6
  IX - apatite, 309
  IX - aragonite, 313
  X - amphibole, 217–18
  XII - perovskite, 285
coordination number, CN, 45
coordination polyhedra, 45–51, 95–9
  distorted, 95–9
  Pauling's rules, 45–6
  radius ratio minima, 51
  regular, 51, 97
corundum-like layer, 174
Coulomb forces, 42
covalent bond, 22, 27–41
  s–s, *p–p*, *s–p*, 27
  single, double, triple, 27–9
covalent radii of atoms, 29–33
  d^2sp^3 hybrid octahedral, 39
  sp^3 single bond tetrahedral, 34
crossed polars, 121
cruciform twin, 169
crystal class, 74–5
crystal field substitution, 103
crystal field theory, 89–94
  Δ_0, 90
  e_g level, 89–94
  t_{2g} level, 89–94
crystallographic interaxial angles, $\alpha = b \wedge c$, $\beta = a \wedge c$, $\gamma = a \wedge b$, 75
crystal systems, 73–5
cubic close packing, CCP, 47–9
cube-octahedron, 52
cyclosilicate, 149, 187–99

Debye–Scherrer camera, 74
defect substitution, 102
density
  calculated, 108, 111
  measurement, 109
  observed, 108
  optical, 108
  optical vs. observed, 123
  range, common minerals, 109
d glide, $a/4 + c/4$
  diamond, 112
  garnet, 159
  spinel, 299
  zircon, 156
diadochy, 100–3
diamagnetism, 38
diamond, crystal chemistry
  C atom locations, 31
  model, 31
  physical properties, 2, 35
  sp^3 hybrid bonds, 30–1, 95
  space group and unit cell, 31, 112
diamond, optical, Na light
  polarizability, 137
  refractive index, 119
  specific refractivity, 131
  velocity, wavelength, frequency, 119
diaspore-like, 170
dipole moment, 18, 134
displacive transformation, 233
distorted polyhedra, 46, 92–4, 95–9
double chain silicates, 215–22

electric vector, 119
electromagnetic radiation, 119
electron affinity, 18–21
electron capture, 5
electron cloud, 6–7
electron density, 6
electron density paths, 140–1
electronegativity
  chemical bond criterion, 22
  concept, 18–21
  ionization potential, 19–20
  Little and Jones values, 23
  Pauling values, 23
  periodicity, 21–3
electron orbitals, 6–7
electron shells, 10
electron spin, 17
electron volts, ev, 18–24
electron wave function, ψ, 6–7
electrons per quantum shell, 10
electrostatic bond strength, EBS, 61–8
exsolution, 209–10
  amphibole, 221–2, 224
  feldspar, 244–9
  pyroxene, 209–14
exsolution angles, augite, 210–13
  pig "001" $\wedge$ *c*, 210
  pig "100" $\wedge$ *c*, 210, 211
  pig "100" $\wedge$ pig "001", 211

exsolution lamellae, augite host, 209–13
C-pigeonite, 210
hypersthene, 212
P-pigeonite, 210
exsolution lamellae, orthopyroxene host, 211–14
augite, 211–13
calcic plagioclase, 212–13
garnet, 212
ilmenite, 213
external symmetry, 71–5

face-centered lattice, *F*, 37
*Fd*3*m* symmetry, 112
feldspars, 235–49
Al—Si ordering, 239–44
density, 237
exsolution, 244–9
iridescence, 247–9
lattice parameters, 237
optical properties, 237
solid solution, 237–9
structure, 241–4
T-site ordering, 240–1
voids, 242–3
*Fm*3*m* symmetry, 112
formula weight, 108

garnet group
CN sites, 156–9
electrostatic bond strength, 62–3
optical properties, 158
screw axes, 159
synthetic analogues, 105, 158
YAG, YIG, 105, 156–9
Gladstone and Dale formula, 108, 122
glide planes, 78–81, 112
Goldschmidt's principles, 101
graphite
comparison with diamond, 1–2, 35–6
σ, π, sp^2 hybrid bonds, 35–6, 95
van der Waals bonds, 36
ground (neutral) state, 15
gypsum plate, 122

half-life, 5
halite structure, 73, 112, 267–8
halogens, 10–11
Hermann–Mauguin symbols, 75
hexagonal close packing, HCP, 47–9
heteropolar (ionic) bond, 42
high energy *d* orbitals, 89–94
homopolar (covalent) bond, 42
Huttenlocher lamellae, 246–9
hybrid bond orbitals, 30–1, 95
d^2sp^3 (pyrite), 36–8, 95
dsp^2 (cooperite), 39, 40, 95
sp (cuprite), 40
sp^2 (graphite), 35–6, 95
sp^3 (diamond), 35–6, 95

hydrogen bonds, 142
hydrogen burning, 11

I-beam structure, 208–9
*Im*3*m* symmetry, 112
independence of polyhedra, 45–6
index of refraction, 118, 120–1
variation with density, 131–2
indicatrix, optical, 121
interatomic distance, 42
interference figures, optical, 121–2
internal symmetry, 74–87
interplanar spacing, 76
inversion axes
$\bar{3}$ calcite, 313
$\bar{4}$ åkermanite, 181–2
$\bar{4}$ SiF_4, 300
$\bar{6}$ bastnaesite, 316
$\bar{6}$ wadeite, 192
ionic bond, 22, 42–69
ionic bonding % vs. Δ_χ, 43
ionic potential, 18–21
ionic radii, 52–8
ionic resonance energy, Δ, 23–4
ionization potential, 18
inosilicates, 149, 201–22
inverted pigeonite, 211–12
iridescence, 249
irrational planes, 211
isobars, 3–4
isoelectronic series, 18
isogyre, 122
isomorphism, 100–3
isostructural, 63, 100–4
isotones, 3–4
isotopes, 3–4
isotropy, 120–1

Jahn–Teller distortion, 90–4

K–Ar decay, 5
k-constants of oxides, 124–9
general use, Jaffe, 124–5
general use, Mandarino, 129
rock-forming minerals, Jaffe, 126–8
kyanite-like zone, 170

lanthanides, 2, 10–11, 315
larvikite, 249
Lewis notation, 27–9
ligand, 89–90
lone pair orbitals, H_2O, 34–5
Lorentz–Lorenz equation, 136–8

M-site, feldspars, 238
M-1, M-2 site
olivine, 95–7
pyroxene, 206–8
M-1, -2, -3, -4 sites
amphibole, 217
ludwigite, 258–9

mass number
  atomic particles, 3
  atoms, 2
mesons, π and μ, 3, 5
mesoperthite, 244–5
metallic bond, 22, 36–8, 320–4
metallic orbital, 322–3
metamict minerals, 5, 103
mica-like segment, 209, 215
mica polymorphs, 225–6
microperthite, 244–5
minimum radius ratio, 52
mirror plane, 74–5
miscibility gap, 239
molar refractivity, 136
molecular (band) spectra, 17–18
molecular orbital, 35–6, 94, 323

n (av), 122
nanometer, nm, 109
nesocarbonate, 305
nesophosphate, 305
nesosilicate, 149–74
nesosulfate, 305
neutron number, 2
n-glide, $a/2 + c/2$
  andalusite, 165
  cuprite, 301
  forsterite, 81
  kamacite, 112
  SnF_4, 301
ninefold CN site
  Ca, apatite, 310–11
  Ca, aragonite, 313–14
  Ce, bastnaesite, 315–16
  K, sanidine, 242
  Na, tourmaline, 195
nuclear stability, 4–5
nucleon, 2
nuclide, 2–4
  even–even, 4
  even–odd, 4
  odd–even, 4
  odd–odd, 4
neutrons, 2

octahedral CFSE energies, Δ_0, 91
octahedral chains
  anatase, 280–2
  andalusite, 152, 165
  brookite, 282
  columbite, 294
  diopside, 206
  epidote, 152, 184–5
  kyanite, 152, 166
  lawsonite, 152, 178–9
  ludwigite, 258–9
  sillimanite, 152, 163
  staurolite, 171
  rutile, 278–9
octahedral distortion, 92–4
octahedral layer, 207–8
  amphibole, pyroxene, 207–8
  chlorite, clays, micas, 223–9
octahedron, 52
octet rule, 27
omniaxial, 120
optical indicatrix, 120–1
optical mineralogy, 118–22
optic angle, 2V, 121
optic axis, OA, 120–1
optic plane, OP, 120
orbital degeneracy, 89
orbital energy levels, 8
orbital filling sequence, 12
orbital overlap, H_2, Cl_2, 22
orbital shapes, 7
orbitals per quantum shell, 8
ordering, 1, 43, 239–41, 297
overbonded anions, 67–9
oxygen atoms per unit cell, 111
oxygen isotopes, 3

packing index, 114–16
paramagnetism, 38, 98
parsimony of polyhedra, 47
Pauling's rules, 45–7, 61–9
periodic classification, 9
  long form table, 2
  spiral form table, 11
periodic irregularities, 13
peristerite, 246–9
peroxy type silicate, 152
petrographic microscope, 118–22
phase diagrams
  Al silicates, 167
  feldspar exsolution, 246
  feldspar solid solution, 239
  $SiO_2 \cdot H_2O$, 233
phase transformation, reversible
  α-, β-quartz, 235
  P-, *C*-pigeonite, 210–13
phyllosilicates, 223, 225–9
  basal spacings, *hkl*, 226–9
  covalent bonds, 227
  dioctahedral, 227–9
  d-spacings, 227, 229
  formulae, 227, 229
  H-bond, 228
  ionic bond, 228
  Pauling layer thickness, 226
  structure, 223–7
  trioctahedral, 226–8
  X-ray data, 226–8
pi (π) bond, 28–38
pig "001", "100" lamellae, 210–12
point group, 74–5
polar mineral structures
  bromellite, 50
  pyrargyrite, 324
  quartz, 236

polar mineral structures (*cont.*)
  tourmaline, 199
  wurtzite, 288
  zincite, 286
polarizability, 18, 134–43
polarizability constants
  electronic, ions, 139
  molecular, minerals, 137
  molecular, oxides, 138
polarizer, 118
polyhedral sharing, 45–6
polyhedra with a common corner, 47
polarizing power, 118
polymerization, 45, 148–9, 254
polymorphism
  $Al_2O\ [SiO_4]$, 166–9
  C, 29–36
  $CaCO_3$, 312–14
  Fe_2O_3, 102, 271, 298
  $KAlSi_3O_8$, 239–41
  Mg_2SiO_4, 81, 97, 177, 296
  $NaAlSi_3O_8$, 237–40
  SiO_2, 233–5
  TiO_2, 276–83
  ZnS, 31–2, 288
primitive lattice, *P*, 37
pyrite
  chemical bonding, 36–9
  physical properties, 37–8
  structure, 38
pyroxene-like segment, 209
pyroxenes
  Al-rich orthopyroxene, 212
  amphibole unit cell comparison, 216
  cleavage, I-beam structure, 208–9
  exsolution, 209–14
  M-1, M-2 sites, 204–6
  optical properties, 204–5, 215
  unit cell data, 204–5, 216

quantum number
  angular momentum, 7
  azimuthal, 7
  magnetic, 7
  principal, 7
  spin, 7
quartz, handedness, 235–6

radiant visible energy, 15–18
radiation damage, 156
radioactive heat, 5
radioactivity, 5, 103, 155, 272
radiogenic substitution, 103
radius ratio, 42, 47–52
  coordination number, 47–51
rare earth garnets, 105
rare or noble gas, 10–11
Rb–Sr decay, 5
reconstructive phase transformation, 233
refraction, 119
refractivity variation
  coordination, 130
  covalency, 131
  density, 131–3
  electronegativity, 131
  valency, 131
resonance, 44
retardation, 119, 122
rotation axes
  crystal classes, 75
  crystals, 73
  symbols, 78
rotation axes, 2-, 3-, 4-, 6-fold
  2, alpha quartz, 231
  2, bastnaesite, 316
  2, sanidine, 238, 241
  3, calcite, 314
  3, CCP assemblages, 48–9
  3, HCP assemblages, 48–9
  3, pyrargyrite, 323
  3, pyrite, 37
  4, halite, 73–4, 112–13
  6, beryl, 188

sawhorse penetration twin, 169
scattering of X-radiation, 74
Schrödinger electron, 6
screw axes
  symbols, 78
  translations, 77
screw axes, 2_1
  åkermanite, 182
  azurite, 318
  datolite, 161
  epidote, 183–5
  sanidine, 238, 242, 243
  wadeite, 193
screw axes, 3_1, 3_2
  quartz, 235–6
screw axes, 4_1, 4_3
  anatase, 281
  diamond, 31, 112
  garnet, 159
  scheelite, 309
  spinel, 299
  zircon, 155
screw axes, 4_2
  cooperite, 290
  gold, 112
  halite, 112
  kamacite, 112
  PtO, 40
  rutile, 278–9
screw axes, 6_3
  apatite, 311
  niccolite, 322
  wadeite, 192
  zincite, 287–8
shared-edge shortening, 46, 96–9, 156, 189, 206–7, 269–70, 276–80

sharing of faces
 corundum, 269–70
 niccolite, 322
 perovskite, 285
sigma (σ) bond, 28
silicate classification, 148–9
silicoaluminates, 151
silicon carbide, 29–33
 isostructural with diamond, 29
single chain repeat patterns, 203
single chain silicates, 201–8
single crystal analysis, 74
site occupancy, 48–50
site occupancy vs. cell dimensions
 amphiboles, 222
 phyllosilicates, 227
sodium light, 17, 119
solid solution, 105, 153
sorosilicate, 149
sp^3 bonds, B-group elements, 32–4
space groups, tabulation, 81–7
specific refractivity, *K*, 108, 122, 130
spectra, emission
 atomic, ionic, molecular, 17
 line multiplicity, 16–17
 quantum theory, 15
 stellar temperature, 18
sphalerite, ZnS
 ionic vs. covalent models, 31–2, 58
 relation to diamond structure, 31–2
spherical electron density, 44
spheroids of revolution, 121
spinel-like layer, 174
square antiprism, 52, 95
spin pairing and π bonds, 93–4
square planar arrays
 azurite, 317–19
 cooperite, 289–90
 CrF_2, CuO, 97
 libethenite, 164
 $PdCl_2$, PdO, PtO, PtS, 39–40
stable nuclide, 4–5
subsolidus, 209
symmetry, 71–87
symmetry axes of crystals, 73

tektosilicate, 149
tetrahedral layer, 208
tetrahedron, 52
tetraoxy type silicate, 152
transition metal elements, 10–11, 89–94
translational symmetry, 74–8
triangular dipyramid, 163–5
triangular dodecahedron, 154–5, 157
triaxial ellipsoids, 121
trigonal antiprism, 196
trigonal prismatic CN, 322

uniaxial + and −, 120–1
underbonded anions, 67–9
unique O^{2-} anion, 63
unit cell, 71, 73
unit cell volume, 108–10
unit cube, 72–4, 112
unstable nuclide, 4–5
uranium and thorium isotopes, 5, 156
uranium decay, 156, 272

vacancy, lattice
 amphibole, 102, 201, 219
 anatase, 282
 andalusite, 165
 beryl, 187–9
 cordierite, 189
 datolite, 161–2
 feldspar, 242
 maghemite, 102, 298
 mica, 219
 pyrochlore, 102, 291–3
 talc, 201, 225–6
 tourmaline, 199
vacancy substitution, 102
valence bands, 324
valence electron, 15–16
van der Waals bond, 22, 36, 228, 300
velocity of light, 119
vibration direction, velocity, 118
visible region, 17, 119

water, H_2O
 index of refraction, 137
 lone pair orbitals, 34–5
 polarity of molecule, 35
 polarizability, 137
 s–p overlap of O and H, 35
 sp^3 hybrid bond, 35
 specific refractivity, 137
 structure, 35
wüstite-like zone, 170

X-radiation
 angle of incidence, 76–8
 crystal identification, 74
 diffraction, 76–8
 scattering, 74
 target anode, 76–8
 wavelength, 76
X-ray diffractometer, 74
X-ray powder diffraction pattern, 74

YAG and YIG, 105

Zachariasen's bond strength, 68–9
zircon geochronometer, 156
zoned zircon, 156

A CATALOG OF SELECTED

DOVER BOOKS

IN SCIENCE AND MATHEMATICS

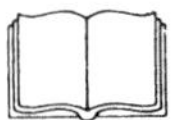

A CATALOG OF SELECTED

DOVER BOOKS

IN SCIENCE AND MATHEMATICS